Fundamentals in Air Pollution

Bruno Sportisse

Fundamentals in Air Pollution

From Processes to Modelling

Springer

Dr. Bruno Sportisse
INRIA
Domaine de Voluceau
Rocquencourt
78153 Le Chesnay CX
France
bruno.sportisse@inria.fr

This work is a translation of the book in French "Pollution atmosphérique; Des processus á la modélisation", B. Sportisse, ISBN 978-2-287-74961-2, Springer, 2008.

ISBN 978-90-481-2969-0 e-ISBN 978-90-481-2970-6
DOI 10.1007/978-90-481-2970-6
Springer Dordrecht Heidelberg London New York

Library of Congress Control Number: 2009938058

Cover illustration: Smog rising from factory (photos.com, item # 4284681)
Cover design: deblik

Printed on acid-free paper

Springer is part of Springer Science+Business Media (www.springer.com)

Preface

This book is a translation of the French book "Pollution atmosphérique. Des processus à la modélisation", published by Springer France (2007).

The content is mainly derived from a course devoted to air pollution I taught at École nationale des ponts et chaussées (ENPC; one of the foremost French high schools, at ParisTech Institute of Technology and University Paris-Est) during the decade 1997–2006. This book has of course been deeply influenced by my research activity at CEREA, the Teaching and Research Center for Atmospheric Environment, a joint laboratory between ENPC and the Research and Development Division of Electricité de France (EDF R&D), that I created and then headed from 2002 to 2007.

I want to thank many of my colleagues for discussions, help and review. Thanks to Vivien Mallet for his careful review, his availability and his pieces of advice (both for the content and the form of this book). Thanks to Marc Bocquet, Karine Sartelet-Kata, Irène Korsakissok for their help in reviewing chapters. I want also to thank a few colleagues for having provided me illustrations from their research work. Thanks to Bastien Albriet, Marc Bocquet, Édouard Debry, Irène Korsakissok, Hossein Malakooti, Denis Quélo, Yelva Roustan, Karine Sartelet, Christian Seigneur and Marilyne Tombette. Thanks also to the American family, Céline and Julien, for their review of the introduction.

I want also to thank the Paris air quality monitoring network, Airparif (Stéphanie Fraincart and Philippe Lameloise), Bénédicte Dousset (Geomer Laboratory and University of Hawaii) and Annie Gaudichet (CNRS and Universities Paris-XII and Paris-VII) for having provided me a few images.

I want to thank Petra Van Steenbergen (English version and invitation to write a book, following SIAM Geosciences 2005) and Nathalie Huilleret (French version and initial project) for their support.

Last, this book is, to some extent, a K project, that was mainly written during night-time. Thanks to my wife, Myriam, and my children, Aude, Marine and Thibaut, for their patience and understanding.

Paris *Bruno Sportisse*

Contents

Introduction 1
Greenhouse Effect, Ozone Hole and Air Quality 1
Brief History 1
Accidents, Impacts and Regulatory Context 4
A Multiplayer Game 8
Role of Scientific Expertise 10
Atmospheric Dilemma 12
Book Objectives and Organization 14
Bibliography 16

1 Primer for the Atmospheric Composition 17
1.1 Atmospheric Chemical Composition 17
1.1.1 Trace Species 17
1.1.2 Gases, Aerosols and Water Drops 20
1.1.3 A Few Species 21
1.1.4 Primary and Secondary Species 21
1.2 Atmospheric Vertical Structure 22
1.2.1 Atmospheric Layers 22
1.2.2 Atmospheric Pressure 25
1.2.3 Vertical Distribution of Species 27
1.3 Timescales 30
1.3.1 Timescales of Atmospheric Transport 30
1.3.2 Atmospheric Residence Time for a Trace Species 32
Problems Related to Chap. 1 36

2 Atmospheric Radiative Transfer 45
2.1 Primer for Radiative Transfer 46
2.1.1 Definitions 46
2.1.2 Energy Transitions 48
2.1.3 Emissions 50
2.1.4 Absorption 52
2.1.5 Scattering 55

2.1.6 Radiative Transfer Equation 59
2.1.7 Additional Facts for Aerosols 60
2.1.8 Albedo 62
2.2 Applications to the Earth's Atmosphere 63
2.2.1 Solar and Terrestrial Radiation 63
2.2.2 Radiative Budget for the Earth/Atmosphere System 68
2.2.3 Greenhouse Effect 71
2.2.4 Aerosols, Clouds and Greenhouse Effect 77
2.2.5 Atmospheric Pollution and Visibility 84
Problems Related to Chap. 2 87

3 Atmospheric Boundary Layer 93
3.1 Meteorological Scales 94
3.2 Atmospheric Boundary Layer 96
3.2.1 Background 96
3.2.2 Classification 97
3.3 Thermal Stratification and Stability 98
3.3.1 A Few Useful Concepts 99
3.3.2 Stability 101
3.3.3 Moist Air 103
3.3.4 Daily Variation of the ABL Stability 105
3.4 ABL Turbulence 106
3.4.1 Background 107
3.4.2 Scale Range and Averaging 108
3.4.3 Turbulent Kinetic Energy 110
3.4.4 Mixing Height and Turbulence Indicators 111
3.5 Fundamentals of Atmospheric Dynamics 113
3.5.1 Primer for Fluid Mechanics 113
3.5.2 ABL Flow 117
3.6 A Few Facts for the Urban Climate 125
3.6.1 Thermal Forcing and Urban Breeze 125
3.6.2 Energy Budget 126
3.6.3 Urban Heat Island 127
3.6.4 Urban Boundary Layer 129
Problems Related to Chap. 3 130

4 Gas-Phase Atmospheric Chemistry 133
4.1 Primer for Atmospheric Chemistry 134
4.1.1 Background for Chemical Kinetics 134
4.1.2 Photochemical Reactions 138
4.1.3 Atmosphere as an Oxidizing Reactor 142
4.1.4 Chemical Lifetime 144
4.1.5 Validity of Chemical Mechanisms 149
4.2 Stratospheric Chemistry of Ozone 150
4.2.1 Destruction and Production of Stratospheric Ozone 150

4.2.2 Ozone Destruction Catalyzed by Bromide and Chloride Compounds . . . 154
4.2.3 Antarctic Ozone Hole . . . 156
4.3 Tropospheric Chemistry of Ozone . . . 159
4.3.1 Basic Facts for Combustion . . . 159
4.3.2 Photostationary State of Tropospheric Ozone . . . 162
4.3.3 Oxidation Chains of VOCs . . . 163
4.3.4 NO_x-Limited Versus VOC-Limited Chemical Regimes . . . 165
4.3.5 Emission Reduction Strategies for Ozone Precursors . . . 167
4.3.6 Example of Photochemical Pollution at the Regional Scale: Case of Île-de-France Region . . . 170
4.3.7 Transcontinental Transport . . . 171
4.4 Brief Introduction to Indoor Air Quality . . . 172
Problems Related to Chap. 4 . . . 174

5 Aerosols, Clouds and Rains . . . 179
5.1 Aerosols and Particles . . . 180
5.1.1 General Facts . . . 180
5.1.2 Residence Time and Vertical Distribution . . . 186
5.1.3 Aerosol Dynamics . . . 188
5.1.4 Parameterizations . . . 193
5.2 Aerosols and Clouds . . . 202
5.2.1 Primer for Clouds . . . 202
5.2.2 Saturation Vapor Pressure of Water, Relative Humidity and Dew Point . . . 203
5.2.3 Condensation Nuclei . . . 204
5.2.4 Mass Transfer Between the Gaseous Phase and Cloud Drops 210
5.3 Acid Rains and Scavenging . . . 212
5.3.1 Acid Rains . . . 213
5.3.2 Wet Scavenging . . . 218
Problems Related to Chap. 5 . . . 222

6 Toward Numerical Simulation . . . 231
6.1 Reactive Dispersion Equation . . . 232
6.1.1 Dilution and Off-Line Coupling . . . 232
6.1.2 Advection-Diffusion-Reaction Equations . . . 232
6.1.3 Averaged Models and Closure Schemes . . . 234
6.1.4 Boundary Conditions . . . 238
6.1.5 Model Hierarchy . . . 239
6.2 Fundamentals of Numerical Analysis for Chemistry-Transport Models . . . 245
6.2.1 Operator Splitting Methods . . . 245
6.2.2 Time Integration of Chemical Kinetics . . . 249
6.2.3 Advection Schemes . . . 254
6.3 Numerical Simulation of the General Dynamic Equation for Aerosols (GDE) . . . 260

6.3.1 Size Distribution Representation 260
6.3.2 Coagulation 263
6.3.3 Condensation and Evaporation 263
6.4 State-of-the-Art Modeling System 265
6.4.1 Forward Simulation 265
6.4.2 Uncertainties 265
6.4.3 Advanced Methods 266
6.4.4 Model-to-Data Comparisons 274
6.4.5 Applications 275
6.5 Next-Generation Models 278
Problems Related to Chap. 6 279

Appendix 1 Units, Constants and Basic Data 283

References 285

Index 293

Introduction

Greenhouse Effect, Ozone Hole and Air Quality

The term of *air pollution* is often used in a misleading way. Actually, air pollution covers many phenomena which are driven by distinct processes and sometimes coupled:

- greenhouse effect due to the so-called greenhouse gases (e.g. carbon dioxide and methane) and the resulting climate change;
- destruction of stratospheric ozone (especially over the South Pole, "ozone hole") catalyzed by chlorofluorocarbons (CFCs);
- air quality with topics ranging from photochemical pollution (ozone, nitrogen oxides and volatile organic compounds[1]) to particulate pollution, acid rains (due to sulfur dioxide and sulfate aerosols), more generally transboundary pollution;
- impact of accidental releases (chemical and biological species, radionuclides) into the atmosphere.

All these topics have in common their strong link to the *chemical composition of the atmosphere* and to *atmospheric dispersion* of pollutants. The emission of *trace species*, with very low concentrations, may strongly alter the atmospheric behavior and the life conditions at the Earth's surface. Considering the pollutant properties, and the space and time characteristic timescales of the processes which govern their atmospheric "fate" makes it possible to *classify* these topics.

Brief History

Air pollution is mentioned in very old texts, even if not named as such. Since Antiquity, a few authors, such as the Chinese philosopher Lao Tzu, were concerned by the impact of anthropogenic activities on environment (especially air). A Roman

[1]In the following, NO_x will stand for nitrogen oxides, VOCs for volatile organic compounds, SO_2 for sulfur dioxide and O_3 for ozone.

B. Sportisse, *Fundamentals in Air Pollution*,

lawyer regulates emissions from a number of activities in York (UK) in the IVth century (Table 0.1).

Historical studies usually focus on the works of the physician and philosopher Moses Maimonides (1135–1204), as giving a precise description of air quality: "*the air becomes stagnant, turbid, thick, misty and foggy*" (using the modern translations, [122]).

Regulatory rules against the use of *sea coal* in the vicinity of the King's Castle are contained in an edict of Edouard I ("*whosoever shall be found guilty of burning coal shall suffer the loss of his head*"). At a larger scale, Richard II regulates the use of coal in London ([143]).

John Evelyn's book, *Fumifugium or the Inconvenience of the Aer and Smoak of London Dissipated* (Fig. 0.1), is published in 1648 while Europe and England both had many other concerns. This book is often presented as the first one which is specifically devoted to air pollution. Actually, the historical British context is a bit more complicated (namely the Restoration of King Charles II, which lowers the environmental focus of the book, [34, 70]). Nevertheless, this book is a good illustration of the starting "industrial prerevolution" with an increasing use of coal for industries and heating, and of the resulting environmental damages (see the astonishing book of Peter Brimblecombe, *The Big Smoke: A History of Air Pollution in London since Medieval Times*, [20]).

While the previous texts were mainly focused on the description of *sanitary* effects, the investigation of the atmospheric chemical composition really starts with Robert Boyle in his book *General history of the Air* (1692), in which *nitros et salinos-sulphurus spiritus* is described. Stephen Hales (*Vegetables Statics*, 1727)

Table 0.1 A brief history

−500	Lao Tzu describes the impact of anthropogenic activities on environment.
300	Local regulation in York (UK, Roman empire).
1200	Moses Maimonides describes air pollution.
1272	Edouard I forbids the use of *sea coal* in the vicinity of his castle.
1390	Richard II regulates the use of coal in London.
1648	*Fumifugium* of John Evelyn.
1692	*A general history of the Air* of Robert Boyle.
1727	Stephen Hales observes the acidity of dew (*Vegetable Statics*).
1840	Christian Schönbein identifies ozone.
1852	Robert Angus Smith distinguishes different pollution regions.
1872	Robert Angus Smith writes *Air and Acid Rain*.
1905	Harold Antoine des Vœux introduces the term of *smog*.
1930	Sidney Chapman formulates a mechanism for stratospheric ozone.
1950s	Arie Jan Haagen-Smit studies the photochemical *smog* of Los Angeles.
1970s	Mechanisms for stratospheric ozone (Crutzen, Rowland, Molina).
1980s	Understanding of the processes driving the stratospheric "ozone hole".
1990s	Convergence of topics related to atmospheric chemistry, greenhouse effect and climate.

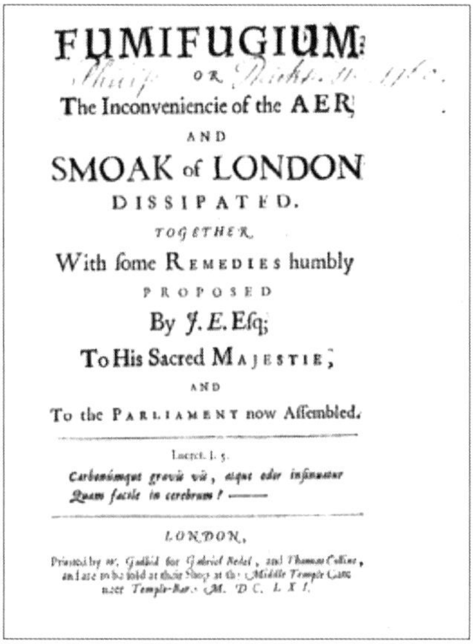

FUMIFUGIUM:

OR

The Inconveniencie of the AER

AND

SMOAK of LONDON

DISSIPATED.

TOGETHER

With some REMEDIES humbly

PROPOSED

By J. E. Esq;

To His Sacred MAJESTIE,

AND

To the PARLIAMENT now Assembled.

Lucret. l. 5.

Carbonúmque gravis vis, atque odor insinuatur
Quam facile in cerebrum! ——

LONDON,

Printed by W. Godbid for Gabriel Bedel, and Thomas Collins, and are to be sold at their Shop at the Middle Temple Gate neer Temple-Bar. M. DC. LXI.

Fig. 0.1 Fumifugium or the Inconvenience of the Aer and Smoak of London dissipated together with some remedies humbly proposed by J. Evelyn Esq. to His Sacred Majesty, and to the Parliament now assembled (1648). A key historical reference (see [34, 70] for the historical context)

studies the acidity of dew on vegetables: "*the air is full of acid and sulphurus particles*".

All through the XIXth century, pollution fogs characterize London. Charles Dickens describes the *London particular* and *pea soupers* in his novels. Claude Monet, in the early XXth century, paints a series of oils in London, with a focus on the Parliament buildings, which illustrates the persistence of fog. These paintings can even provide elements to investigate *a posteriori* the atmospheric conditions over London in this period ([12]).

In 1852, Robert Angus Smith gives a description of pollution over Great Britain in a very precise way, on the basis of observational data and with particularly *modern* words. He notices that the pollution type may differ, depending on its distance from the emission sources:

> [...] we may therefore find easily three kinds of air, [...], that with carbonate of ammonia in the fields at a distance, [...], that with sulfate and ammonia in the suburbs, [...] and that with sulphuric acid, or acid sulphate, in the town (from [44]).

The concept of *acid rain* is the subject of his book *Air and Acid Rain: the Beginnings of a Chemical Climatology* (1872). As a General Inspector in charge of the application of the *Alkaly Act*, he organizes an extended monitoring network, which can be viewed as a "precursor" of the modern air quality monitoring networks.

In 1905, the scientist Harold Antoine des Vœux introduces the term of *smog* to describe "*a fog intensified by smoke*" (there are possibly earlier uses). This term is widely used after his study of the pollution event over Glasgow (autumn 1909).

At the scientific level, the accelerating advances in physics and chemistry result in a finer and finer understanding of atmospheric processes. Meanwhile, the increase in anthropogenic emissions, due to growing industrial activities and birth of the automobile era, contributes to the emergence of environmental concerns.

Ozone is measured in the second half of the XIXth century (following its identification by Christian Schönbein, due to its characteristic odor). Atmospheric chemical mechanisms are formulated all through the XXth century to explain the atmospheric chemical composition. In the early 1930s, Sidney Chapman proposes the first chemical mechanism for stratospheric ozone. Arie Jan Haagen-Smit describes the possible composition of the photochemical smog over Los Angeles in the early 1950s, namely a mixture of ozone, nitrogen oxides and volatile organic compounds.

New topics are added to these "classical" pollutions (London and Los Angeles smogs) from the 1960s: acid rains, transboundary pollution, stratospheric ozone destruction, greenhouse effect (and resulting climate change), and more generally the study of the atmospheric chemical composition.

Viewing the atmosphere as a chemical reactor is definitely accepted after the works of P. J. Crutzen, M. J. Molina and F. S. Rowland, among many other scientists, sharing the Nobel prize in 1995 "*for their work in atmospheric chemistry, particularly concerning the formation and decomposition of ozone*".

Accidents, Impacts and Regulatory Context

Simultaneously with this increasing understanding, the pollution manifestations have sometimes resulted in spectacular impacts (Table 0.2). When specific emission and meteorological conditions are met, a few pollution events may result in hundreds or thousands of deaths in a few days (*Great Smog*, also referred to as *Big Smoke*, of 1952 in London: 4000 deaths from 5 to 9 December, Fig. 0.2).

The impacts of air pollution are not limited to health impacts. The works of Arie Jan Haagen-Smit were initially focussed on the impact of photochemical smog on agriculture. In the 1960s and 1970s, acid rains are indirectly observed by their impact on ecosystems (forests, lake eutrophication, soil acidity). Interactions between

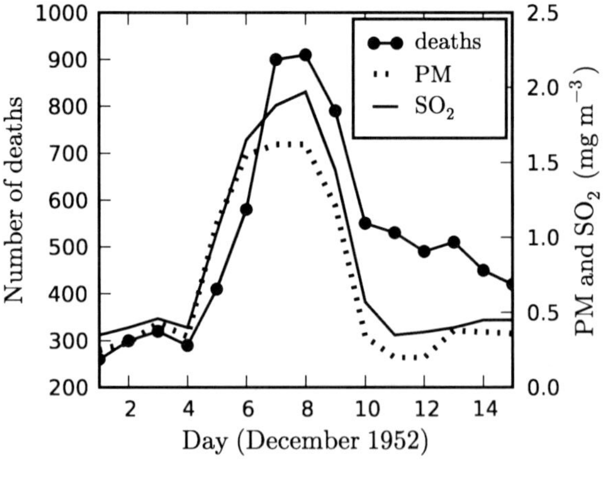

Fig. 0.2 *Great Smog* (London, December 1952): evolution of mortality, sulfur dioxide concentration (SO_2) and smoke ("PM" stands for particulate matter). The concentration unit is mg m^{-3} (a scaling factor up to 100, as compared to the "modern" concentrations!). Sources: [20] and [156]

Table 0.2 A few historical "accidents" related to air pollution. For example, see [98] for the study of the Meuse Valley smog

1873	London	1000 deaths (?)	London smog
1909	Glasgow	1000 deaths	London smog
1930	Meuse Valley (Belgium)	60 deaths	London smog
1948	Donora (USA)	20 deaths	London smog
1952	London *Great Smog*	4000 deaths	London smog
1962	London	750 deaths (?)	London smog
1966	New York (24–30 November)	168 deaths	London smog
1984	Bhopal (India)	2000 deaths	chemical accident
1986	Chernobyl (USSR)	?	nuclear accident

SO_2, particulate matter and atmospheric water result in the black alteration of building surfaces.

As a consequence, a regulatory corpus (in a more systematic way than the aforementioned cases) is established (Table 0.4). Local rules may originate in the Middle Ages: they often focus on chimney heights. In Great-Britain, there is a growing initiative to regulate smoke emissions (*smoke abatement*) in the first half of the XIXth century. The so-called Mackinnon committee (including the scientist M. Faraday) is actually the *Committee for Means and Expediency of preventing the Nuisance of Smoke arising from Fires or Furnaces* (1843). Several regulatory texts are proposed but are subject to strong opposition of industrialists. As a result, only a "dampened" text is added to the *Public Health Act* in 1846. In 1853, more constraints are detailed in the *Smoke Nuisance Abatement Metropolis Act*. Other amendments will be added in the *Public Health Act* of 1875. A specific focus is put on the saponification industry, which emits chloride compounds, with the *Alkaly Act* of 1863 (it will result in a decrease of about 95% of chloride emissions).

In 1895, the United States of America start to regulate the emissions related to …automobiles to decrease "*the showing of visible vapor as exhaust from steam automobiles*".

The increasing number of smog events over Los Angeles results in the creation of the first modern air quality monitoring network in 1947 (*Los Angeles Air Pollution Control District*). Following the Great London Smog, the *British Clean Air Act* (CAA) is enacted in 1956 ([157] for an historical perspective). A similar regulation is taken in 1963 by the USA, with a specific part for traffic-induced emissions in 1965. While air quality monitoring was previously mainly in charge of states, a few amendments (CAAA, *Clean Air Act Amendments*) establish in 1970 the role of a federal agency, the *Environmental Protection Agency* (US EPA), and define federal guidelines for six pollutants (NAAQS, *National Ambient Air Quality Standards*).

The assessment of the acid rain impacts in North America results in a regulation devoted to sulfur dioxide emissions (a specific item is included in the 1970 CAAA). Persistency of acid rains on a few sites (as evaluated by NAPAP, the *National Acid Precipitation Assessment Program*) leads to the establishment of an emission trading market of dioxide sulfur emissions (title IV of the *US Clean Air Act*, 1990, and

Table 0.3 CLRTAP protocols. EMEP is the technical center in charge of evaluation, measurements and modeling (*Co-operative Programme for Monitoring and Evaluation of the Long-range Transmission of Air pollutants in Europe*)

1984	Long-term funding of EMEP.
1985	Reduction of sulfur dioxide emissions of 30%.
1988	Control of NO_x emissions and transboundary fluxes.
1991	Control of VOC emissions and transboundary fluxes.
1994	Supplementary reduction of sulfur dioxide emissions.
1998	Persistent organic pollutants (POPs).
1998	Heavy metals.
1999	Acidification, eutrophization and ozone.

Acid Rain Program). North-American electric companies, strongly based on coal combustion, are mainly concerned.

In Europe, from 1967, the Swedish scientist Svante Oden investigates the impacts of sulfur dioxide emissions on rain acidity. In spite of an initial skepticism toward the possible long-range impact of emissions, transboundary pollution is recognized as a key concern in the 1970s:

> [...] air quality in any European country is measurably affected by emissions from other European countries ... [and] if countries find it desirable to reduce substantially the total deposition of sulphur within their borders, individual national control programmes can achieve only a limited success (OECD Convention on Long-Range Transboundary Air Pollution, 1977).

In 1979, the *Convention on Long-Range Transboundary Air Pollution* (CLRTAP, see Table 0.3) is established by the United Nations. A few key concepts stem from this framework, such as the *critical load*, defined as "a quantitative estimate of an exposure to one or more pollutants below which significant harmful effects on specified sensitive elements of the environment do not occur according to present knowledge".

Many directives of the European Union will be issued during the following years: for sulfur dioxide in 1980 (80/779/EEC), for nitrogen oxides in 1985 (85/203/EEC), for ozone in 1992 (92/72/EC), etc. A global policy devoted to air quality control is initiated with the framework directive of 1996 (96/62/EC), which results in many "daughter" directives: particulate matter, sulfur, lead and nitrogen oxides in 1999 (99/30/EC), carbon monoxide and benzene in 2000 (2000/69/EC), ozone in 2002 (2002/3/EC), heavy metals, mercury and PAH (polycyclic aromatic hydrocarbons) in 2004 (2004/107/EC).

One of the most spectacular consequences of this intensive regulatory activity is related to lead. The decrease by a scaling factor greater than 2 of the authorized lead content in gasoline, in 1985 (directive 85/210/EEC), quickly results in a similar decrease in the air concentrations. The use of lead for gasoline will be forbidden in 2000.

Overcoming the fragmented approach which follows the 1979 convention (a few protocols devoted to specific pollutants, Table 0.3), the Göteborg Protocol, in 1999, adopts a global approach with a multipollutant and multimedia (water, air and soil) focus. The European Union initiates thereafter a process to decrease the regulated

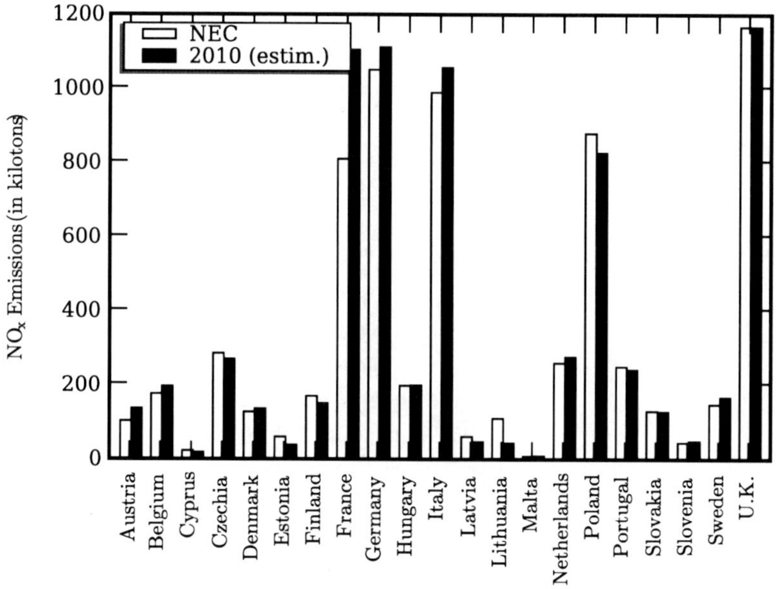

Fig. 0.3 NO_x national emission ceilings for 2010: comparison between the value of the NEC directive and the estimations (2006) of a few national plans. Source: [37]

concentrations (CAFE, *Clean Air For Europe*). In 2001, the NEC directive (National Emissions Ceilings, 2001/81/EC) defines for each country emission ceilings for 2010, for four pollutants: NO_x, SO_2, VOCs and ammonia (NH_3). As an illustration, Fig. 0.3 shows the evaluation, in early 2007, of the ability of countries to achieve the targets for NO_x (the most challenging issue, especially for France and Germany, because it is related to traffic-induced emissions).

At the global scale, the understanding of the chemical mechanism of stratospheric ozone destruction and the observation of the antarctic "ozone hole" in 1985 (a decrease by a factor 2 of the ozone column as compared to the 1960s) result in a series of international conferences to address this issue. The decision of reducing emissions of a few pollutants (e.g. CFCs) is taken by the Montreal Protocol in 1987. The extension to other species and to more drastic reductions is carried out by the London (1990), Copenhagen (1992) and Vienna (1995) protocols. The noticeable fact is that there are *only* a few years from the understanding of the adverse role of CFCs (on stratospheric ozone destruction) to the regulatory consequence (namely the progressive CFC emission ban).

Meanwhile, a strong increase in the atmospheric CO_2[2] is measured, especially by Charles Keeling (Hawai, Mauna Loa) in the 1960s. The possible resulting perturbation in the radiative behavior of the atmosphere ("additional" greenhouse effect)

[2]More generally of a few greenhouse gases, defined as gases which absorb terrestrial infrared radiation (Chap. 2).

Table 0.4 A brief history of air quality reglementation

1853	Smoke Nuisance Abatement Metropolis Act.
1863	Alkaly Act (Great Britain).
1895	Regulation of automobile exhaust smoke (USA).
1947	Los Angeles Air Pollution Control District.
1956	British Clean Air Act.
1963	US Clean Air Act (US CAA).
1965	Title II US CAA (Motor Vehicle Air Pollution Control Act).
1970	Clean Air Act Amendments and creation of the US EPA (USA).
1979	Convention on long-range transboundary air pollution (Geneva).
1980	SO_2 directive (European Union).
1987	Montreal Protocol (stratospheric ozone).
1990	Title IV US Clean Air Act (acid rains).
1992	Ozone directive (European Union).
1996	Framework directive for air quality (European Union).
1997	Kyoto Protocol.
1999	Göteborg Protocol (multipollutants, multimedia).
2001	NEC directive (National Emissions Ceilings; European Union).

and in the climate becomes a major concern in the 1990s. Following a cycle of international conferences, the Kyoto protocol (1997) determines emission reductions for a few countries. At the same time, the IPCC works (*Intergovernmental Panel on Climate Change*, for example [106]) result in a better understanding of the underlying processes and a finer evaluation of the possible impacts.

A Multiplayer Game

This framework drives the issues to be addressed and the strategies to be taken by the different "players" (public authorities and emission sectors).

A key question for the public authorities, at national and international levels, is the appropriate choice of emission reductions: how to define emission ceilings for a transboundary pollution, how to allocate emission reductions per country and per emission sector? Once an emission reduction is fixed, the issue of *monitoring* (namely of the monitoring networks to be deployed) becomes a prevailing issue. What pollutants should be measured (when possible)? How to reduce the cost of monitoring networks (trade-off between a large number of "coarse" stations and a smaller number of fine "supersites")? For example, Fig. 0.4 shows the evolution from 1991 to 2001 of the French monitoring network devoted to ozone observation (a "continental" pollutant); meanwhile, the number of measurement stations for less classical pollutants has not significantly increased.

As expected, the issues are quite different for the emitting sectors and, as a result, the corresponding industries. How to forecast and to apply regulatory constraints

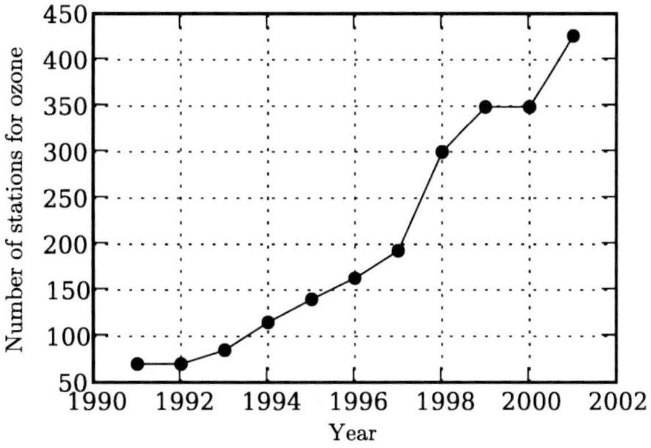

Fig. 0.4 Evolution from 1991 to 2001 of the number of measurement stations for ozone in France. Source: ADEME (French environmental protection agency)

Table 0.5 Evolution (in %) of SO_2 emissions from 1980 to 2000 in Europe. The estimated range of uncertainties is indicated between brackets. Source: [1]

Country	Evolution (in %)	Country	Evolution (in %)
Austria	−90	Netherlands	−[85, 90]
Denmark	−90	Poland	−[60, 65]
France	−80	Switzerland	−[80, 85]
Germany	−90	Sweden	−[85, 90]
Italy	−75	Great Britain	−90

Table 0.6 Evolution (in %) of NO_x, VOCs and CO emissions from 1990 to 2002, in Europe and in the USA. Source: [72]

Region	NO_x	VOCs	CO
Europe (EU25)	−31	−39	−45
USA	−16	−16	+14

which have been dramatically increased for many pollutants? What are the consequences for investments and the choice of R&D projects to initiate? Table 0.5 details the decrease between 1980 and 2000 of SO_2 emissions for a few European countries. The corresponding industrial sector is that of electric power production (accounting for up to 80%). For example, in the case of thermal power plants for electric production, the innovative techniques which make it possible for such a decrease in emissions are FGD (for *fuel gas desulfurization*) and SRC (*selective reduction catalysis*; see e.g. [103]). Illustrative costs are about 100 millions of euros and 50 millions of euros for a power plant of 600 megawatts. As an illustrative case, Poland spent more than 8 billions of euros in the 1990s to reduce its annual emissions of SO_2 and NO_x of 800 kt (kilotons) and 300 kt, respectively ([71]).

Similarly, another most impacted sector is the automobile sector, which comprises both car and oil industries. The evolution of reglementations devoted to unitary emissions (emissions for a given vehicle) has impacted gasoline quality and the design of car engines. Figure 0.5 shows the evolution of the so-called Euro norms from 1993 to 2005 for a gasoline-fueled vehicle. In spite of the increase in traffic,

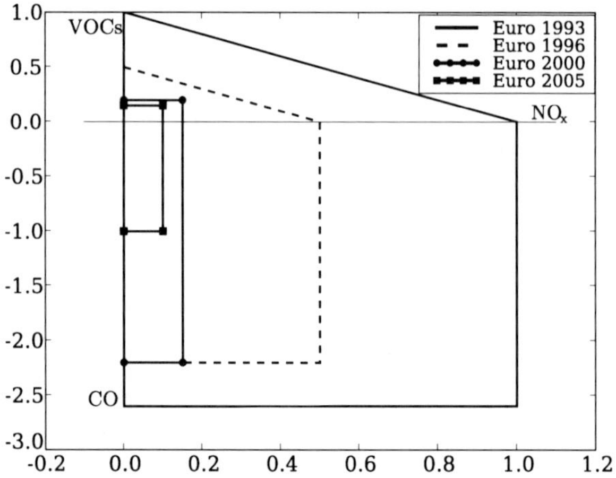

Fig. 0.5 Evolution of the European reglementation for unitary emissions of gasoline vehicles (Euro 1993–2005 norms). The values are dimensionless. The polygon of "regulatory constraints" is defined by NO_x for positive abscissae, by VOCs for positive ordinates and by CO for negative ordinates

this results in a strong decrease in the emissions of ozone precursors (nitrogen oxides and volatile organic compounds). We can refer to Table 0.6, which indicates a reduction ranging from 30 to 40% between 1990 and 2002 in Europe.

Reductions of CO_2 emissions for cars is another example. In 1998, the ACEA (European Automobile Manufacturers' Association) signed an agreement with the European Union to reach a mean CO_2 emission of $140\,g\,km^{-1}$ for new cars in 2008. Such an effort underlies many changes in the automobile sector (increase in number of diesel vehicles, introduction of new fuels, technological improvement of engines). Note the strong differences among manufacturers (Table 0.7). ACEA predicts that the target of the European Union for 2012 ($130\,g\,km^{-1}$) cannot be achieved with only technological approaches (position paper dated 7 June 2007, [6]). Moreover, another tough point for the evaluation of the 1998 agreement is related to the impact of the so-called *external factors* (changes in the automobile market, regulatory framework, etc.).

A key point is the difference between an emission reduction and an atmospheric concentration reduction due to the long-range transport of pollutants and the formation of secondary pollutants through chemical and physical processes (Fig. 0.6, Chap. 5).

Role of Scientific Expertise

In such a context, scientific and technical expertise plays a leading role. Classically, this concerns:

- understanding of the underlying phenomena for the adverse effects on health and environment to evaluate the contribution of anthropogenic activities;

Table 0.7 CO_2 emissions for new gasoline and diesel vehicles in the European Union with 15 countries (EU–15). Value in 2003 (in g km^{-1}) and evolution from 1995 to 2003, depending on the manufacturer origin (ACEA for European manufacturers, JAMA for Japanese manufacturers and KAMA for Korean manufacturers). Source: [2]

	Global		Gasoline		Diesel	
	2003	1995–2003	2003	1995–2003	2003	1995–2003
EU–15	164	−11.8%	171	−9.5%	157	−12.3%
ACEA	163	−11.9%	171	−9.0%	154	−12.5%
JAMA	172	−12.2%	170	−11.0%	177	−25.9%
KAMA	179	−9.1%	171	−12.3%	201	−35.0%

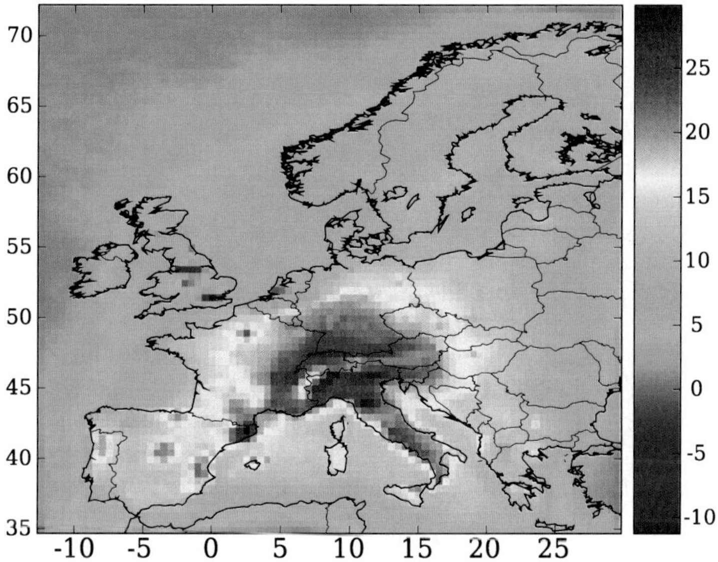

Fig. 0.6 Contribution (in %) of traffic-induced emissions to the ozone peaks (Europe, summer 2001). The estimation is carried out by comparing a reference simulation with a simulation without traffic-induced emissions. The simulation configuration does not take into account the nonlinear effects of photochemistry (Chap. 4). Simulation with the POLYPHEMUS system. Credit: Yelva Roustan, CEREA

- definition of appropriate monitoring networks to supply regulatory decisions or to improve scientific knowledge (satellite observation of the atmospheric chemical composition).

During the last decade, numerical models have become a decisive tool, with many applications:

- *process studies* (to improve scientific understanding);

- *environmental forecasting*: how to forecast a photochemical pollution event, how to estimate the dispersion of an accidental release (Fig. 0.7 for the assessment of the Chernobyl accident)?
- *impact studies*: how to assess the impact of emission reduction scenarios at the European scale (national emission ceilings; Fig. 0.6) or at the local scale (impact of changes in traffic management)?
- *long-term climate studies* of the atmospheric chemical composition (greenhouse effect and climate change);
- *inverse modeling of emission fluxes*: how to estimate poorly accurate emissions (possibly regulated) from observational data of atmospheric concentrations?

Atmospheric Dilemma

In many cases, environmental policy decisions face a dilemma because the improvement of a given criterion may result in a *disbenefit* of another. Investigating all possible consequences of changes in emissions requires therefore scientific expertise, as illustrated by the four following examples.

Ozone concentration depends on the emissions of precursors, (VOCs and NO_x) in a complicated (nonlinear) way. Depending on the *chemical regime*, emission reduction may lead to an increase in ozone concentration (see the North-American case in the 1980s, Chap. 4)!

Similarly, the introduction of a new engine or of a new fuel for car traffic may result in adverse effects, whose prior evaluation is challenging. *Mass* reduction of *emitted* particles may result in an increasing number of fine particles which are *formed* in the atmosphere (the most adverse ones at the sanitary level, Problem 5.4).

Another example is provided by the introduction of biofuels (ethanol) whose prior motivation is the decrease in emitted fossil carbon. Based on a lifecycle analysis, the resulting impact should be positive by reducing the net budget of greenhouse gas emissions. However, there are at least two concerns. First, the impact on air quality could be negative, similarly to the first previous example (Exercise 4.7). Second, the extension of agro-biofuels is expected to generate concomitant emissions, especially of nitrogen peroxide (N_2O, Exercise 4.7), a strong greenhouse gas, which could negate the expected gain.

Last, the improvement of air quality, due to the decrease in sulfate particulate pollution in the Northern Hemisphere during the last two decades, results in a reduction of the cooling effect of particles with respect to solar radiation (Chap. 2), which can be viewed as the annealing of the counterbalance to the greenhouse effect.

This latter example provides a typical case of links between scientific expertise and decision-making. It also illustrates the temptation of *atmospheric engineering* (*geo-engineering*). A classical case in meteorology (whose effects are controversial) is *cloud seeding* (particles are used to initiate precipitations, Chap. 5). In the context of atmospheric chemistry, P.J. Crutzen questions the possible injection of

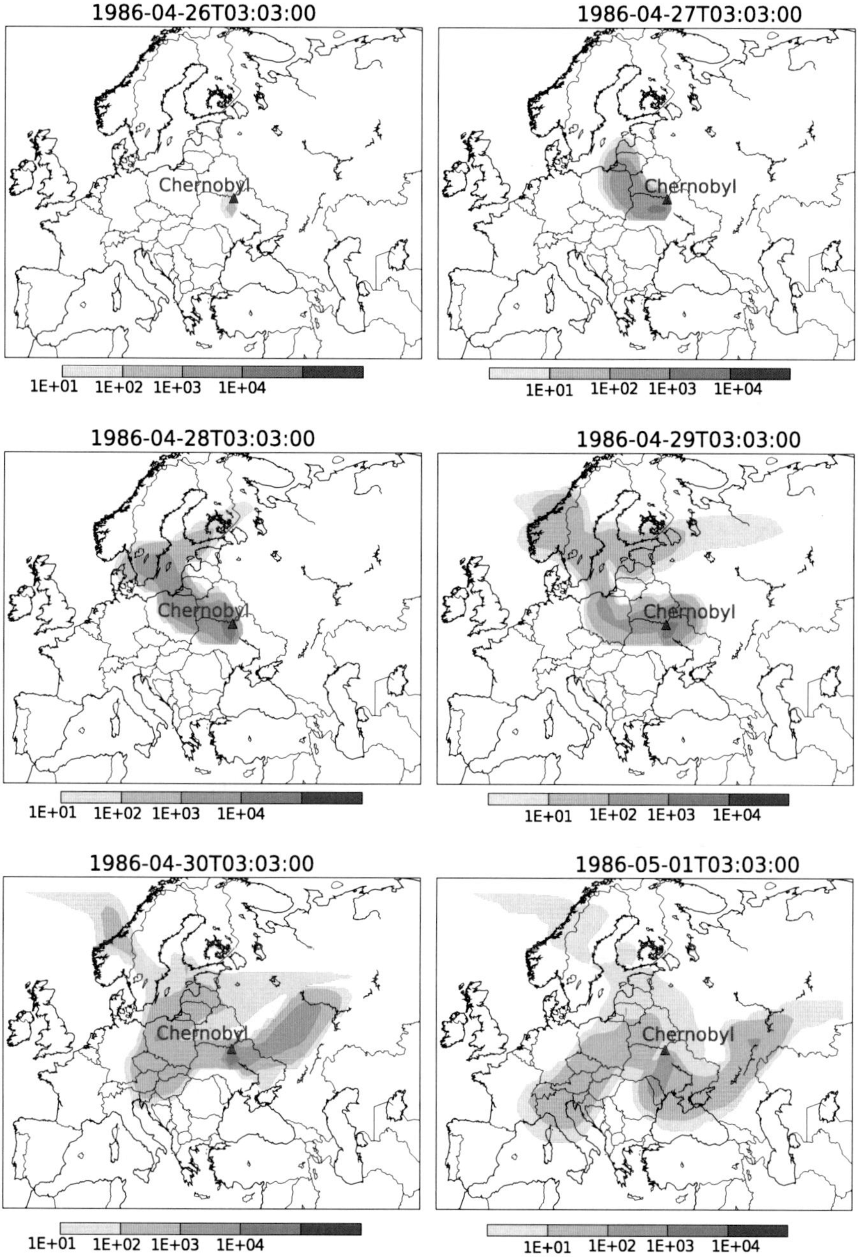

Fig. 0.7 Simulated evolution of the radionuclide plume over Europe following the Chernobyl accident (marked by a *triangle*). From *left to right* and from *top to bottom*: field of cesium 137 (in becquerel) at 3:00 (TU) from 26 April to 1 May 1986. Simulation with the POLYPHEMUS system. Credit: Denis Quelo, CEREA/Institute of Radiological Protection and Nuclear Safety. Source: [115]

sulfate particles into the stratosphere to increase the planetary albedo[3] and, thus, to compensate the reduction in particulate burden (Exercise 2.9):

> [...] this can be achieved by burning S_2 or H_2S, carried into the stratosphere on ballons and by artillery guns to produce SO_2 [...and this has to be viewed] as an escape route against strongly increasing temperatures ... ([26]).

Apart from a "general" position toward such projects, resulting negative effects of such projects have of course to be carefully estimated (e.g. possible increasing destruction of stratospheric ozone).

Book Objectives and Organization

This book aims at giving the key elements to understand *atmospheric pollutions* (Table 0.8).

The objective of this book is not to give a global and comprehensive overview of issues, which would require the knowledge of many scientific fields (fluid mechanics, atmospheric chemistry, radiative transfer, aerosol and cloud physics, etc.). Reference textbooks, more or less easy to read, are available (see the bibliographical references at the end of this chapter). This book is based on these references, especially for a few exercises.

Complementary to these comprehensive monographs, this books aims at giving a few "rules of the (scientific) game", beyond the rule of the thumb.

Table 0.8 Classification of atmospheric pollutions. The *regional* scale corresponds to the meteorological meso scale (from a big city to the continental scale)

Pollution	Historical peak	Species	Scale	Regulation
London smog	London 1952	SO_2	local	CAA (1956)
Photochemistry	Los Angeles (1940s)	ozone, NO_x, VOCs	local, regional	O_3 directive (1992)
Acid rains	USA (1960s)	SO_2	regional	US CAA (1970s)
Transboundary pollution	Europe (1970s)	sulfates, nitrates		CLRTAP (1979), Göteborg (1999)
Statospheric ozone	antarctic hole (1980s)	CFCs	global	Montreal (1987)
Greenhouse effect and climate change	1990s	CO_2, CH_4	global	Kyoto (1997)

[3]That is to say the ability of the Earth/atmosphere system to reflect solar radiation back to space.

The book organization is detailed in Table 0.9.

The fundamentals are given in Chap. 1, to be viewed as a short primer for the atmospheric chemical composition. Magnitudes of a few characteristic scales are calculated. The different atmospheric pollution types are classified by considering the impact scales.

Chapter 2 reviews the radiative issues with a focus on the atmospheric energy budget. The interaction between solar and terrestrial radiations and the atmospheric matter (gases, aerosols, cloud liquid water) is investigated. This provides an introduction to the greenhouse effect issue.

Atmospheric dynamics is briefly summarized in Chap. 3, with a focus on the *atmospheric boundary layer* (let us say the first kilometer just above the Earth's surface). Starting from bases of fluid mechanics, a few key meteorological models are presented. Attention is paid to the role of meteorological conditions in the development of a pollution event (stability and vertical mixing of pollutants).

Chapter 4 should be viewed as an introduction to gas-phase atmospheric chemistry, with applications to stratospheric ozone and to photochemical smog ("ozone peaks"). A few issues related to the oxidizing power of the atmosphere are also presented.

Multiphase processes are detailed in Chap. 5 with a focus on aerosols (atmospheric particles). The fundamentals of aerosol dynamics are given to understand their atmospheric evolution, the interactions with gas-phase species and with clouds.

Table 0.9 Questions, scientific fields and keywords for each chapter

Chap.	Issues
1	How to classify atmospheric pollutions? What are the characteristic scales? *Keywords*: bases of atmospheric sciences, emissions, residence time.
2	What is the impact of atmospheric chemistry on the atmospheric energy budget? What is the connection between air pollution and visibility degradation? *Keywords*: radiative transfer, greenhouse effect.
3	To what extent do meteorological conditions govern pollution? What are the urban specificities? *Keywords*: atmospheric boundary layer.
4	What are the main cycles of atmospheric chemistry? What is the genesis of a photochemical pollution event? What is the efficiency of emission reduction strategies? *Keywords*: gas-phase atmospheric chemistry.
5	What is the role of atmospheric particles (aerosols)? By what processes are their evolution governed? What is acid rain? *Keywords*: microphysics, aerosol dynamics.
6	What are the current state-of-the-science models? What are the applications and the limitations? *Keywords*: chemistry-transport models, numerical simulation.

Applications are related to acid rains, particulate pollution and scavenging by precipitations.

Last, numerical simulation is briefly introduced in Chap. 6 with the presentation of the *chemistry-transport models* (CTMs). Applications and current challenging issues are also illustrated.

Each chapter includes not only exercises for direct applications but also problems for more realistic issues. Constants, units, etc. may be found in the Appendix 1. A complete list of References, including the selected Bibliography below, and a comprehensive Index complete the book.

Bibliography

At least two comprehensive books give a nice general presentation of atmospheric chemistry and physics:

- J. Seinfeld and S. Pandis, *Atmospheric Chemistry and Physics*, Wiley-Interscience, 1998
- G. Brasseur, J. Orlando, and G. Tyndall, *Atmospheric Chemistry and Global Change*, Oxford University Press, 1999

For radiative transfer theory, classical references are:

- R. Goody and Y. Yung, *Atmospheric Radiation. A Theoretical Basis*, Oxford University Press, 1986
- K. Liou, *Radiation and Cloud Processes in the Atmosphere*, vol. 20, Oxford Monograph on Geology and Geophysics, 1992
- G. Thomas and K. Stamnes, *Radiative Transfer in the Atmosphere and Ocean*, Cambridge University Press, 1999

The study of the atmospheric boundary layer is detailed in:

- J. Holton, *An Introduction to Dynamic Meteorology*, Academic Press, 1992
- R. Pielke, *Mesoscale Meteorological Modelling*, Academic Press, 1984
- R. Stull, *An Introduction to Boundary Layer Meteorology*, Kluwer Academic Publishers, 1988
- J. Garrat, *The Atmospheric Boundary Layer*, Cambridge University Press, 1992

Numerical simulation is investigated in

- M. Z. Jacobson, *Fundamentals of Atmospheric Modeling*, Cambridge University Press, New York, 1998
- B. Sportisse and B. Mallet, *Introduction to Computational Atmospheric Chemistry: From Fundamentals to Advanced Applications of Chemistry Transport Models*, Springer Verlag, 2008

To end these references, a *marvelous* book, for its clarity and its concision, is:

- D. Jacob, *Introduction to Atmospheric Chemistry*, Princeton University Press, 1999

Chapter 1
Primer for the Atmospheric Composition

The objective of this chapter is to give the fundamentals of atmospheric composition and of air pollution. The terminology is also determined and a few key data are given.

This chapter is organized as follows. The atmospheric composition is investigated in Sect. 1.1. We introduce the concept of *trace species*. The main *primary* (emitted) and *secondary* (formed in the atmosphere) species are also presented. Emission inventories at global scale are given. Section 1.2 details the vertical structure of the atmosphere. The characteristic timescales of the major atmospheric species are defined in Sect. 1.3. A comparison between the timescales of atmospheric transport and atmospheric residence time makes it possible to classify the different pollutions.

1.1 Atmospheric Chemical Composition

1.1.1 Trace Species

Mixing Ratio The *molar fraction* of species X, C_X, is defined as the ratio of the mole number of X to the mole number of air. It is also referred to as the *mixing ratio*.

In the following, *atmosphere* will refer to the *terrestrial atmosphere*. Air is mainly composed of molecular nitrogen (N_2) and of molecular oxygen (O_2). In the atmosphere, $C_{N_2} = 0.78\,\text{mol}\,\text{mol}^{-1}$ and $C_{O_2} = 0.21\,\text{mol}\,\text{mol}^{-1}$. When ranked according to abundance, argon is the third species with $C_{Ar} = 0.0093\,\text{mol}\,\text{mol}^{-1}$.

The mixing ratio of water (H_2O) has a strong variability, ranging from 10^{-6} to 10^{-2} according to humidity. Except water, the remaining species are *trace species*. Their molar fractions are indeed measured in ppmv ($10^{-6}\,\text{mol}\,\text{mol}^{-1}$, *part per million of volume*), ppbv ($10^{-9}\,\text{mol}\,\text{mol}^{-1}$, *part per billion of volume*) or pptv ($10^{-12}\,\text{mol}\,\text{mol}^{-1}$, *part per trillion of volume*). The notation v is often omitted in the following.

B. Sportisse, *Fundamentals in Air Pollution*,

Table 1.1 Chemical composition of dry air (2000). Units: 1 ppmv for 10^{-6} mol mol^{-1} and 1 ppbv for 10^{-9} mol mol^{-1}

Species	Symbol	Mixing ratio
Nitrogen	N_2	780 000 ppmv
Oxygen	O_2	210 000 ppmv
Argon	Ar	9300 ppmv
Carbon dioxide	CO_2	365 ppmv
Ozone	O_3	1 ppbv–10 ppmv
Methane	CH_4	1.8 ppmv
Nitrogen protoxide	N_2O	314 ppbv

Table 1.1 summarizes the chemical composition of dry air (with some of the major atmospheric species).

For instance, $C_{CO_2} \simeq 365$ ppmv for carbon dioxide, $C_{O_3} \simeq 5$ ppmv for ozone (actually from 10 to 100 ppbv in the troposphere) and $C_{CH_4} \simeq 1.8$ ppmv for methane. Note that the mixing ratios of a few species have dramatically increased since the "preindustrial" times (Table 1.2 and Exercise 1.5).

Ideal Gas Law Air can be viewed as an ideal gas, satisfying the relation

$$P = NRT, \tag{1.1}$$

with P the pressure (in Pa), T the temperature (in K), $R = 8.314\,\mathrm{J\,mol^{-1}\,K^{-1}}$ the universal gas constant and N the mole number per air volume. An alternative formulation is

$$P = nk_BT, \tag{1.2}$$

with n the molecule number per air volume and $k_B = 1.38 \times 10^{-23}\,\mathrm{J\,K^{-1}}$ the Boltzmann constant, defined as $R/\mathcal{A}_v$ where $\mathcal{A}_v$ is the Avogadro number (6.02×10^{23} molecule mol^{-1}).

An equivalent equation is, for dry air,

$$P = \rho r_{air} T, \tag{1.3}$$

with ρ the air density (in kg m^{-3}) and $r_{air} = R/M_{air}$, where M_{air} is the molar mass (molecular weight) of dry air (Exercise 1.1).

Exercise 1.1 (Molar Mass of Air) Compute the molar mass of dry air.
Data: $M_{N_2} = 28\,\mathrm{g\,mol^{-1}}$, $M_{O_2} = 32\,\mathrm{g\,mol^{-1}}$ and $M_{Ar} = 40\,\mathrm{g\,mol^{-1}}$.
Solution:
Let M_{X_i} be the molar mass of species X_i. With $M_{air} = \sum_i C_{X_i} M_{X_i}$, by keeping the three most abundant species, we obtain $M_{air} \simeq 28.9\,\mathrm{g\,mol^{-1}}$.

The *standard thermodynamic conditions* are defined by $P = 1.013 \times 10^5$ Pa (1013 hPa or 1 atm, mean value at sea level) and $T = 273.15$ K (0 °C).

Table 1.2 Estimation of the evolution of a few chemical species from preindustrial times to 1998. Take care about the units: 1 pptv for 10^{-12} mol mol^{-1}. Source: [106]

Species (unit)	Preindustrial (1750)	1960	1980	1990	1998
CO_2 (ppmv)	278	315	325	352	365
CH_4 (ppmv)	0.7	1.27	1.57	1.68	1.745
N_2O (ppbv)	270	300	303	310	314
CFC-11 (pptv)	0	11	173	275	268
CFC-12 (pptv)	0	33	297	468	533

Moist Air The extension to wet (moist) air is challenging because the molar mass of wet air depends on the water vapor concentration, which has a large variability. The *virtual temperature*, T_v, is defined as the temperature that dry air should have to keep the density of wet air, with a constant pressure, that is to say:

$$P = \rho r_{air} T_v. \tag{1.4}$$

Moist air may be characterized by the ratio of the water vapor mass to the dry air mass, w. This ratio ranges from 1 to 10 g kg^{-1}. The *specific humidity*, q_s, is defined as the ratio of the water vapor mass to the wet air mass. It is easy to check that $q_s = w/(1+w)$.

The virtual temperature can then be computed by (Exercise 1.2)

$$T_v \simeq T(1 + 0.62w). \tag{1.5}$$

Mass Concentration For species X, the mass concentration, ρ_X, is defined as the species mass per unit of air volume. Let M_X be the molar mass of species X. The mole number of X per unit of air volume is therefore ρ_X/M_X. It is connected to the mixing ratio of X by

$$C_X = \frac{\rho_X}{M_X}\frac{1}{N} = \frac{\rho_X}{M_X}\frac{RT}{P}. \tag{1.6}$$

If the mass concentration is expressed in µg m^{-3}, the mixing ratio in ppb and the molar mass in kg mol^{-1}, respectively, no conversion factor is required.

For ozone ($M_{O_3} = 48$ g mol^{-1}), under standard thermodynamic conditions, we obtain ρ_{O_3}(in µg m^{-3})$/C_{O_3}$(in ppb) $\simeq 2$. This means that a mixing ratio of 50 ppb for ozone corresponds to a mass concentration of about 100 µg m^{-3}.

For species such as mercury or heavy metals, the magnitude of the massic concentrations is the nanogram (10^{-9} g) or the picogram (10^{-12} g) per cubic meter of air.

Exercise 1.2 (Virtual Temperature) Prove (1.5).

Solution:

In this exercise, d (*dry*) and v (*vapor*) are related to dry air and to water vapor, respectively (except T_v). For instance, $w = m_v/m_d$ with m_v the mass of water vapor and m_d the mass of dry air. It is easy to get:

$$P = (N_d + N_v)RT = \left(\frac{\rho_d}{M_d} + \frac{\rho_v}{M_v}\right)RT$$

$$= (\rho_d + \rho_v)\frac{R}{M_d}T\left(\frac{\rho_d}{\rho_d + \rho_v} + \frac{\rho_v}{\rho_d + \rho_v}\frac{M_d}{M_v}\right)$$

$$= \rho r_{air} T\left(\frac{1}{1+w} + \frac{1}{1+1/w}\frac{1}{\varepsilon}\right),$$

where $\varepsilon = M_v/M_d = 18/29 \simeq 0.62$ (with $M_d = M_{air}$). This yields

$$T_v = T\frac{w+\varepsilon}{\varepsilon(1+w)}.$$

Up to first order in w, with $1/(1+w) \simeq 1 - w$, we get $T_v \simeq T(1 + w(1-\varepsilon)/\varepsilon)$. We conclude with the value of ε.

Molecule Concentration The molecule concentration, defined as the number of molecules per air volume, can be computed with

$$n_X = \mathcal{A}_v C_X \frac{P}{RT} = C_X \frac{P}{k_B T}. \tag{1.7}$$

If C_X is in ppmv, n_X is in molecule cm^{-3}.

Exercise 1.3 (Loschmidt Number) Compute the number of air molecules in a cubic meter under the standard thermodynamic conditions. This defines the so-called *Loschmidt number*.

Solution:

Using the ideal gas law (see (1.2)), $P/k_B T = 2.69 \times 10^{25}$ molecule cm^{-3}.

1.1.2 Gases, Aerosols and Water Drops

The atmospheric matter is composed of gases and of condensed matter. The latter comprises liquid (cloud and rain drops) and solid forms (snow and graupel) of water, and the so-called *aerosols* (liquid and solid particles in suspension).

The size of the atmospheric bodies plays a leading role for many processes, such as the interaction with radiation and the wet scavenging by precipitations. The diameter of a gaseous molecule is about 1 Angström (0.1 nm), the diameter of a cloud drop ranges from a few micrometers to about 100 µm, and the diameter of a rain drop is above 0.1 mm (Table 1.3).

In a first approximation, an aerosol can be assumed to be a sphere (see Exercise 5.3 for the case of soots). The diameters range from a few nanometers to tens of micrometers for the mineral particles (formed from dust). The diameter of the "urban" aerosol is about one micrometer.

Table 1.3 Characteristic size of different atmospheric bodies

Atmospheric "body"	Size
Gaseous molecule	0.1 nm
Fine aerosol	1 nm–1 μm
Coarse aerosol	10–20 μm
Cloud drop	10–50 μm
Rain drop	0.1 mm

1.1.3 A Few Species

The atmospheric pollutions can be classified with respect to the trace species to be investigated. We can cite:

- sulfur dioxide (SO_2), in the gaseous phase, related to fossil combustion (fuel and coal);
- gas-phase photochemical compounds: ozone (O_3), nitrogen oxides (NO and NO_2) and volatile organic compounds (VOC);
- heavy metals (lead, cadmium, zinc), related to industrial emissions, in the particulate phase;
- mercury (Hg) in gaseous and particulate phase;
- aerosols (particulate matter), composed of a mixture of sulfate (SO_4^{2-}), ammonium (NH_4^+), nitrate (NO_3^-), organic matter, dust, sea salt and liquid water (when not solid), ...;
- radionuclides, related to natural emissions (radon, Rn), to atmospheric nuclear tests (strontium, Sr), to accidental releases in nuclear power plants (iodine, I, and cesium, Cs) or to processes in the nuclear industry (krypton Kr);
- greenhouse gases, such as carbon dioxide (CO_2), methane (CH_4), nitrogen protoxide N_2O, etc.;
- carbon monoxide (CO);
- persistent organic pollutants (POP), defined as long-lived organic compounds (pesticides, dioxine).

Each pollution is characterized by a specific set of species with given properties: emission sources, chemical and physical properties, characteristic timescales, etc.

1.1.4 Primary and Secondary Species

The emitted species define the so-called *primary* species while the species that are formed in the atmosphere are said to be *secondary* species. For example, in the case of photochemistry, nitrogen oxides are emitted (primary species; especially NO) while ozone is produced by chemical reactions (secondary species).

The emission sources are usually classified into two categories:

Table 1.4 Emissions of NO_x into the troposphere (2000), in Tg(N) yr^{-1} (atomic oxygen is not taken into account in the emitted mass). The teragram is defined as $1\ Tg = 10^{12}$ g. Source: [106]

Source	Annual flux
Fossil combustion	33
Biomass burning	7.1
Aviation	0.7
Soil	5.6
Lightnings	5
Total	51.4

Table 1.5 Emissions of VOC into the troposphere (2000), in Tg(C) yr^{-1}. The emissions related to vegetation are highly uncertain (other estimations may be as much as twice higher). Source: [106]

Source	Annual flux
Fossil combustion	161
Biomass burning	33
Total for biogenic emissions (vegetation)	377
incl. isoprene (C_5H_8)	220
incl. monoterpene ($C_{10}H_{16}$)	127
incl. others	30
Total	571

- *biogenic* emissions, related to natural processes, such as volcanic emissions, Aeolian erosion of dust, sea salt emissions, VOC emissions due to photosynthesis, etc.
- *anthropogenic* emissions, induced by human activities (transports, energy production, industries, agriculture, heating, etc.).

Depending on the primary species, the fraction of anthropogenic sources may be more or less important. For example, the anthropogenic fraction of NO_x emissions is high, especially because of fossil fuel combustion (and traffic-induced emissions). For methane, the anthropogenic part is supposed to be up to $3/4$.

On the other hand, the emissions related to vegetation dominate the VOC emissions, when taken into account as a whole (Table 1.5). Note that the anthropogenic part can be important for given species among the VOC (Table 1.6).

1.2 Atmospheric Vertical Structure

1.2.1 Atmospheric Layers

The vertical distribution of temperature can be used for classifying the different atmospheric layers (Fig. 1.1):

Table 1.6 Standard *speciation* of VOC (in % of the mass fraction) as a function of the emitted sectors. TMB stands for trimethylbenzene. The highest contribution is boxed for a given sector. Source: [142]

Compounds	Cars	Evaporation (fuel)	Solvents	Heating (gas)
ethylene	16.4			
toluene	16	2.8	53.3	8.5
meta-para-xylene	9.4	1.1	12.2	
isopentane	8.1	26.3		
benzene	6.9	1.5	0.1	15.3
propene	6.8			
acetylene	6.5			
1,2,4-TMB	4.1		4.8	
ortho-xylene	3.7		8.1	
ethylbenzene	3.2		4.2	
1,3-butadiene	2.5			
ethane	1.9			
n-pentane	1.8	8.6		23.7
n-butane	1.8	35.8		33.9
1-butene	1.6			
1,3,5-TMB	1.2		1.9	
1,2,3-TMB	1.			
n-heptane	0.9		1.6	
trans-2-pentene	0.8			
trans-2-butene	0.8	1.3		
isobutane	0.7	12.3		
isooctane	0.6			
1-pentene	0.6			
cis-2-butene	0.6			
n-hexane	0.6	1.8	12.6	3.4
n-octane	0.5		1.3	
cis-2-pentene	0.5			
isoprene	0.4			
propane	0.1	3.4		15.3
1-hexene	0.1			

- the *troposphere* for heights below 8 km above the polar regions, and below 18 km above the Equator.
  The temperature is a decreasing function of altitude, down to 220 K above the polar regions and to 190 K above the Equator. The averaged temperature gradient is about $-6.5\,\mathrm{K\,km^{-1}}$.
- then the *stratosphere* for heights up to 50 km.

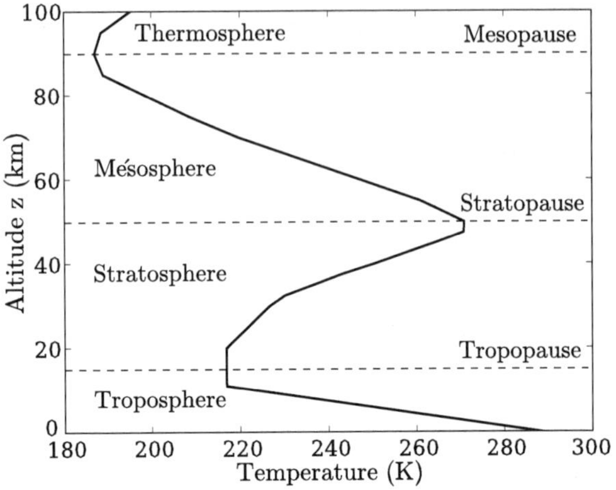

Fig. 1.1 Vertical distribution of temperature (standard atmosphere USA 1976)

The temperature is first constant and then an increasing function of altitude, up to about 270 K. This heating is directly related to the absorption of the ultraviolet solar radiation (UV) by ozone (O_3) and by molecular oxygen (O_2) (Exercise 2.3 and Exercise 4.5). This inversion layer is a specific property of the Earth's atmosphere.

- then the *mesosphere* up to 85–90 km.
  The temperature is a decreasing function of altitude down to 170 K (the coldest atmospheric temperature), due to the rarefaction of ozone and of oxygen.
- and then the *thermosphere* and the *ionosphere* (up to about 150 km).
  The temperature increases and is more and more dependent on solar activity. The UV radiations dissociate N_2 and O_2 and gas-phase molecules are ionized (Chap. 2, Sect. 2.2.1.2). Air becomes a rarefied gas: the air density is about 10^{19} molecule m^{-3} at 100 km, to be compared with 10^{25} molecule m^{-3} at sea level (Exercise 1.3).
- Beyond, the Earth's attraction can be neglected. In the *exosphere* (at about 500 km), atomic hydrogen can escape from the atmosphere (Remark 1.2.1 and Exercise 1.7).

There are two *inversion* layers in the atmosphere, characterized by a positive gradient of temperature: in the stratosphere and in the ionosphere. Part of the solar radiation is absorbed by a few gas-phase species in these layers, playing a filtering role, which results in an increasing temperature. The vertical distribution of these gas-phase species determines the vertical distribution of temperature.

Similarly, the atmospheric layers can be classified with respect to other properties.

The atmospheric dynamics can also differ from one layer to another. In inversion layers, the atmosphere is stable: warm air parcels are subject to uprising motions while cold air parcels can be blocked by the warm layers above them (Chap. 3).

Above the mesosphere, the gravitational effects play a leading role for the vertical distribution of species: the "light" species are at higher altitudes than the "heavy"

Fig. 1.2 Balance between weight and gradient-pressure force

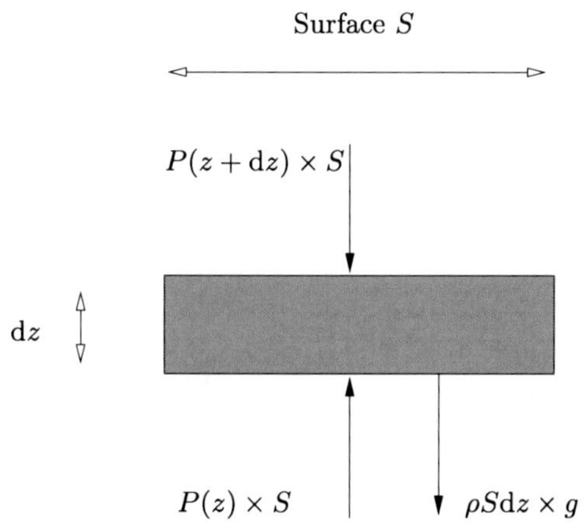

species. This defines the so-called *heterosphere*; the atmosphere below is sometimes referred to as the *homosphere* (Sect. 1.2.3).

1.2.2 Atmospheric Pressure

The atmospheric pressure P is defined as the force exerted by the atmosphere on a unit area of surface. Let $P(z)$ be the pressure at altitude z. For a volume of vertical thickness dz, of horizontal area S, at altitude z, the balance between the exerted forces (Fig. 1.2) is

$$[P(z+\mathrm{d}z)-P(z)]S+\rho S\,\mathrm{d}z\,g=0 \tag{1.8}$$

with g the acceleration of gravity and ρ the air density. Remember that g can be computed at altitude z by

$$g=\left(\frac{R_t}{R_t+z}\right)^2 g_0 \tag{1.9}$$

with $R_t = 6400$ km the radius of the Earth and $g_0 = 9.81\ \mathrm{m\,s^{-2}}$ the acceleration of gravity at ground.

Rearranging (1.8) with d$z \to 0$ yields the so-called *hydrostatic equation*,

$$\frac{\mathrm{d}P}{\mathrm{d}z}=-\rho g. \tag{1.10}$$

The ideal gas law is $P=\rho r_{air}T$. In a first approximation, the temperature and the air molecular weight can be supposed to have constant values. Upon integration, we obtain

$$P(z)=P(0)\exp\left(-\frac{z}{H}\right)\quad \text{with } H=\frac{r_{air}T}{g}, \tag{1.11}$$

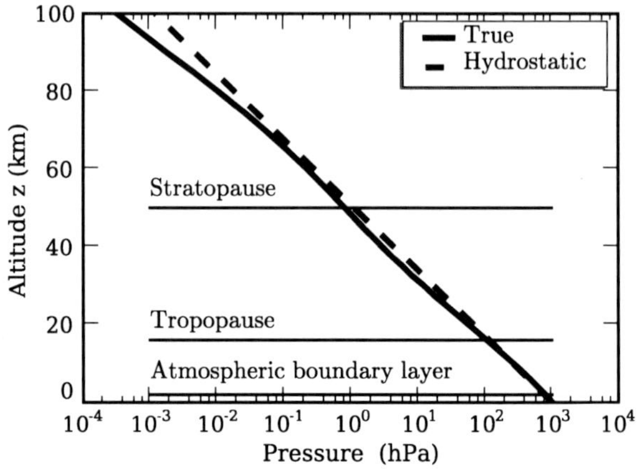

Fig. 1.3 Comparison of the vertical pressure profile between an hydrostatic atmosphere (with a scale height $H = 7.3$ km) and a real atmosphere (standard atmosphere USA 1976)

where H is referred to as the *scale height*. With a mean atmospheric temperature of 250 K, we calculate $H \simeq 7.3$ km. The comparison between the hydrostatic profile and the "real" profile is shown in Fig. 1.3.

Actually, the temperature, the air molecular weight and the acceleration of gravity depend on altitude. The exact calculation yields

$$P(z) = P(0)\exp\left(-\int_0^z \frac{dz'}{H(z')}\right) \quad \text{with } H(z) = \frac{RT(z)}{M_{air}(z)g(z)}. \tag{1.12}$$

For a mean atmospheric pressure at the Earth's surface of 1000 hPa, we can calculate the pressure:

- at the top of the atmospheric boundary layer (Chap. 3): $P(2\,\text{km}) \simeq 760$ hPa;
- at the tropopause: $P(16\,\text{km}) \simeq 110$ hPa;
- at the stratopause: $P(50\,\text{km}) \simeq 1$ hPa.

Pressure is a strongly decreasing function of altitude. The same holds for the vertical distribution of the atmospheric mass (Exercise 1.4): up to 90% of total atmospheric mass is in the troposphere, 75% in the atmospheric boundary layer (below 2 km).

Exercise 1.4 (Atmospheric Mass) Calculate the total atmospheric mass. What are the tropospheric and stratospheric masses? Results are to be compared with the Earth's mass (about 10^{25} kg) and to the mass of the oceans (about 10^{21} kg).
Data: $P(0) = 984$ hPa (mean pressure at the surface of the Earth) and $R_t = 6400$ km (radius of the Earth).
Solution:
At ground, the force exerted by the atmosphere is $m_{atm}g = 4\pi {R_t}^2 \times P(0)$ with m_{atm} the atmospheric mass. We deduce $m_{atm} \simeq 5 \times 10^{18}$ kg. With obvious notations, the force exerted on the troposphere by the remaining part of the atmosphere is $(m_{atm} - m_{tropo})g = 4\pi {R_t}^2 \times P(16)$, thus $m_{tropo}/m_{atm} = 1 - P(16)/P(0) \simeq 89\%$. Similarly, $m_{strato}/m_{atm} = [P(16) - P(50)]/P(0) \simeq 11\%$. The mass of the atmosphere above the stratopause is only 0.1% of the total atmospheric mass.

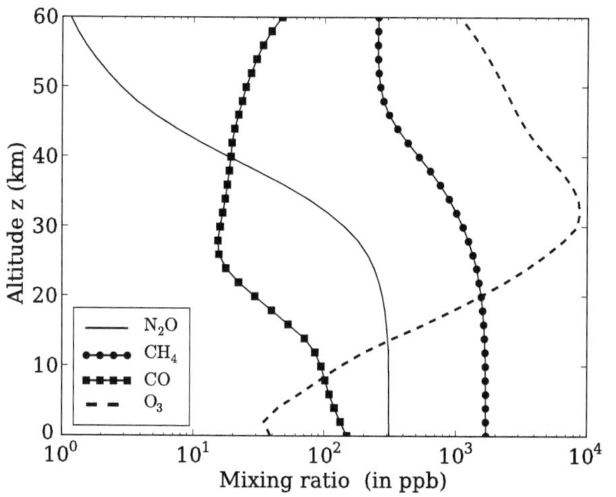

Fig. 1.4 Typical vertical distribution for a few species (mixing ratio expressed in ppb), at latitude 30° North, in March. Source: [18]

Exercise 1.5 (Emitted Mass of CO_2 since Preindustrial Times) Calculate the emitted mass of carbon since preindustrial times (evolution of C_{CO_2} from 278 ppmv to 365 ppmv, Table 1.2). Assume that the molecular weight of air and the atmospheric mass have not been modified.
Data: $M_C = 12\,\text{g}\,\text{mol}^{-1}$.
Solution:
The total carbon mass associated to CO_2 is $m_C = n^{CO_2} M_C$ where n^{CO_2} is the total number of CO_2 moles. By definition, $n^{CO_2} = C_{CO_2} \times n^{atm}$ with $n^{atm} = m_{atm}/M_{air}$ the total number of atmospheric moles. Thus,

$$\Delta m_C = (M_C/M_{air}) m_{atm} \Delta C_{CO_2} \simeq 1.8 \times 10^{14}\,\text{kg}.$$

1.2.3 Vertical Distribution of Species

The typical vertical distribution (in the first 60 kilometers above the Earth's surface) is shown in Fig. 1.4 for a few species (at latitude 30° North, in March). For example, most of the ozone is located in the stratosphere (Exercise 1.6 for the integrated ozone column).

Exercise 1.6 (Ozone Column) The vertical profile shown in Fig. 1.4 is associated to an integrated ozone column of about $\bar{n}_{O_3} = 7.5 \times 10^{22}$ molecule m^{-2}. Calculate the ozone column thickness if it were brought to the Earth's surface under standard thermodynamic conditions. This defines the so-called *Dobson unit* (DU), with 1 DU = 0.01 mm for this virtual column.
Solution:
For a unit area of surface, the thickness Δz satisfies $P\Delta z = \bar{n}_{O_3} k_B T$. Thus $\Delta z \simeq 2.8$ mm or 280 DU.

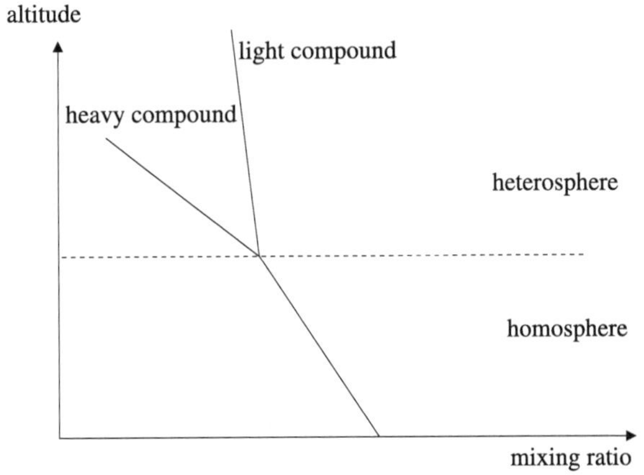

Fig. 1.5 Homosphere versus heterosphere

In the first part of the atmosphere (in the *homosphere*), turbulent and molecular mixing plays a leading role. In the *heterosphere*, the species are segregated according to their own scale height, directly related to their molecular weight. For a species X_i, the scale height is $H_i = RT/(M_i g)$ with M_i the molecular weight. The lightest compounds (with large values of the scale height) are therefore above the heaviest compounds, as expected (Fig. 1.5).

The altitude of the heterosphere bottom can be estimated by comparing the eddy diffusivity, K_z (typically $10\,\mathrm{m^2\,s^{-1}}$; Chap. 6) with the molecular diffusion coefficient ν. From the kinetic theory of gases, for a given trace species,

$$\nu \propto \frac{\sqrt{T}}{n}, \tag{1.13}$$

with n the molecular density. The multiplying factor depends only on the species. With the ideal gas law, this yields $\nu \propto T^{1.5}/P$. The effects related to molecular diffusion become more and more important with an increasing altitude since the variations of temperature can be neglected with respect to those of pressure (a decreasing exponential function of altitude). Thus, $\nu(z) \simeq [P(z)/P(0)]\nu_{air}$, where $\nu_{air} = 1.6 \times 10^{-5}\,\mathrm{m^2\,s^{-1}}$ at the Earth's surface. The altitude at which molecular diffusion effects have to be taken into account is obtained by using the hydrostatic profile for pressure (with a scale height H). The resulting altitude is $H \ln(K_z/\nu_{air}) \simeq 100\,\mathrm{km}$.

Remark 1.2.1 (Thermal Escape) One can wonder if a few trace species can escape from the Earth's atmosphere. This process is referred to as *thermal escape* or *Jeans' escape*. It takes place in the upper region of the atmosphere, the so-called *exosphere*.

In the Earth's atmosphere, the only species affected by thermal escape is atomic hydrogen (Exercise 1.7). The loss flux is estimated to be about one hundred tons per day. This corresponds to the hydrogen contained in 1800 tons of water. In comparison to the total mass of water for the Earth (1.5×10^{18} tons), this shows that the escape flux for the Earth/atmosphere system can be neglected, even for long-term studies.

Exercise 1.7 (Thermal Escape)

1. The escape velocity, u_{esc}, is defined as the velocity required to escape from the Earth's gravitational attraction. Give an expression of the escape velocity. Hint: use the conservation of total energy for a body of mass m, of velocity v, at a distance r from the Earth's center. The total energy is $mv^2/2 - mM_tG/r$ with G the universal gravity constant and M_t the Earth's mass.
2. The escape takes place at the bottom of the exosphere, at the so-called *exobase*. The exobase is defined as the altitude at which the air mean free path (distance between two collisions) is equal to the scale height. The exobase altitude ranges from 400 to 500 km. Calculate u_{esc}.
3. The flux of atoms or molecules subject to escape is specific to the species. We label by i the escaping atom or molecule. We assume that the velocity distribution is given by the Maxwell distribution, namely in spherical coordinates for the three-dimensional velocity (v, θ, ϕ) by

$$P_i(v,\theta,\phi)\,\mathrm{d}\theta\,\mathrm{d}\phi\,\mathrm{d}v = \left(\frac{1}{\pi U_i^2}\right)^{3/2} \exp\left(-\frac{v^2}{U_i^2}\right) v^2 \sin\theta\,\mathrm{d}\theta\,\mathrm{d}\phi\,\mathrm{d}v$$

where $U_i = \sqrt{2RT/M_i}$ is the most probable velocity and M_i is the molar mass.

Calculate F_i, the average flux of escaping atoms or molecules per unit area of surface and per unit of time. We denote by n_i the atom or molecule density. The vertical component of the velocity is $v\cos\theta$. Prove Jeans' formula (1925, [69]),

$$F_i = \frac{n_i U_i}{2\sqrt{\pi}}(1 + Y_i)\exp(-Y_i),$$

with Y_i, the *escape parameter*, given by $r/H_i = (u_{esc}/U_i)^2$.
4. Give the values of Y_i for which the thermal escape is significant. Calculate Y_H and Y_{H_2} for an exobase temperature $T \in [800, 1600]$K. Conclude.

Data: $G = 6.67 \times 10^{-11}\,\mathrm{m^3\,kg^{-1}\,s^{-2}}$, $M_t = 6 \times 10^{24}$ kg, $M_H = 1\,\mathrm{g\,mol^{-1}}$ and $M_{H_2} = 2\,\mathrm{g\,mol^{-1}}$.

Solution:

1. Far from the Earth ($r \to \infty$), the velocity is null. The escape takes place at the distance r defined by

$$u_{esc} = \sqrt{\frac{2M_tG}{r}} = \sqrt{2g(r)r},$$

with $g(r) = GM_t/r^2$ the acceleration of gravity at this distance.
2. This yields $u_{esc} \simeq 11\,\mathrm{km\,s^{-1}}$. Note that this value does not depend on the species. Hint: do not forget to add the Earth's radius to the altitude!
3. The average flux per unit area of surface and per unit of time is obtained by multiplying the vertical component $v\cos\theta$ by the density n_i at the given altitude. Upon integration over the "appropriate" part of the distribution (e.g. $v \geq u_{esc}$), we obtain

$$F_i = n_i \int_0^{\pi/2}\int_0^{2\pi}\int_{u_{esc}}^{\infty} P_i(v,\theta,\phi)\,v\cos\theta\,\mathrm{d}\theta\,\mathrm{d}\phi\,\mathrm{d}v,$$

by using $\int_0^{\pi/2} \sin\theta \cos\theta \, d\theta = [\sin^2\theta]_0^{\pi/2}/2 = 1/2$ and $\int_0^{2\pi} d\phi = 2\pi$. With the new variable $y = v/U_i$, we integrate by parts,

$$\int_{\sqrt{Y_i}}^{\infty} y^3 \exp(-y^2) dy = \left[-\frac{1}{2}\exp(-y^2)y^2\right]_{\sqrt{Y_i}}^{\infty} + \int_{\sqrt{Y_i}}^{\infty} y \exp(-y^2) dy$$
$$= \left[-\frac{1}{2}\exp(-y^2)y^2\right]_{\sqrt{Y_i}}^{\infty} + \left[-\frac{1}{2}\exp(-y^2)\right]_{\sqrt{Y_i}}^{\infty}$$
$$= \frac{1+Y_i}{2}\exp(-Y_i).$$

It is easy to conclude for the multiplying factor.

4. The function $(1+Y)\exp(-Y)$ is a decreasing function of Y. Its value is 5×10^{-2} for $Y = 5$, 5×10^{-4} for $Y = 10$ and 5×10^{-6} for $Y = 15$. Thus, the thermal escape can be neglected for $Y \leq 10$ or 15. The only species affected by thermal escape are the lightest bodies (atomic and molecular hydrogen, helium). With the range of variation of T, we calculate $Y_H \in [4.5, 9]$ and $Y_{H_2} \in [9, 18]$. The only body to take into account in the Earth's atmosphere is therefore atomic hydrogen.

To know more ([132]):

F. Selsis, *Évaporation plantaire*. In *Formation plantaire et exoplantes*, École CNRS de Goutelas XXVIII, 2005. Édité par J.L. Halbwachs, D. Egret et J.M. Hameury

1.3 Timescales

The comparison between the atmospheric residence time and the characteristic timescales for atmospheric transport determines the impact scale of a given species (primary or secondary).

If the magnitude of the residence time is about that of the transport at the continental scale (e.g. over Europe), then the impact is at least continental. If the magnitude of the residence time is about that of the exchange time from troposphere to stratosphere, then the species may reach the stratosphere, etc.

1.3.1 Timescales of Atmospheric Transport

The timescales of atmospheric transport can be derived from a dimensional analysis. Another powerful approach is the study of *atmospheric tracers*, defined as trace species involved only in *linear* processes (e.g. radionuclides, such as krypton, radon, strontium, etc.). For example, this makes it possible to estimate the characteristic timescales of the exchanges between the "atmospheric reservoirs" (troposphere, stratosphere, hemispheres). A few examples are given in Problem 1.1 and 1.3.

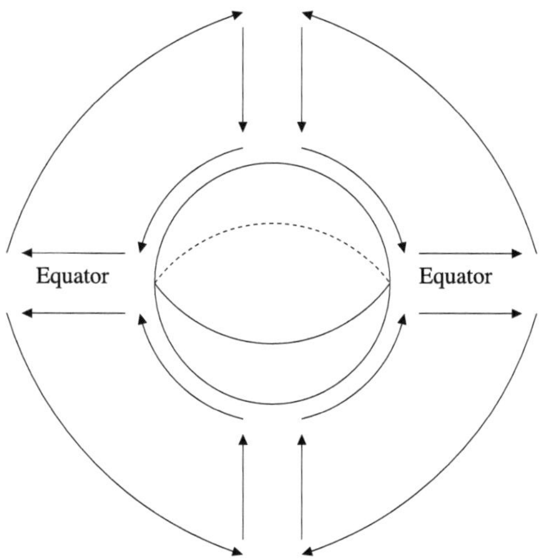

Fig. 1.6 Simplified model for the atmospheric general circulation (Hadley's circulation, XVIIIth century): updrafts of warm air in the Equatorial region and subsidence in the polar regions, which defines the so-called *Hadley's cells*. Actually, other *cells* can be defined in each hemisphere: *Ferrel's cell* between the Equatorial zone and the 30° latitude, *polar cell* between the 30° latitude and the 60° latitude

1.3.1.1 Horizontal Transport

The horizontal transport is driven by the wind fields. A characteristic wind velocity for the *zonal transport* (West/East) is $U \simeq 10\,\mathrm{m\,s^{-1}}$. For the *meridional* wind velocity (South/North), the value is lower, let us say $U \simeq 1 - 2\,\mathrm{m\,s^{-1}}$.

It is then straightforward to estimate the characteristic timescales for the transport of a trace species emitted at mid-latitudes (e.g. in Europe or in North-America). Let L be the characteristic spatial scale. As the corresponding timescale is $\tau = L/U$, we obtain:

- a few days for continental transport;
- from one to two weeks for transcontinental transport (transatlantic, transpacific);
- from one to two months for hemispheric mixing (in the Northern Hemisphere or in the Southern Hemisphere);
- from one to two months for transport to the Equatorial region or to the polar regions.

A key point is that the interhemispheric mixing is not favored by the general atmospheric circulation. In the equatorial region, the *Inter-Tropical Convergence Zone* (ITCZ) is characterized by strong updrafts of warm and wet air masses (Fig. 1.6). As a result, the timescale for the interhemispheric exchange is about 1 year (Problem 1.1). The interhemispheric mixing mainly occurs through the seasonal meridional motions of the ITCZ or through local breaks of the ITCZ (e.g. due to monsoon).

1.3.1.2 Vertical Transport

The vertical component of the wind fields is much lower than the horizontal components. Moreover, the molecular diffusion can be neglected, except in a thin layer

Table 1.7 Timescales for atmospheric transport

Transport	Characteristic timescale
Continental	1 week
Transcontinental	2 weeks
Hemispheric	1 month
Interhemispheric	1 year
Atmospheric boundary layer	1 hour–1 day
Free troposphere ($\simeq 5000\,\mathrm{m}$)	1 week
Troposphere	1 month
Exchange from troposphere to stratosphere	from 5 to 10 years
Exchange from stratosphere to troposphere	from 1 to 2 years

just above the Earth's surface (Chap. 3). The dynamic viscosity, ν_{air}, is about $10^{-5}\,\mathrm{m}^2\,\mathrm{s}^{-1}$.

A much more efficient mechanism is related to the *buoyancy effects*: a warm air parcel is "lighter" and can be implied in updrafts. The vertical distribution of temperature governs the vertical motion (at least in the lower part of the atmosphere), deeply related to the atmospheric *turbulence* (see Chap. 3). The strength of the vertical motion is given by a vertical turbulent diffusion coefficient, $K_z \simeq 10\text{–}20\,\mathrm{m}^2\,\mathrm{s}^{-1}$.

For a spatial scale L (now in the vertical direction), the characteristic timescale is $\tau = L^2/K_z$. Thus:

- the timescale for transport in the atmospheric boundary layer ranges from a few hours to one day: it strongly depends on the *atmospheric stability*, which defines the magnitude of K_z (Chap. 3);
- the timescale for tropospheric mixing (up to the tropopause) is about one month.

At the tropopause, the transport from the troposphere to the stratosphere is not favored by inversion of the vertical temperature profile (Fig. 1.1). The characteristic timescale for this exchange is estimated to range from 5 to 10 years. The timescale for the reverse exchange (from the stratosphere to the troposphere) ranges from 1 to 2 years (Problem 1.3).

The timescales for the atmospheric transport are summarized in Table 1.7.

1.3.2 Atmospheric Residence Time for a Trace Species

The atmospheric residence time is fixed by the *loss* processes that affect the trace species:

- chemical or photochemical reactions;
- radioactive decay (for radionuclides);
- dry deposition at the Earth's surface;

Table 1.8 Atmospheric residence time of a few CFCs. Source: [106]

Species	Symbol	Residence time (in years)
CFC 11	CCl_3F	45
CFC 113	CCl_2FCClF_2	85
CFC 12	CCl_2F_2	100
CFC 114	$CClF_2CClF_2$	300
CFC 13	$CClF_3$	640
CFC 115	CF_3CClF_2	1700

- wet scavenging for soluble gas-phase species and aerosols (by clouds or precipitations);
- gravitational settling (sedimentation) for the coarse fraction of aerosol distributions (Exercise 5.1 and Problem 5.3).

Note that the loss flux for the Earth/atmosphere system to space can be neglected (see Remark 1.2.1). As a result, we suppose that the Earth/atmosphere system is closed.

The above processes can be parameterized by first-order kinetic laws. This is valid for linear processes and this is a linearized approximation for chemical kinetics. The time evolution of the mass concentration of species X, ρ_X, is governed by

$$\frac{d\rho_X}{dt} = -k\,\rho_X \tag{1.14}$$

where t stands for time and k is similar to a first-order kinetic rate (Chap. 4). This defines the characteristic timescale $\tau = 1/k$.

Let us consider an ensemble of processes, parameterized by kinetic rates k_i (equivalently by timescales $\tau_i = 1/k_i$). The time evolution is then given by

$$\frac{d\rho_X}{dt} = -\left(\sum_i k_i\right)\rho_X. \tag{1.15}$$

The global characteristic timescale is then defined as

$$\frac{1}{\tau} = \sum_i \frac{1}{\tau_i}. \tag{1.16}$$

The magnitude of the atmospheric residence time for a few species is shown in Fig. 1.7. The residence time is fixed by the leading processes for each trace species.

- The atmospheric residence time of most of the chemical species is governed by the *atmospheric oxidizing power*, mainly related to the dydroxyl radical OH. We refer to Chap. 4 (devoted to atmospheric chemistry) for a deeper understanding (Problem 4.3). In a first approximation, $k = k_{OH}[OH]$, with k_{OH} the oxidation rate associated to OH and [OH] the OH concentration (in the appropriate units).

- The characteristic timescale associated to dry deposition can be computed from the dry deposition velocity (Exercise 1.8).
- For radioelements, radioactive decay is characterized by the half-lifetime $\tau_{1/2}$, corresponding to the time at which the initial concentration is divided by two. It is easy to derive a relation with the residence time (as defined above): as $\rho_X(t) = \exp(-t/\tau)\rho_X(0)$, we obtain $\tau_{1/2} = \tau \times \ln 2$.

Exercise 1.8 (Characteristic Timescale for Dry Deposition) This exercise aims at defining a characteristic timescale for dry deposition.

1. Dry deposition is characterized by the so-called *dry deposition velocity*, v_{dep} (usually about $1\,\mathrm{cm\,s^{-1}}$). Define a characteristic timescale for the deposition in an atmospheric layer of height H.
2. The vertical diffusion of a passive tracer in the column $[0, H]$ is given by (Chap. 3)

$$\frac{\partial \rho_X}{\partial t} + \frac{\partial}{\partial z}\left(K_z(z)\frac{\partial \rho_X}{\partial z}\right) = 0$$

with $K_z(z)$ the vertical turbulent diffusion coefficient.

At ground ($z = 0$), the boundary condition reads $-K_z(z)\partial\rho_X/\partial z = E - v_{dep}\rho_X$ with E an emission flux. At the top of the column ($z = H$), $-K_z(z)\partial\rho_X/\partial z = 0$.

Derive the time evolution of the column-averaged value ρ_X, supposing that the column is well mixed. Conclude for the characteristic timescale of dry deposition.

Solution:

1. With a dimensional analysis, $\tau = H/v_{dep}$.
2. We define $\bar{\rho}_X = 1/H \times \int_0^H \rho_X(z)\,\mathrm{d}z$. Upon integration by parts, $\mathrm{d}\bar{\rho}_X/\mathrm{d}t = (E - v_{dep}\rho_X(z=0))/H$. If the column is supposed to be well mixed, $\rho_X(z=0) = \bar{\rho}_X$, yielding the characteristic timescale $\tau = H/v_{dep}$.

Consider a species X emitted with a mass rate E_X. The global mass budget for X is then

$$\frac{\mathrm{d}m_X}{\mathrm{d}t} = E_X - \frac{m_X}{\tau_X} \tag{1.17}$$

with m_X the total species mass. At equilibrium, $m_X = \tau_X E_X$ (see Exercise 1.9 for methane).

Exercise 1.9 (Atmospheric Residence Time of Methane) The global annual emission of methane ([106]) is estimated to be 598 Tg (including 70% related to anthropogenic emissions). Calculate the atmospheric residence time of methane, assuming equilibrium.

Solution:

As $C_{CH_4} = 1.7$ ppmv, the total mass is $1.7 \times 10^{-6} \times (M_{CH_4}/M_{air})m_{atm}$, namely 4700 Tg

(m_{atm} is the mass of the atmosphere, Exercise 1.4). Thus, the residence time is 4700/535 $\simeq$ 8 years.

It is now easy to classify the impacts of the main atmospheric trace species:

- CFCs are highly stable and have a lifetime up to tens of years (Table 1.8 and Problem 1.5): the impact is then global and statospheric;
- mercury has a lifetime ranging from 1 to 2 years: the impact is global and tropospheric (Problem 1.2 with the different forms of mercury);
- tropospheric ozone has a lifetime of a few days: the appropriate scale for the study of photochemistry is then at least the continental scale, with a possible role of transcontinental transport (Chap. 4, Sect. 4.3.7).
- the chemical timescale of SO_2 is about 2 days (oxidation to sulfuric acid H_2SO_4): the direct impact of SO_2 emissions is then relatively local. The lifetime of H_2SO_4 is governed by wet scavenging and is about 5 days: the impact scale of acid rains is then the continental scale (Exercise 5.10);
- tropospheric aerosols have a residence time of about 10 days (Problem 1.4);
- persistent organic pollutants (POP) have a high residence time in the atmosphere (at least a few months). They can accumulate in different ecosystems through deposition. Moreover, they are subject to reemissions. One key concern is related to the high concentrations measured in the arctic region (due to the so-called *cold condensation* process, [52, 151]).

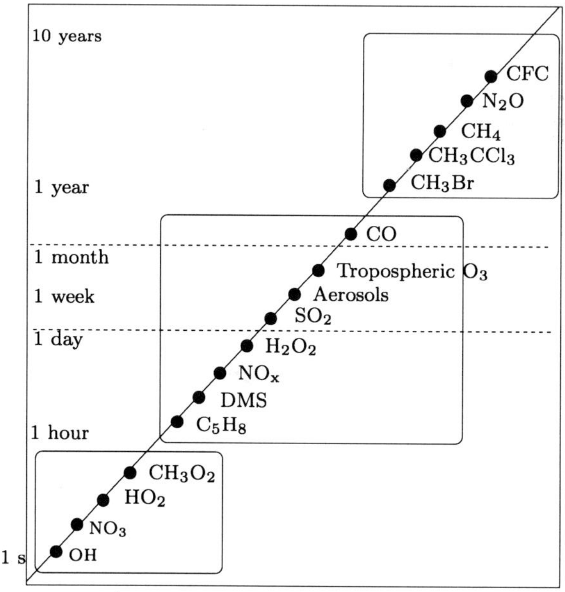

Fig. 1.7 Atmospheric residence time of the main atmospheric species. Source: [18]

Problems Related to Chap. 1

Problem 1.1 (Interhemispheric Exchange) Krypton (^{85}Kr) is a noble radioactive gas that can be emitted during a few industrial processes in the nuclear power plants. Due to the location of the plants, it is mostly emitted in the Northern Hemisphere. Its lifetime is $\tau = 15.5$ years. In 1983, the emission rate was estimated to be $E = 15\,\mathrm{kg\,yr^{-1}}$. On the basis of observational data, the mass was likely 93 kg in the Northern Hemisphere and 86 kg in the Southern Hemisphere, respectively.

Calculate the interhemispheric exchange timescale, assuming equilibrium.

Solution:

Let m_n and m_s be the mass in the Northern and Southern Hemispheres, respectively. Let τ_h be the interhemispheric exchange timescale. The time evolution is

$$\begin{cases} \frac{dm_n}{dt} = -\frac{m_n - m_s}{\tau_h} - \frac{m_n}{\tau} + E \\ \frac{dm_s}{dt} = \frac{m_n - m_s}{\tau_h} - \frac{m_s}{\tau}. \end{cases}$$

Thus,

$$\frac{\mathrm{d}}{\mathrm{d}t}(m_n - m_s) = -\left(\frac{2}{\tau_h} + \frac{1}{\tau}\right)(m_n - m_s) + E.$$

At equilibrium, $\tau_h = 2(E/(m_n - m_s) - 1/\tau)^{-1} \simeq 1$ an.

Other estimations provide higher values (of a few months). Note that other atmospheric tracers can be used, such as SF_6 (sulfur hexafluoride), with a lifetime of 800 years, mainly emitted by electricity production. Moreover, finer approaches (with general circulation models) can estimate the seasonal variation of the interhemispheric exchange (Fig. 1.8). The exchanges are much faster from November to April, when the ITCZ moves toward North, which favors the injection of air masses from the Northern Hemisphere into the Southern Hemisphere.

To know more ([60, 86]):

D. JACOB, M. PRATHER, S. WOFSY, AND M. MCELROY, *Atmospheric distribution of ^{85}Kr*

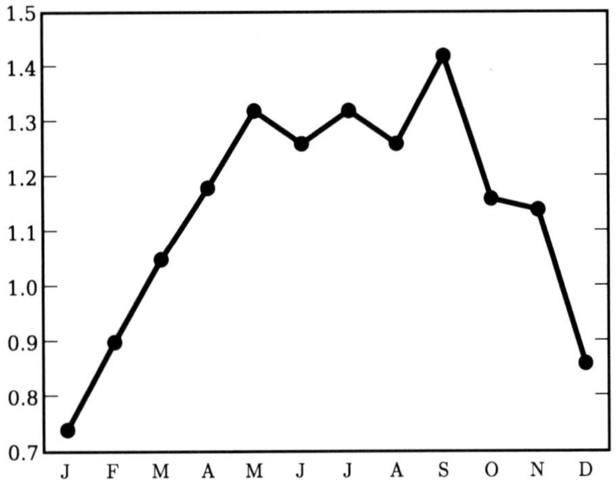

Fig. 1.8 Evolution of the timescale of interhemispheric exchange (in yr), as a function of the month. Estimation derived from the study of krypton (for the year 1987). Source: [86]

simulated with a general circulation model, J. Geophys. Res., **22** (1987), pp. 6614–6626
I. LEVIN AND V. HESSHAIMER, *Refining of atmospheric transport model entries by the globally observed passive tracer distributions of* ^{85}*krypton and sulfur hexafluoride* (SF_6), J. Geophys. Res., **101** (1996), pp. 16745–16755

Problem 1.2 (Atmospheric Dispersion of Mercury) Mercury is a toxic pollutant that impacts environmental media such as lakes. More than one third of its emissions is related to industrial sources. The main data for mercury (emissions and atmospheric burden) are shown in Table 1.9.

1. Estimate the atmospheric residence time and the impact scale of mercury.
2. Elemental mercury, Hg(0), is in the gaseous phase. It can be deposited and is oxidized by the gas-phase oxidants. What are the processes that govern the fate of elemental mercury? Assume that it is not a soluble gas (in a first approximation). Derive the evolution equation for the concentration of elemental mercury. Calculate the global residence time. Comment on the assumption relative to the solubility.
   Hint: use Exercise 1.8.
   Data (take care about the units!):
   - Dry deposition velocity for elemental mercury: $v_{dep} \simeq 0.01\,\mathrm{cm\,s^{-1}}$.
   - Average concentration for tropospheric OH: 10^6 molecule cm^{-3}.
   - Tropospheric mixing ratio of O_3: 40 ppbv.
   - Kinetic rates for the oxidation by OH and O_3: $k_{OH} \simeq 8 \times 10^{-14}$ and $k_{O_3} \simeq 2.8 \times 10^{-20}$ (in cm^3 molecule^{-1} s^{-1}).
   - Take the height of the atmospheric boundary layer as $H \simeq 1$ km.
3. High mercury concentrations can be measured in ecosystems in the vicinity of emissions. Comment.
4. Actually, mercury can also be emitted in an oxidized form, Hg(II) (sometimes referred to as *particulate mercury*). Its dry deposition velocity is then 0.5 cm s^{-1}. Moreover, particulate mercury can be scavenged by precipitations. Calculate the residence time for particulate mercury and conclude.

Note: I thank Christian Seigneur for having provided me with most of the data required for this problem.

Table 1.9 Atmospheric budget of mercury, as estimated by three authors. 1 Mg = 10^6 g. Source: [131]

	[96]	[129]	[131]
Total emissions (Mg an^{-1})	7000	6107	7000
anthropogenic	4000	2104	2200
Atmospheric burden (Mg)	5000	6900	5360

Solution:

1. Let E be the annual mass emissions and m the atmospheric burden of mercury, respectively. Let τ be the atmospheric residence time. The time evolution of m is governed by
$$\frac{dm}{dt} = E - \frac{m}{\tau}.$$
At equilibrium, $\tau = m/E$. Using the data, τ ranges from 0.71 to be 1.13 yr. The impact scale is therefore at least hemispheric. The upper value can indicate that interhemispheric exchange may occur.
2. The loss processes are dry deposition, wet scavenging, and oxidation by the main atmospheric oxidants, namely OH and O_3. With obvious notations, the equation becomes
$$\frac{dm}{dt} = E - (\lambda_{dep} + \lambda_{less} + \lambda_{OH} + \lambda_{O_3})m,$$
yielding the residence time $\tau = 1/(\lambda_{dep} + \lambda_{less} + \lambda_{OH} + \lambda_{O_3})$.
The timescale related to dry deposition is $1/\lambda_{dep} = H/v_{dep} \simeq 115$ jours.
For an oxidation reaction with an oxidant Ox, we obtain $\lambda_{Ox} = k_{Ox} \times C_{Ox}$ with C_{Ox} expressed in molecule cm^{-3}. The characteristic timescale is about 145 days for the oxidation by OH and 400 days for the oxidation by O_3.
Finally, the residence timescale is $\tau \simeq 2$ months, which is coherent with the value obtained from the mass budget.
If mercury were soluble, the residence time would be much lower, due to wet scavenging (about 10 days).
3. The observational data indicate that there is either deposition or scavenging in the vicinity of emissions. This cannot be related to elemental mercury since its timescale for dry deposition is 115 days and it is not a soluble species. This is therefore related to another form of mercury, namely the oxidized form.
4. For particulate mercury, the timescale of dry deposition is about 2 days. Moreover, particulate mercury is soluble and, as a result, can be scavenged. This explains the observational data.

A key uncertainty for the modeling of mercury is the partitioning of emissions between the elemental and particulate forms, which strongly affects the impact scale.
To know more ([131]):
N. SELIN, D. JACOB, R. PARK, R. YANTOSCA, S. STRODE, L. JAEGLE, AND D. JAFFE, *Chemical cycling and deposition of atmospheric mercury: global constraints from observations*, J. Geophys. Res., 112 (2007)

Problem 1.3 (Troposphere/Stratosphere Exchanges) This problem aims at estimating the timescales for the transfer between the troposphere and the stratosphere. We focus on strontium (^{90}Sr), a radionuclide that was emitted by nuclear tests in the 1950s and 1960s.

1. Assuming the equilibrium of the tropospheric mass (supposed to have transfer fluxes only with the stratosphere), prove that there is an asymmetry in the exchange timescales: the timescale for the transfer from troposphere to stratosphere is lower than the timescale for the reverse transfer.
Hint: use Exercise 1.4.

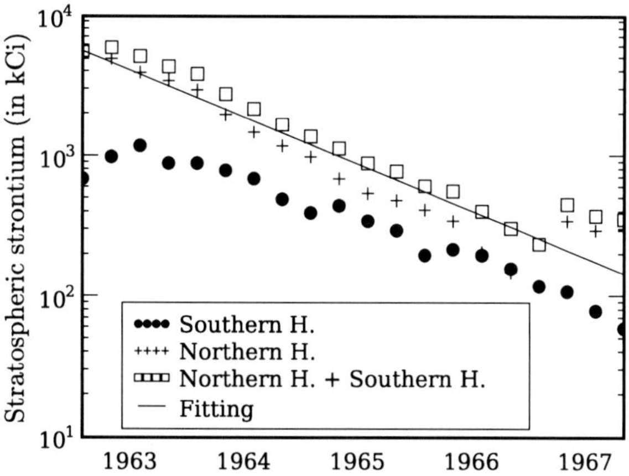

Fig. 1.9 Evolution of the strontium concentration in the stratosphere (Northern and Southern Hemispheres), measured in kilocivert ($1\ \mathrm{Ci} = 3.7 \times 10^{10}$ Bq for strontium). The "fitting line" defines a timescale of 1.3 yr. Source: [152]

2. The half-life time of strontium is 28.5 years. Strontium can also be scavenged in the troposphere, when combined with other compounds (to form salts). The observational data in the stratosphere, during the period 1963–1967, have shown that the radioactivity of ^{90}Sr (β radiation, expressed in Becquerel, Bq) had an exponential decay with a characteristic timescale of 1.30 yr. Moreover, the difference between both hemispheres (Northern and Southern Hemispheres) had a characteristic timescale of 0.76 yr (Fig. 1.9). Conclude.

Solution:

1. Let $\tau_{t \to s}$ and $\tau_{s \to t}$ be the timescales for the transfer from the troposphere to the stratosphere and for the reverse transfer, respectively. At equilibrium, with the notations used in Exercise 1.4 for the tropospheric and stratospheric masses, we obtain

$$m_{tropo}/\tau_{t \to s} = m_{strato}/\tau_{s \to t}.$$

Using the results of Exercise 1.4, we deduce $\tau_{t \to s}/\tau_{s \to t} \simeq 8$ (coherent with the data shown in Table 1.7).

2. Let a_n and a_s be the radioactivity in the Northern Hemisphere and in the Southern Hemisphere, respectively. The processes to be taken into account are the transfer from the stratosphere to the troposphere (with a rate k_{ST}), the stratospheric interhemispheric exchanges (with a rate k_S; as the masses of the stratospheric hemispheres are supposed to be equal, there is no assymetry for the interhemispheric exchange) and the radioactive decay (with a rate k). The transfer from the troposphere to the stratosphere are neglected due to the tropospheric wet scavenging of strontium (and following Exercise 1.3). Thus,

$$\begin{cases} \frac{da_s}{dt} = -k_{ST} a_s - k_S(a_s - a_n) - k a_s \\ \frac{da_n}{dt} = -k_{ST} a_n - k_S(a_n - a_s) - k a_n. \end{cases}$$

Let us define $\Sigma = a_n + a_s$ and $\Delta = a_n - a_s$. Upon integration,

$$\Sigma \sim \exp(-(k_{ST} + k)t), \qquad \Delta \sim \exp(-(k_{ST} + 2k_S + k)t),$$

namely $k_{ST} = 1/1.30 - \ln 2/28.5 \simeq 0.74\,\mathrm{yr}^{-1}$ and $k_S = (1/0.76 - \ln 2/28.5 - 0.74)/2 \simeq 0.28\,\mathrm{yr}^{-1}$. This defines the timescale $\tau_{s \to t} \simeq 1.35$ yr (1/0.74), to be compared with the values given in Table 1.7. Note that the timescale for the stratospheric interhemispheric exchange is about 3.6 years (1/0.28).

To know more ([78, 120]):

P. KREY AND B. KRAJEWSKI, *Comparison of atmospheric transport model calculations with observations of radioactive debris*, J. Geophys. Res., **75** (1970), pp. 2901–2908

E. REITER, *Stratospheric-tropospheric exchange processes*, Rev. Geophys. Space Phys., **13** (1975), pp. 459–474

Problem 1.4 (Aerosol Lifetime and Radon Radioactive Decay) The study of the atmospheric tracers "bound" to aerosols provides an estimation of the aerosol atmospheric residence time. The loss processes to be described for the atmospheric tracers also include the loss processes for aerosols (represented by the global residence time for aerosols, τ). The focus is often put on long-lived radionuclides, implied in the radioactive chain of radon (^{222}Rn). The emissions of radon are biogenic (the spatial distribution is highly variable).

Let λ_F be the kinetic rate describing the radioactive decay from a *father* radioelement F. The kinetic rate is connected to the half-life time $\tau_{1/2}$ with $\lambda\tau_{1/2} = \ln 2$. For the radioactive chain $P \to D$ (D is the *daughter* radionuclide), the time evolution of the concentration of F, C_F, is $\mathrm{d}C_F/\mathrm{d}t = -\lambda_F C_F$.

We will write $X_F = \lambda_F C_F/(\lambda_{Pb} C_{Pb})$ the ratio of the activity of F to that of lead (isotopic ratio).

1. Assuming equilibrium for the concentration of F, derive an equation for τ as a function of the isotopic ratio.
2. A first example is provided by the radioactive chain ^{210}Pb→ ^{210}Bi→ ^{210}Po (lead, bismuth, polonium). The half-life times are 22.3 years and 5.1 days for lead and bismuth, respectively. The observational data show that $X_{Bi} \in [0.48, 0.68]$. Estimate the aerosol residence time.
3. Consider the radioactive chain ^{210}Po→ ^{206}Pb (that ends the chain since ^{206}Pb is stable). The half-life time of polonium is 138 days. Assuming equilibrium for polonium, express τ as the root of a quadratic equation, whose coefficients depend on X_{Po}.

   Estimate τ from the observational data $X_{Po} \in [0.054, 0.092]$. Comment.

Solution:

1. The evolution of C_F is given by

$$\frac{\mathrm{d}C_F}{\mathrm{d}t} = \lambda_F C_F - \left(\lambda_F + \frac{1}{\tau}\right) C_F.$$

At equilibrium, it is easy to show

$$\frac{\lambda_F C_F}{\lambda_F C_F} = \frac{1}{1 + \frac{1}{\lambda_F \tau}},$$

which provides an equation for τ.

2. For the radioactive chain of lead,
$$X_{\text{Bi}} = \frac{\lambda_{\text{Bi}} C_{\text{Bi}}}{\lambda_{\text{Pb}} C_{\text{Pb}}} = \frac{1}{1 + \frac{1}{\lambda_{\text{Bi}} \tau}},$$
which gives the following equation for τ:
$$\tau = \frac{1}{\lambda_{\text{Bi}}} \frac{X_{\text{Bi}}}{1 - X_{\text{Bi}}}.$$
With the observational data, $\tau \in [6.8, 15.6]$ days, which is coherent with the expected values (the aerosol residence time is governed by wet scavenging, Chap. 5).
3. For the radioactive chain of bismuth, we obtain
$$\frac{\lambda_{\text{Po}} C_{\text{Po}}}{\lambda_{\text{Bi}} C_{\text{Bi}}} = \frac{1}{1 + \frac{1}{\lambda_{\text{Po}} \tau}},$$
and finally,
$$X_{\text{Po}} = \frac{\lambda_{\text{Po}} C_{\text{Po}}}{\lambda_{\text{Pb}} C_{\text{Pb}}} = \frac{1}{1 + \frac{1}{\lambda_{\text{Po}} \tau}} \frac{1}{1 + \frac{1}{\lambda_{\text{Bi}} \tau}}.$$
The quadratic equation is therefore
$$\tau^2 (1 - X_{\text{Po}}) - X_{\text{Po}} \left(\frac{1}{\lambda_{\text{Bi}}} + \frac{1}{\lambda_{\text{Po}}} \right) \tau - \frac{X_{\text{Po}}}{\lambda_{\text{Po}} \lambda_{\text{Bi}}} = 0.$$
As $X_{\text{Po}} \in [0, 1]$, it is easy to prove that there exists a unique positive root (the product of roots is negative).
Solving the quadratic equation yields $\tau \in [16.8, 26.5]$ days. The estimation of the residence time is too high. This is induced by an underestimation of the polonium concentration in the atmosphere. Actually, there are other sources of polonium to be added to the production by the radioactive chain of radon (volcanic emissions and mineral aerosols). The fraction of the "terrestrial" polonium (emitted polonium) can be up to two thirds of the total concentration, yielding $\tau \in [7.4, 10.8]$ days.

To know more ([107]):
C. PAPASTEFANOU, *Residence time of tropospheric aerosols in association with radioactive nuclides*, Applied Radiation and Isotopes, **64** (2006), pp. 93–100

Problem 1.5 (Montreal Protocol) The Montreal Protocol (1987) is the first international treaty devoted to a strong reduction of the CFC emissions. We focus on CFC12, a trace species with an atmospheric residence time $\tau \simeq 100$ years. This problem, partially taken from [58], aims at illustrating that delaying the emission reduction is a hazardous strategy. We refer to Sect. 4.2 for a description of the physical and chemical processes to be taken into account.

Let $m(t)$ be the trace species mass at the year t. In 1989, the annual emission rate was estimated to be $E = 4 \times 10^8\ \text{kg}\,\text{yr}^{-1}$ and the mass in the atmosphere was likely $m(1989) = 10^{10}$ kg.

1. Express the evolution equation for $m(t)$. Calculate the stationary state.
2. It is desired to investigate two strategies for reducing the emissions:

- the first strategy is based on a reduction of 50% from 1989. Calculate the resulting stationary state and the mass in 2050.
- the second strategy is based on a delayed reduction, to be applied from 1999. Calculate the resulting stationary state and the mass in 2050. Comment.

3. We want to generalize the conclusion of the previous question (impact of a delayed reduction).
Consider two strategies. The first one is a reduction of emissions to $E/2$ from 1989. The second one is a reduction down to $\alpha E/2$ from the year $T \geq 1989$ (delayed reduction, e.g. at $T = 1999$). The emissions are supposed not to be modified during the period from 1989 to T.
Calculate the concentration at the year $t \geq T$ (e.g., $t = 2050$) for both strategies. Show that the reduction must be stronger for the second strategy ($\alpha < 1$) if it is desired to reach the same concentration as for the first strategy. Is it possible to reach all "targets"?

Solution:

1. The evolution of $m(t)$ is given by
$$\frac{dm}{dt} = -\frac{m}{\tau} + E, \qquad m(0) = m_0.$$
The stationary state is $\tau E = 4 \times 10^{10}$ kg. The transient solution is
$$m(t) = \tau E(1 - e^{-t/\tau}) + m_0 e^{-t/\tau}.$$
2. For the first strategy, the stationary state is $\tau E/2 = 2 \times 10^{10}$ kg; in 2050, $m = 1.46 \times 10^{10}$ kg. For the second strategy, the stationary state is the same one. In 1999, $m = 1.29 \times 10^{10}$ kg. In 2050, $m = 1.57 \times 10^{10}$ kg. There is a negative impact of the delayed reduction.
3. For the first strategy, the solution is $m(t) = \tau E/2(1 - e^{-t/\tau}) + m_0 e^{-t/\tau}$. For the second strategy, $m(t) = \tau\alpha E/2(1 - e^{-(t-T)/\tau}) + m_1 e^{-(t-T)/\tau}$, with the initial condition at T, $m_1 = \tau E(1 - e^{-T/\tau}) + m_0 e^{-T/\tau}$. After a few algebraic manipulations, it is easy to show that both concentrations are equal if and only if
$$\alpha = 1 - \frac{e^{T/\tau} - 1}{e^{t/\tau} - e^{T/\tau}}.$$
Note that $\alpha < 1$ (as $t \geq T$). Moreover, $\alpha \geq 0$ if
$$T \leq \tau \ln\left(\frac{e^{t/\tau} + 1}{2}\right).$$
In other words, delaying the reduction is hazardous. First, it has "to be paid": the delayed reduction has to be greater than the reduction initally planned ($\alpha < 1$). Second, the target can even not be reached (conditional existence of $\alpha \geq 0$).

Figure 1.10 shows the evolution of CFC-11 and CFC-12 mixing ratios, from 1955 to 2005 (as given by the observational data) and the estimation to 2100. Figure 1.11 shows the evolution from 1980 to 2004 of the emissions (in kiloton yr^{-1}).

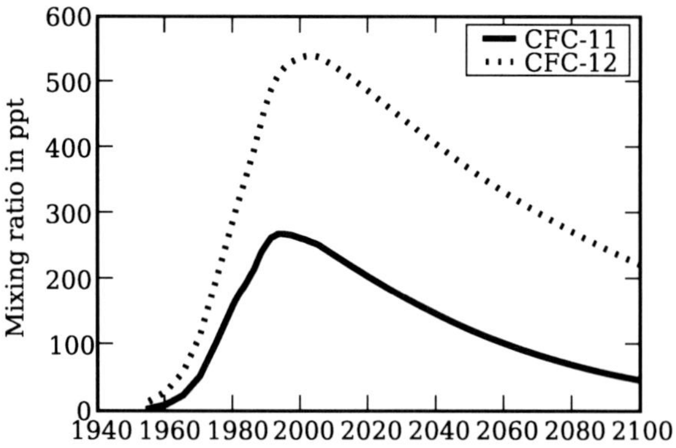

Fig. 1.10 Evolution from 1955 to 2100 of CFC-11 and CFC-12 mixing ratios (expressed in ppt). The mixing ratios are based on observational data for the period before 2005. Source: [5]

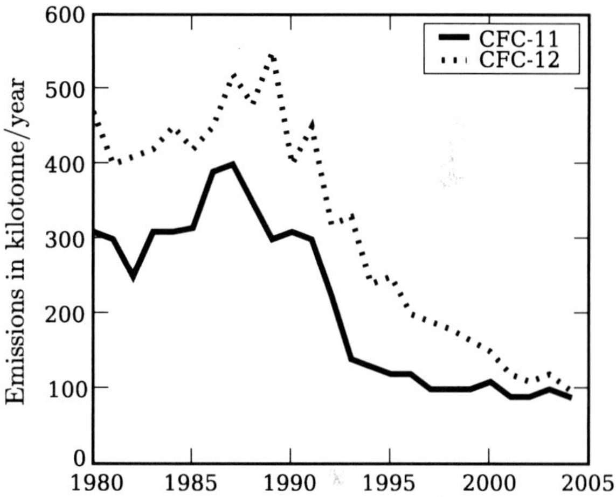

Fig. 1.11 Evolution from 1980 to 2004 of CFC-11 and CFC-12 emissions (in kiloton year^{-1}). Source: [5]

To know more ([5]):
Scientific assessment of ozone depletion: 2006. Chapter 8: Halocarbon Scenarios, Ozone Depletion Potentials, and Global Warming Potentials, 2007. World Meteorological Organization

Chapter 2
Atmospheric Radiative Transfer

The interaction between atmospheric matter and solar and terrestrial radiation plays a leading role for life conditions at the Earth's surface:

- stratospheric ozone filters the solar ultraviolet radiation;
- the absorption by a few gas-phase species (e.g. water, methane or carbon dioxide) of the terrestrial radiation defines the so-called *greenhouse effect*, which results in a surface temperature greater than 273 K (otherwise, the Earth would be a "white planet");
- more generally, the state of the atmosphere is determined by its energy budget, comprising the radiative fluxes (solar and terrestrial radiation) and the latent and sensible heat fluxes.[1]

The radiative properties and the concentrations of the atmospheric *trace species* determine the general behavior of the atmosphere. For example, a few species, emitted by anthropogenic activies, play a decisive role by increasing the greenhouse effect or by decreasing the filtration of ultraviolet radiation (by taking part in the consumption of stratospheric ozone).

At first glance, the energy budget of the Earth/atmosphere system should not be perturbed by anthropogenic activities (Table 2.1). The Earth and the atmosphere absorb 235 $\mathrm{W\,m^{-2}}$ of solar radiation, to be compared with 0.087 $\mathrm{W\,m^{-2}}$ of internal energy flux (geothermy) and with the energy directly related to anthropogenic activies, 0.025 $\mathrm{W\,m^{-2}}$. However, there is an indirect anthropogenic contribution to this energy budget, which is deeply related to pollution. Emission of greenhouse gases will indeed imply much stronger effects: the radiative forcing due to greenhouse gases since pre-industrial times is estimated to be about 2.5 $\mathrm{W\,m^{-2}}$, which means that the scaling factor (for the direct contribution) is 100!

This chapter is organized as follows. A primer for radiative transfer is given in Sect. 2.1, with the description of radiative emission, absorption and scattering. Section 2.2 illustrates the application to the atmosphere, with a focus on the greenhouse effect. The main sources of uncertainties, namely the role of clouds and aerosols,

[1] Latent heat is associated to the condensation of water vapor, an exothermic process; the sensible heat is associated to turbulence.

B. Sportisse, *Fundamentals in Air Pollution*,

Table 2.1 Energy fluxes for the Earth/atmosphere system. Source: [25]

Solar energy absorbed by the Earth and the atmosphere	$235\,\mathrm{W\,m^{-2}}$
Internal energy flux (geothermy)	$0.087\,\mathrm{W\,m^{-2}}$
Anthropogenic energy production	
average	$0.025\,\mathrm{W\,m^{-2}}$
urban area	$\simeq 50\,\mathrm{W\,m^{-2}}$
Radiative forcing due to greenhouse gases since the preindustrial times	$\simeq 2.5\,\mathrm{W\,m^{-2}}$

are also detailed. The link between particulate pollution and visibility reduction is briefly presented at the end of this chapter.

2.1 Primer for Radiative Transfer

The atmosphere filters the energy received from the Sun and from the Earth. Radiative transfer describes the interaction between radiation and matter (gases, aerosols, cloud droplets). The three key processes to be taken into account are:

- *emission*;
- *absorption* of an incident radiation by the atmospheric matter (which corresponds to a decrease of the radiative energy in the incident direction);
- *scattering* of an incident radiation by the atmospheric matter (which corresponds to a redistribution of the radiative energy in all the directions).

2.1.1 Definitions

2.1.1.1 Radiation

There are two possible viewpoints for describing electromagnetic radiations. A radiation is composed of particles, the so-called *photons*. Similarly, it can be viewed as a wave propagating at the speed of light ($c \simeq 3 \times 10^8\,\mathrm{m\,s^{-1}}$ in a vacuum, a similar value in air). It is then characterized by its frequency ν (in $\mathrm{s^{-1}}$ or in hertz) or, equivalently, by its wavelength $\lambda = c/\nu$ (usually expressed in nm or in µm).

As shown in Fig. 2.1, the radiation spectrum can be divided into wavelength regions: γ-ray region ($\lambda \leq 0.1\,\mathrm{nm}$), X-ray radiation ($0.1\,\mathrm{nm} \leq \lambda \leq 10\,\mathrm{nm}$), ultraviolet radiation ($10\,\mathrm{nm} \leq \lambda \leq 380\,\mathrm{nm}$), visible radiation ("light", from blue to red: $380\,\mathrm{nm} \leq \lambda \leq 750\,\mathrm{nm}$), infrared radiation ($750\,\mathrm{nm} \leq \lambda \leq 10\,\mu\mathrm{m}$), microwave region, etc.

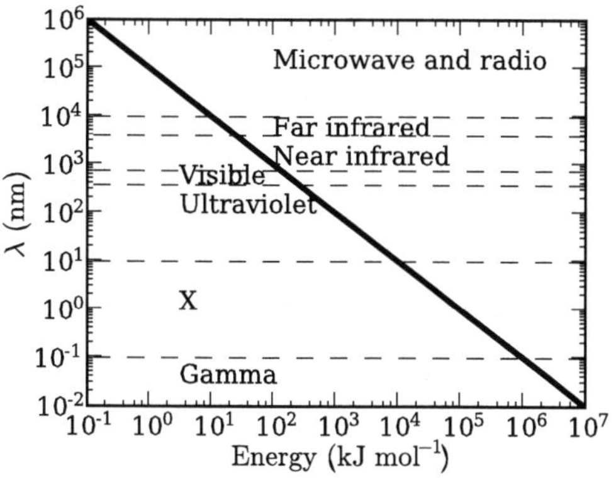

Fig. 2.1 Decomposition of the electromagnetic spectrum. *Abscissa*: energy (in $kJ\,mol^{-1}$); *ordinate*: wavelength (in nm)

2.1.1.2 Solid Angle

The concept of *solid angle* is used for quantifying the solar radiation received by a surface. Let σ be a surface element on a sphere of radius r centered at point O. A solid angle Ω is then defined as the ratio of σ to the square of r: $\Omega = \sigma/r^2$.

In spherical coordinates, the differential surface element $d\sigma$ is generated by the variations of the *zenithal* angle $d\theta$ and of the *azimuthal* angle $d\phi$ (Fig. 2.2). The differential solid angle is then given by

$$d\Omega = \frac{d\sigma}{r^2}. \tag{2.1}$$

Since $d\sigma = r\,\sin\theta\,d\phi \times r d\theta$, this yields

$$d\Omega = \sin\theta\,d\phi\,d\theta. \tag{2.2}$$

A solid angle is measured in *steradian* (sr). For a sphere, as $\theta \in [0, \pi]$ and $\phi \in [0, 2\pi]$, we obtain upon integration $\Omega = 4\pi$.

2.1.1.3 Radiance and Irradiance

We consider a differential surface area dA (Fig. 2.2). Let dE_λ be the radiative energy (expressed in joules) intercepted by dA for incident photons of wavelength in $[\lambda, \lambda + d\lambda[$, during a time interval dt, in the solid angle $d\Omega$. The *monochromatic radiance* is defined as the energy that is propagated through the surface dS generated by dA in a direction perpendicular to the incident direction ($dS = \cos\theta \times dA$), namely

$$dI_\lambda = \frac{dE_\lambda}{dS\,d\lambda\,dt\,d\Omega}. \tag{2.3}$$

The radiance is usually expressed in $W\,m^{-2}\,nm^{-1}\,sr^{-1}$ (with $1\,W = 1\,J\,s^{-1}$).

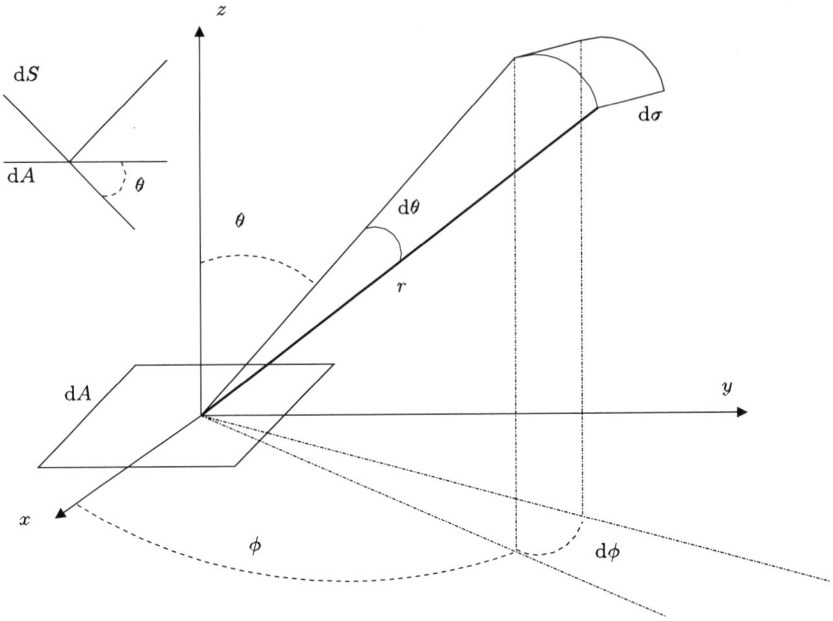

Fig. 2.2 Definition of a solid angle. In the *left part* of the figure, the surface elements $\mathrm{d}A$ and $\mathrm{d}S$ are represented by their projection in a vertical plane

Monochromatic irradiance is defined as the component of the monochromatic radiance that is normal to $\mathrm{d}A$, upon integration over the whole solid angle (the normal direction is given by the vertical axis in Fig. 2.2),

$$F_\lambda = \int_\Omega \cos\theta \, I_\lambda \, \mathrm{d}\Omega . \tag{2.4}$$

Upon integration over all the wavelengths, the *irradiance* is defined by $F = \int_\lambda F_\lambda \, \mathrm{d}\lambda$. It is expressed as a (received) power per unit area of surface (in $\mathrm{W\,m^{-2}}$).

2.1.2 Energy Transitions

Quantum mechanics describes the energy levels of a given molecule. The key point is that the energy levels are given by a *discrete* sequence, let us say $(E_n)_n$, specific of the *spectroscopic properties* of the molecule.

The simplest illustration is detailed in Exercise 2.1, with the case of a particle supposed to be "trapped in a well".

Exercise 2.1 (Discrete Energy Levels) Consider a particle "trapped in a well" (the well is actually defined by an energy potential). For convenience, we suppose that the well corresponds to a one-dimensional interval, let us say [0, 1]. Let x be the spatial variable.

The particle location is described by a *probability density function*, $p(x)$, which can be computed from the wavefunction $f(x)$ as $p(x) = |f(x)|^2$. The wavefunction is governed by the Schrödinger equation

$$-\frac{h^2}{2m}\frac{\mathrm{d}^2 f}{\mathrm{d}x^2} + V(x)f = Ef,$$

where m is the particle mass, $V(x)$ is the energy potential that defines the well, E corresponds to the particle energy and $h = 6.63 \times 10^{-34}$ J s is Planck's constant.

The particle motion is free inside the well ($V = 0$) but the particle is "trapped" (its probability density function is null at the boundaries). Calculate the possible levels of energy (E).

Solution:

The governing equation for f is

$$\frac{\mathrm{d}^2 f}{\mathrm{d}x^2} = \frac{2mE}{h^2} f, \qquad f(0) = f(1) = 0.$$

The solutions are in the form $f(x) \sim \sin(\sqrt{\frac{2mE}{h^2}}x)$ where $\sqrt{\frac{2mE}{h^2}} = n\pi$ with n a positive integer. This results in a discrete spectrum of possible values for the energy level, $E_n = n^2\frac{h^2}{2m}\pi^2$.

Consider a molecule with a given energy level, let us say E_1. The emission of a photon by this molecule corresponds to a transition from E_1 to a lower energy level, let us say $E_2 < E_1$. On the opposite, the absorption of a photon by this molecule implies the transition from E_1 to a higher energy level, let us say $E_2 > E_1$. The wavelength of the photon is fixed by the energy transition. Planck's law (Fig. 2.3) states that

$$\Delta E = h\nu = \frac{hc}{\lambda}. \tag{2.5}$$

A photon can therefore be absorbed or emitted only if its wavelength corresponds to a possible transition. As a result, for a given molecule, absorption and emission are possible only in specific parts of the radiation spectrum (determined by the spectroscopic properties of the molecule).

If λ has a small value, the energy gap is large: the shortwave radiations (e.g. ultraviolet radiation) are the most energy-containing ones (Fig. 2.1). On the other hand, if the wavelength has a large value, the energy gap is low: the longwave radiations (e.g. infrared radiation) do not contain a lot of radiative energy.

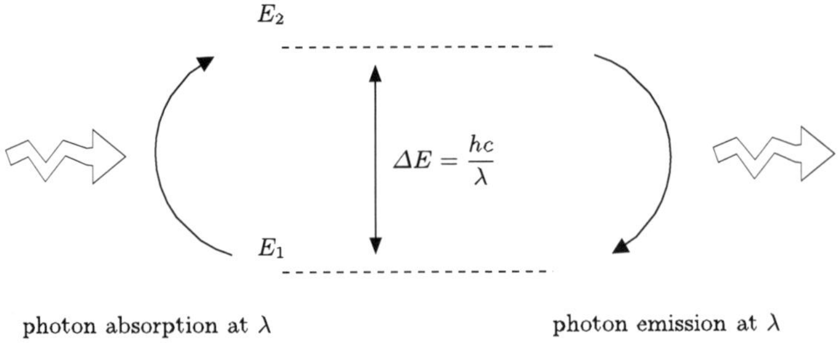

Fig. 2.3 Emission and absorption of a photon: transition between two energy levels E_1 and E_2 ($E_1 < E_2$). The wavelength is given by Planck's law, $\lambda = hc/(E_2 - E_1)$

2.1.3 Emissions

2.1.3.1 Blackbody Emission

Planck's Distribution The radiative energy emitted by a "body" at equilibrium depends on its temperature. Intuitively, it is expected that the higher the temperature is, the higher the emission is.

The maximum of radiative energy that can be emitted per unit area of surface and per time unit defines the so-called *blackbody emission*. For a body at temperature T, the maximum of emitted radiance at wavelength λ is given by the so-called Planck distribution,

$$B_\lambda(T) = \frac{2hc^2}{\lambda^5} \frac{1}{\exp\left(\frac{ch}{\lambda k_B T}\right) - 1}, \tag{2.6}$$

where $k_B = 1.38 \times 10^{-23}\,\mathrm{J\,K^{-1}}$ is Boltzmann's constant. The unit is $\mathrm{W\,m^{-2}\,nm^{-1}}$.

Remark 2.1.1 (Shape of Planck's Distribution) Let us try to justify the qualitative shape of Planck's distribution (following A. Einstein).

Consider a blackbody, defined as a substance that aborbs all incident radiation. The blackbody is supposed to be inside a box, such that the walls do not emit nor absorb radiation. The reflection on the walls are therefore supposed to be perfect. Thus, the blackbody receives an incident radiation, I, that is exactly the emitted radiation.

Let $E_1 < E_2$ be two energy levels of the blackbody. At equilibrium, the probability of having E_i ($i = 1, 2$) is given by the Maxwell-Boltzmann distribution and is equal to $\exp(-E_i/k_B T)$, up to a normalization factor.

The probability $P_{1\to 2}$ that the absorption of radiation leads to a transition from E_1 to a higher value E_2 is proportional to the number of molecules at state E_1, namely

$$P_{1\to 2} = \alpha I \exp\left(-\frac{E_1}{k_B T}\right) \tag{2.7}$$

with α a multiplying factor.

The transition from E_2 to a lower value E_1 is driven by two processes: first, the spontaneous emission of a photon and, second, the emission induced by the absorption of the incident radiation, I. The transition probability $P_{2\to1}$ is therefore composed of two terms: the first one is proportional to the number of molecules at state E_2 while the second one is proportional to I. Thus,

$$P_{2\to 1} = \underbrace{\beta I \exp\left(\frac{-E_2}{k_B T}\right)}_{\text{induced emission}} + \underbrace{\gamma \exp\left(\frac{-E_2}{k_B T}\right)}_{\text{spontaneous emission}}, \tag{2.8}$$

with β and γ two multiplying factors.

At equilibrium, $P_{1\to2} = P_{2\to1}$, so that the number of molecules at a given energy level is constant. The incident radiation is exactly the emitted radiation and satisfies

$$I = \frac{\gamma}{\alpha \exp(-\frac{\Delta E}{k_B T}) - \beta}, \tag{2.9}$$

with $\Delta E = E_2 - E_1$, connected to the wavelength λ by $\Delta E = hc/\lambda$.

This justifies Planck's law, (2.6), if $\alpha = \beta$, namely if the probability of absorption is equal to the probability of induced emission.

Wien's Displacement Law Planck's distribution is an increasing function with respect to temperature (T) and is a concave function of the wavelength (λ). At fixed T, the wavelength of the maximum is given by $\partial B_\lambda/\partial\lambda = 0$. This yields (expressed in nm)

$$\lambda_{max} \simeq \frac{2898 \times 10^3}{T}. \tag{2.10}$$

λ_{max} is inversely proportional to the temperature. This is the so-called Wien's displacement law: a warm body emits shortwave radiations, corresponding to energy-containing radiations.

Stefan-Boltzmann Law Upon integration over the entire wavelength domain, the total emitted radiance of a blackbody at temperature T is

$$B(T) = \int_0^{+\infty} B_\lambda(T)\, \mathrm{d}\lambda = \sigma T^4. \tag{2.11}$$

The unit is $\mathrm{W\,m^{-2}}$.

Admitting that $\int_0^\infty \nu^3 \mathrm{d}\nu/(e^\nu - 1) = \pi^4/15$, the Stefan-Boltzmann constant is[2] $\sigma = 2\pi^5 {k_B}^4/(15c^2h^3)$, namely $\sigma = 5.67 \times 10^{-8}\,\mathrm{W\,m^{-2}\,K^{-4}}$. As expected, the radiative energy emitted by a blackbody is an increasing function of the temperature (proportional to the fourth power of the temperature).

---

[2]The notation could be misleading with respect to that of a differential area used for solid angles.

Table 2.2 Typical values of the emissivity in the infrared region, for different surface types. Source: [67]

Surface	ε_{IR}	Surface	ε_{IR}
Sea	0.95–1	Grass	0.90–0.95
Fresh snow	0.99	Desert	0.85–0.90
"Old" snow	0.80	Forest	0.95
Liquid water clouds	0.25–1	Concrete	0.70–0.90
Cirrus	0.10–0.90	Urban	0.85

2.1.3.2 Emissivity

A realistic medium is not a blackbody. The radiative energy that is actually emitted by a medium at temperature T, for a given wavelength λ, is

$$E_\lambda(T) = \epsilon_\lambda(T) B_\lambda(T), \tag{2.12}$$

where $\epsilon_\lambda(T)$ is the so-called *emissivity* at wavelength λ and at temperature T. By definition, $\epsilon_\lambda \leq 1$ (unitless).

For example, in the case of infrared radiation, the radiative behavior of fresh snow (the "whitest" snow) is similar to a blackbody,[3] with $\epsilon_\lambda \simeq 1$ (Table 2.2). On the opposite, the urban environment has a low emissivity, which will strongly impact the urban climate (Sect. 3.6).

2.1.3.3 Application to the Earth and to the Sun

The Sun can be considered as a blackbody with a temperature of 5800 K. Applying Wien's law justifies that the Sun's emission peaks in the visible region (maximum at 500 nm). The Earth can be considered as a blackbody at 288 K. The maximum is then in the infrared region (10 μm).

Upon a direct application of the Stefan-Boltzmann law, the Sun emits about 20^4 as much as the Earth ($5800/288 \simeq 20$).

The key point is that both emitted radiation spectra can be *split* (Fig. 2.4). The atmosphere will have a different behavior with respect to the solar and terrestrial radiation: in a first approximation, the infrared terrestrial radiation is absorbed while the atmosphere is transparent to the solar visible radiation.

2.1.4 Absorption

2.1.4.1 Beer-Lambert Law

A fraction of the incident radiation is absorbed along the path of propagation in a medium (here the atmosphere). The Beer-Lambert law (also referred to as the Beer-

[3] The terminology has nothing to do with colors!

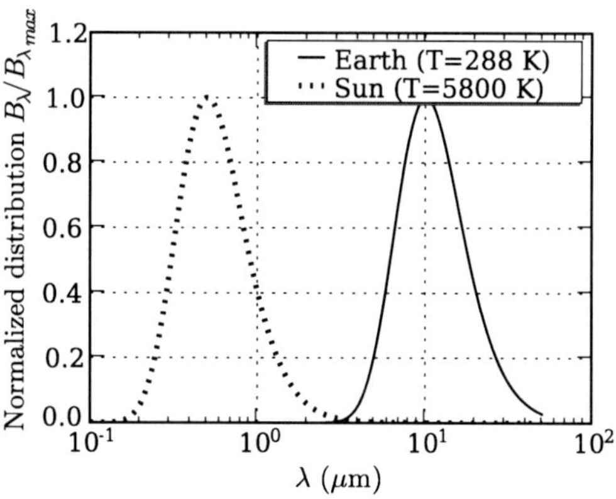

Fig. 2.4 Normalized emission spectrum for the Earth (blackbody at $T = 288$ K) and for the Sun (blackbody at $T = 5800$ K)

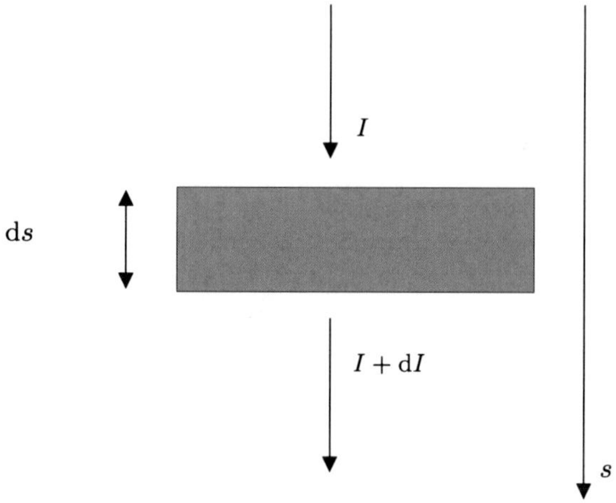

Fig. 2.5 Absorption of an incident radiation traversing a medium (*gray box*)

Lambert-Bouguer law) governs the reduction in the radiation intensity I_λ at wavelength λ (Fig. 2.5). If s stands for the medium thickness (oriented in the direction of propagation), the evolution of the radiation intensity is

$$\frac{\mathrm{d}I_\lambda}{\mathrm{d}s} = -a_\lambda(s)I_\lambda, \tag{2.13}$$

where $a_\lambda(s)$ is the absorption coefficient at wavelength λ (depending on the medium). The unit of a_λ is, for instance, m^{-1} or cm^{-1}. Assuming that the medium is homogeneous, then a_λ has a constant value and

$$I_\lambda(s) = I_\lambda(0) \times \exp(-sa_\lambda). \tag{2.14}$$

Consider a medium composed of p absorbing species, with densities n_i ($i = 1, \ldots, p$), expressed in molecule cm^{-3}. The absorbing coefficient is then obtained by summing over all species. For a given species, the contribution depends on the density and on the so-called *absorption cross section* (the effective cross section resulting in absorption), $\sigma_i^a(\lambda, s)$, usually expressed in cm^2:

$$a_\lambda(s) = \sum_{i=1}^{p} n_i(s)\sigma_i^a(\lambda, s). \tag{2.15}$$

A way to define the absorption cross section is to consider an incident flux of energy per surface, F (in W cm^{-2}). The resulting absorbed energy is then $F_a = \sigma_a \times F$ (expressed in W).

Another classical concept is the so-called *optical depth* τ_λ (unitless), defined for a monochromatic radiation by

$$\mathrm{d}\tau_\lambda = a_\lambda(s)\mathrm{d}s. \tag{2.16}$$

Rewriting the Beer-Lambert law yields

$$\frac{\mathrm{d}I_\lambda}{\mathrm{d}\tau_\lambda} = -I_\lambda. \tag{2.17}$$

2.1.4.2 Kirchhoff's Law

For a given wavelength λ, the *absorptivity* A_λ is defined as the fraction of the incident radiation that is absorbed by the medium. Kirchhoff's law (1859) connects the absorptivity and the emissivity of a medium at thermodynamic equilibrium, namely

$$\epsilon_\lambda = A_\lambda. \tag{2.18}$$

The absorption properties of a medium are therefore directly related to its emission properties.

Note that A_λ can be derived from a_λ. For a medium supposed to be homogeneous, with a thickness Δz (typically a cloud), with an absorbing coefficient a_λ, the ratio of the absorbed intensity to the incident intensity is $A_\lambda = 1 - \exp(-a_\lambda \Delta z)$.

At thermodynamic equilibrium, when taking into account absorption and emission, the evolution of the intensity is then

$$\frac{\mathrm{d}I_\lambda}{\mathrm{d}s} = a_\lambda(s)(B_\lambda(T) - I_\lambda). \tag{2.19}$$

2.1.4.3 Spectral Line Broadening

For a given energy transition ΔE, Planck's law describes only monochromatic absorption or emission, with a unique wavelength λ_0, given by $|\Delta E| = hc/\lambda_0$. This defines the so-called *spectral lines*. In practice, monochromatic radiations are not observed. As shown by the absorption spectrum for a few species (Sect. 2.2), there is a *broadening* of the wavelengths, mainly related to two effects.

Doppler Broadening For moving molecules, the *Doppler effect* implies that the emission and absorption wavelengths are broadened. This is usually described by the so-called Doppler profile, centered at λ_0, given by a Gaussian distribution with respect to the frequency $\nu = c/\lambda$,

$$f_D(\nu) = \frac{S_D}{\alpha_D}\sqrt{\frac{\ln 2}{\pi}}\exp\left(-\frac{(\nu-\nu_0)^2}{\alpha_D{}^2}\ln 2\right), \tag{2.20}$$

where α_D and S_D stand for the half width of the line and the line strength, respectively. The half width of the line is related to the velocity of the molecule in the direction of the incident radiation, and is proportional to $\sqrt{T}$.

This shape is derived from the probability density function of the velocity, given by the Maxwell distribution

$$P(v) = \sqrt{\frac{m}{2\pi k_B T}}\exp\left(-\frac{mv^2}{2k_B T}\right) \tag{2.21}$$

with m the molecule mass. The Doppler effect states that the frequency ν appears shifted as seen by a stationary observer to the frequency $\tilde{\nu} = \nu(1 \pm v/c)$.

Pressure Broadening (Lorentz Effect) The collisions between the molecules contribute to broaden the lines. The distribution function is then

$$f_L(\nu) = \frac{S_L}{\pi}\frac{\alpha_L}{(\nu-\nu_0)^2 + \alpha_L{}^2} \tag{2.22}$$

where α_L and S_L stand for the width and the strength of the line, respectively. The width is related to the collision frequency and is proportional to the product of the molecule density, n, by the velocity (proportional to $\sqrt{T}$). With the ideal gas law, $n \propto P/T$ (P is the pressure), and thus $\alpha_L \propto P/\sqrt{T}$.

The Lorentz effect is a decreasing function of altitude. For a hydrostatic atmosphere (Chap. 1), supposed to be adiabatic (Chap. 3), the vertical profiles of pressure and of temperature are indeed $P(z) \simeq P_0 \exp(-z/H)$ and $T(z) \simeq T_0 - \Gamma z$.

The typical shape of the Doppler and Lorentz profiles is shown in Fig. 2.6. Up to 40 kilometers, the Lorentz effect is the dominant effect (due to high densities), then the Doppler effect and, finally, the joint impact of both effects (described by the so-called *Voigt profile*).

2.1.5 Scattering

Let us consider a gaseous molecule or a particle (aerosol or cloud drop), with a characteristic size. The incident radiation is also *scattered* in all directions. The shape of the scattered intensity strongly depends on the characteristic size.

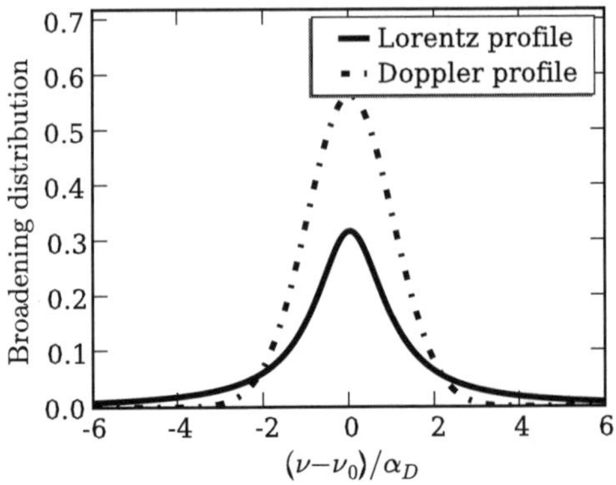

Fig. 2.6 Normalized distribution functions for the Lorentz and Doppler effects

2.1.5.1 Scattering Regimes

The scattering of the incident electromagnetic wave by a gas-phase molecule or by a particle mainly depends on the comparison between the wavelength (λ) and the characteristic size (d). We recall that $d \simeq 0.1$ nm for a gas-phase molecule, $d \in [10\,\text{nm}, 10\,\mu\text{m}]$ for an aerosol and $d \in [10, 100]\mu$m for a liquid water drop (Chap. 1). The wide range covered by the body size will induce different behaviors.

Three scattering regimes are usually distinguished: the Rayleigh scattering (typically for gases), the scattering represented by the optical geometry's laws (typically for liquid water drops) and the so-called Mie scattering (for aerosols).

Rayleigh Scattering If $d \ll \lambda$ (the case for gases), the electromagnetic field can be assumed to be homogeneous at the level of the scattering body. This defines the so-called *Rayleigh scattering* (also referred to as *molecular scattering*).

The scattered intensity in a direction with an angle θ to the incident direction, at the distance r from the scattering body (see Fig. 2.7), for a media of mass concentration C, composed of spheres of diameter d and of density ρ, is then given by ([89])

$$I(\theta, r) = I_0 \frac{8\pi^4}{r^2\lambda^4} \frac{\rho^2 d^6}{C^2} \left(\frac{m^2 - 1}{m^2 + 2} \right)^2 (1 + \cos^2\theta). \tag{2.23}$$

The incident intensity is I_0. m is the complex refractive index, specific to the scattered body: it is defined as the ratio of the speed of light in the vacuum to that in the body, and depends on the chemical composition for aerosols (e.g. $m = 1.34$ for water at $\lambda = 450$ nm, Table 2.3).

This formula is inversely proportional to λ^4: scattering is therefore much stronger for the shortwave radiations (Remark 2.1.2 devoted to the sky color). As a result, the terrestrial longwave radiations are weakly scattered.

Note that the Rayleigh scattering is an increasing function of the size (d) and is a decreasing function of the distance (r). Moreover, Rayleigh scattering is symmetric between the backward and forward directions: $I(\theta, r) = I(\pi - \theta, r)$.

Fig. 2.7 Scattering of an incident radiation (I_0)

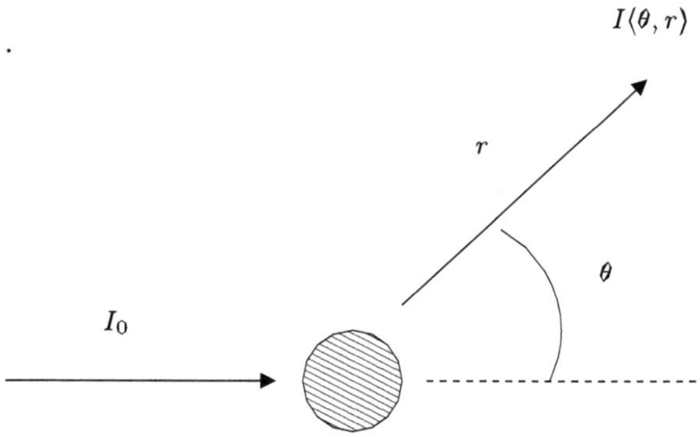

Optical Geometry If $d \gg \lambda$ (this is the case of liquid water drops with respect to the solar radiation), the laws of optical geometry can be applied, leading to the understanding of many physical phenomena (e.g. rainbow formation). The scattering weakly depends on the wavelength.

Mie Scattering If $d \simeq \lambda$ (the case for most of atmospheric aerosols), the simplifications used above are no longer valid. A detailed calculation of the interaction between the electromagnetic field and the scattering body is required: this is given by the *Mie theory*.

The intensity of the scattered radiation in the direction with an angle θ to the incident direction, at a distance r, is ([89])

$$I(\theta, r) = I_0 \frac{\lambda^2 (i_1 + i_2)}{4\pi^2 r^2}, \tag{2.24}$$

where i_1 and i_2 are the intensity Mie parameters, given as complicated functions of d/λ, θ and m. The parameters i_1 and i_2 are characterized by a set of *maxima* as a function of the angle θ. Note that the forward fraction of the scattering intensity is dominant (Fig. 2.8).

Remark 2.1.2 (Sky Color) In a simplified approach, the intensity scattered by aerosols can be parameterized as a function proportional to $\lambda^{-1.3}$. As a result, the scattering does not filter specific wavelengths, which explains why polluted skies (with high concentrations of particulate matter) are gray (Sect. 2.2.5).

On the opposite, for a "clean" sky (sometimes referred to as "Rayleigh sky"), Rayleigh scattering can be applied. The scattering bodies are gas-phase molecules, such as N_2 and O_2, whose characteristic size is about one angström (0.1 nm). The solar radiation is mainly in the visible region (the X-rays and the ultraviolet radiation have been filtered in the ionosphere and in the stratosphere, respectively). The dependence in λ^{-4} peaks the scattered intensity in the smallest wavelengths (those corresponding to the blue color).

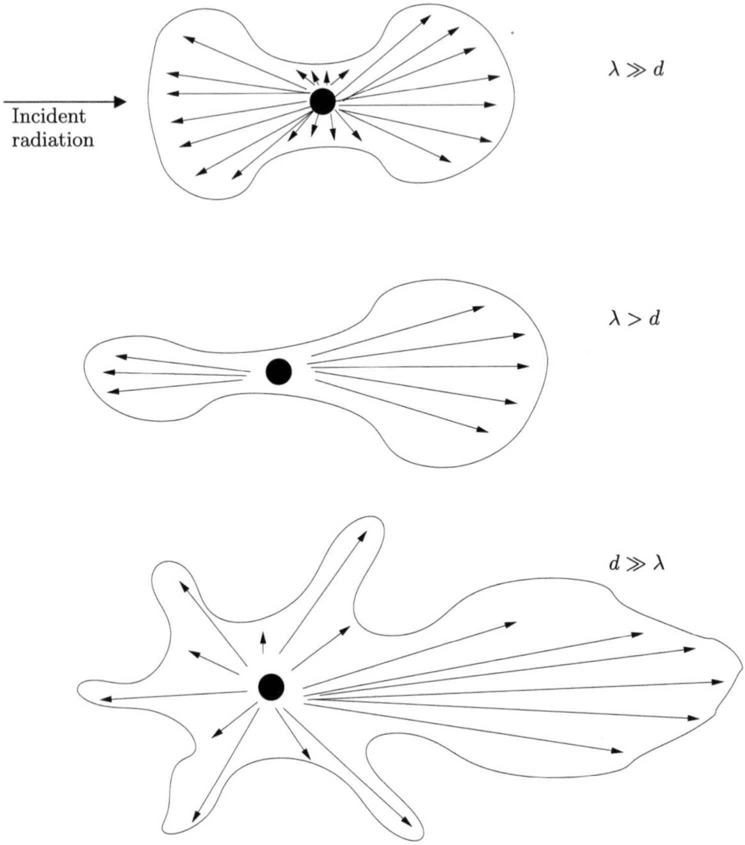

Fig. 2.8 Scattering of an incident radiation of wavelength λ by an aerosol (*gray sphere*) of diameter d. The size of the vectors originating from the aerosol is proportional to the scattered intensity in the vector direction

2.1.5.2 Modeling of Scattering

Modeling the scattering requires us to describe the intensity as a function of not only the medium thickness (s) but also of the solid angle $\Omega = (\theta, \phi)$.

The evolution of the intensity is given by

$$\frac{\mathrm{d}I_\lambda}{\mathrm{d}s} = -d_\lambda(s)I_\lambda + \frac{d_\lambda}{4\pi}\int P(\Omega,\Omega')I_\lambda(\Omega')\,\mathrm{d}\Omega'. \tag{2.25}$$

The first term, similar to the Beer-Lambert law, corresponds to an extinction of the incident radiation. The scattering coefficient $d_\lambda(s)$ (for instance expressed in cm^{-1})

is given, similar to absorption, by

$$d_\lambda(s) = \sum_{i=1}^{p} n_i(s)\sigma_i^d(\lambda, s), \tag{2.26}$$

with $\sigma_i^d(\lambda, s)$ the scattering cross section for species i (expressed e.g. in cm^{-2}). Similar to absorption, the scattering cross section can be defined as the fraction of the incident flux of energy that is scattered. Let F be the incident flux of energy per surface (in $\mathrm{W\,cm}^{-2}$). The scattered energy, F_d (in W), is then given by

$$F_d = \sigma_d \times F. \tag{2.27}$$

The second source term in (2.25) corresponds to the scattering in all the directions. The scattering probability density function, $P(\Omega, \Omega')$, describes the scattering from the solid angle Ω to the solid angle Ω'. It satisfies

$$\frac{1}{4\pi}\int P(\Omega, \Omega')\,\mathrm{d}\Omega' = 1. \tag{2.28}$$

2.1.6 *Radiative Transfer Equation*

The three processes (emission, absorption and scattering) are actually coupled. The radiative transfer equation reads

$$\frac{\mathrm{d}I_\lambda}{\mathrm{d}s} = -(a_\lambda(s)+d_\lambda(s))I_\lambda(s)+a_\lambda(s)B_\lambda(T)+\frac{d_\lambda}{4\pi}\int P(\Omega, \Omega')I_\lambda(\Omega')\,\mathrm{d}\Omega'. \tag{2.29}$$

The sum $a_\lambda + d_\lambda$ defines the *extinction coefficient*, usually written as b_λ^{ext}.

The optical depth determines the opacity of the medium and is defined, similar to (2.16), by

$$\mathrm{d}\tau_\lambda = (a_\lambda(s) + d_\lambda(s))\mathrm{d}s. \tag{2.30}$$

Thus

$$\frac{\mathrm{d}I_\lambda(\tau)}{\mathrm{d}\tau} = -I_\lambda(\tau) + \omega_a\, B_\lambda(T(\tau)) + \frac{\omega_d}{4\pi}\int P(\Omega, \Omega')I_\lambda(\Omega')\,\mathrm{d}\Omega', \tag{2.31}$$

where $\omega_a = a_\lambda/(a_\lambda + d_\lambda)$ and $\omega_d = d_\lambda/(a_\lambda + d_\lambda)$ are the absorption and scattering albedos, respectively.

We investigate two simplified cases: the case of infrared radiation (only absorption and emission are taken into account) and the case of visible radiation (only scattering is described).

2.1.6.1 Infrared Radiation

Scattering can be neglected for the infrared radiation. Moreover, the temperature T is a function of the altitude: we write $T(\tau)$ with τ an increasing function of the altitude (the terrestrial radiation propagates from the bottom to the top of the atmosphere).

Integrating (2.31) yields

$$I_\lambda(\tau) = I_\lambda(0)e^{-\tau} + \int_0^\tau B_\lambda(T(\tau'))e^{(\tau'-\tau)}\,\mathrm{d}\tau'. \tag{2.32}$$

The first term is a pure extinction term while the second term describes the emission from the atmosphere. $I_\lambda(0)$ is the emitted radiation at the Earth's surface.

2.1.6.2 Visible Radiation

For the visible region of the electromagnetic spectrum, we can neglect both absorption and emission in the atmosphere. Thus,

$$\frac{\mathrm{d}I_\lambda(\tau)}{\mathrm{d}\tau} = -I_\lambda(\tau) + \frac{1}{4\pi}\int P(\Omega,\Omega')I_\lambda(\Omega')\,\mathrm{d}\Omega'. \tag{2.33}$$

In this case, the optical depth, τ, is a decreasing function of the altitude (the radiation propagates from the top to the bottom of the atmosphere). The boundary condition $I_\lambda(0)$ corresponds to the solar radiation received at the top of the atmosphere.

There does not exist any analytical solution in the general case. Solving this equation can be performed with the *method of successive orders*. The solution is built by solving successively the systems

$$\frac{\mathrm{d}I_\lambda^{n+1}(\tau)}{\mathrm{d}\tau} = -I_\lambda^{n+1}(\tau) + \frac{1}{4\pi}\int P(\Omega,\Omega')I_\lambda^n(\Omega')\,\mathrm{d}\Omega'. \tag{2.34}$$

Scattering is then applied to the radiation computed in the previous iteration. We can then apply a superposition approach since the radiative transfer equation is linear, yielding $I_\lambda = \sum_{n=0}^{\infty} I_\lambda^n$.

2.1.7 *Additional Facts for Aerosols*

The extinction properties of a particle are determined by its *extinction efficiency*, defined as the ratio of the scattering cross section to the interception surface. For a particle of diameter d, the interception surface is $A = \pi(d/2)^2$. The extinction comprises a part associated to absorption and a part associated to scattering, namely

$$Q_\lambda^{ext} = Q_\lambda^a + Q_\lambda^d = \frac{\sigma_\lambda^a + \sigma_\lambda^d}{\pi(d/2)^2}. \tag{2.35}$$

Table 2.3 Typical values of the complex refractive index ($n_\lambda + jk_\lambda$), for a few aerosol types. The wavelength is $\lambda = 450$ nm, corresponding to visible radiation. Source: [53]

Aerosol type	n_λ	k_λ
Water	1.34	0.
Ammonium	1.53	-5×10^{-3}
Sulfate	1.43	0.
Sea salt	1.5	0.
Soot	1.75	−0.45
Mineral aerosol	1.53	-8.5×10^{-3}
Organic aerosol	1.53	-8.5×10^{-3}

The absorption and scattering efficiencies, Q_λ^a and Q_λ^d respectively, are functions of

- the *size parameter*

$$\alpha_\lambda = \frac{\pi d}{\lambda}, \tag{2.36}$$

  with d the particle diameter (the particle is supposed to be a sphere);
- the properties of the medium defined by the particle (due to its chemical composition), described by its *complex refractive index*

$$m_\lambda = n_\lambda + jk_\lambda, \quad j^2 = -1. \tag{2.37}$$

  The real part n_λ is related to scattering while the imaginary part k_λ is related to absorption.

Actually, m_λ is normalized with respect to the "ambient" medium (here, air, whose refractive index is about 1 for visible radiation).

The refractive index in the visible region of the electromagnetic spectrum is given in Table 2.3 for different particle types (with different chemical composition). Note that soot (*elemental* or *black carbon*) is characterized by a strong absorption.

The extinction coefficient for a particle can be deduced from the extinction efficiency. For a particle density n (expressed in number of particles per volume of air), we obtain

$$b_\lambda^{ext} = \sigma_\lambda^{ext} \times n = \frac{\pi d^2}{4} Q_\lambda^{ext} n. \tag{2.38}$$

The extinction efficiency is a function of the refractive index (m_λ) and of the size parameter (α_λ). For large values of the size parameter ($d/\lambda \gg 1$, typically for cloud drops with visible radiation), Q^{ext} is about 2 (Fig. 2.9), yielding the extinction coefficient

$$b_\lambda^{ext} \simeq \frac{\pi d^2}{2} n. \tag{2.39}$$

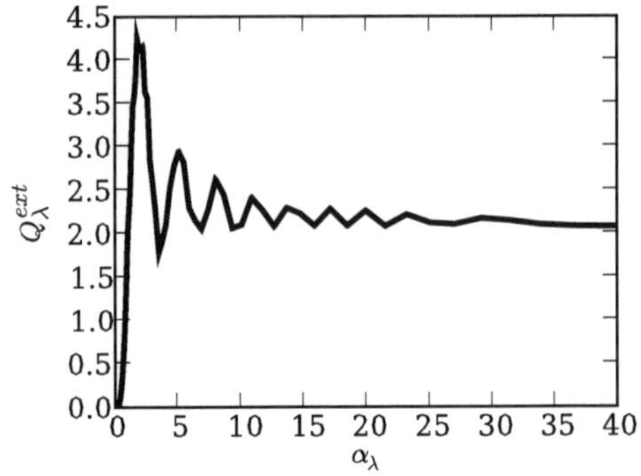

Fig. 2.9 Evolution of the extinction efficiency as a function of the size parameter. Credit: Marilyne Tombette, CEREA

Table 2.4 Typical values of the surface albedo (visible radiation) for different surface types. Source: [67]

Surface	Albedo (visible)	Surface	Albedo (visible)
Liquid water	0.05–0.20	Grass	0.15–0.25
Fresh snow	0.75–0.95	Desert	0.20–0.40
"Old" snow	0.40–0.70	Forest	0.10–0.25
Sea ice	0.25–0.40	Bitume	0.05–0.20
Clouds	0.20–0.90	Urban	0.10–0.27

2.1.8 Albedo

For a given wavelength, the albedo of a surface (by extension of an atmospheric layer) is defined as the fraction of the incident radiation that is scattered backward. As shown in Table 2.4 for infrared radiation, the albedo varies according to the surface type.

We refer to Exercise 2.2 for the albedo of a two-layer atmosphere.

Exercise 2.2 (Albedo of a Two-Layer Atmosphere) We assume that the atmosphere is composed by a layer of albedo A_0 with respect to the solar radiation. Typically, the layer represents clouds. Consider a second layer of albedo A_1 (typically for the sulfate aerosols; Fig. 2.10), above the first layer. The second layer is supposed to be a perturbation of the first layer ($A_0 \gg A_1$). Calculate the global albedo.
Hint: take into account multiple reflection between both layers.
Solution:
Let I be the incident solar radiation. We write R_n the reflected radiation (scattering backward to space), D_n the transmitted radiation from the upper layer to the lower layer, and U_n the reflected radiation from the lower layer to the upper layer, after n pairs of reflection on the layers.

After the first reflection, $R_0 = A_1 \times I$, $D_0 = (1 - A_1) \times I$, $U_0 = A_0 \times D_0$. The fluxes can be iteratively calculated with

$$U_n = A_0 \times D_n, \qquad D_{n+1} = A_1 \times U_n = A_0 A_1 \times D_n, \qquad R_{n+1} = (1 - A_1) \times U_n.$$

Thus, for $n \geq 0$,

$$R_{n+1} = A_0 (1 - A_1)^2 (A_0 A_1)^n \times I.$$

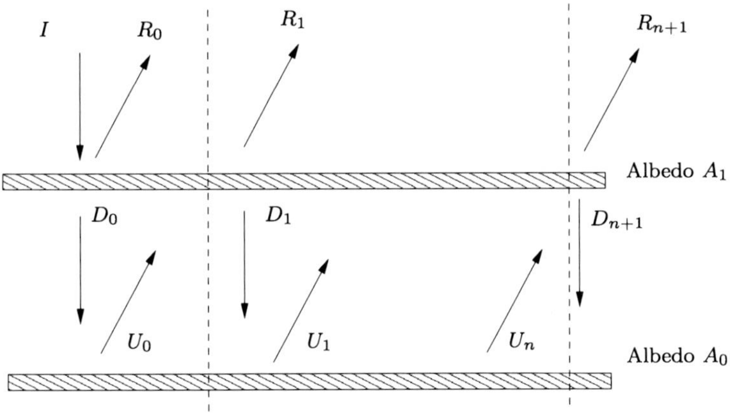

Fig. 2.10 Fate of the incident solar radiation (I) in a two-layer atmosphere

Summing over all the contributions due to reflection yields

$$R_0 + \sum_{n=0}^{\infty} R_{n+1} = \underbrace{\left(A_1 + \frac{A_0(1-A_1)^2}{1-A_0A_1}\right)}_{A} \times I.$$

Conserving the first-order terms in A_1 in an asymptotic expansion leads to

$$A \simeq A_1 + A_0(1-2A_1)(1+A_0A_1) \simeq A_0 + A_1(1-A_0)^2.$$

The global albedo is therefore not the sum of the layer albedos.

2.2 Applications to the Earth's Atmosphere

The concepts presented above are required for investigating the radiative properties of the atmosphere. The main application is the so-called *greenhouse effect.*

2.2.1 Solar and Terrestrial Radiation

2.2.1.1 Absorption Spectra

Calculating the absorption spectrum of the atmospheric compounds (namely of the absorption cross sections $\sigma_i^a(\lambda)$) is the objective of *spectroscopy*. This is based on the possible energy transitions for a given molecule. With the energy jumps ranked in an increasing order:

- the electronic transitions correspond to ultraviolet (UV) and visible radiation;

- the vibration transitions correspond to infrared (IR) radiation;
- the rotation transitions correspond to infrared and radio radiation.

For convenience, we do not cite the rotation-vibration transitions.

The vibration transitions can occur only for molecules presenting asymmetry. The geometrical structure of the molecules is therefore an important property for the interaction with the infrared radiation. This explains why CO_2, H_2O, N_2O or O_3 absorb and emit infrared radiation, on the contrary to O_2 or N_2. These gases are referred to as *greenhouse gases*.

The main absorption bands for the infrared radiation are shown in Table 2.5.

2.2.1.2 Absorption of Solar Radiation

Ionization The X-ray region of the electromagnetic spectrum (the most energetic radiation) is filtered in the ionosphere through the ionization process. Let A be a molecule or an atom. The ionization process reads

$$A + h\nu(\lambda) \longrightarrow A^+ + e^-. \qquad \text{(R 1)}$$

Ionization requires high energies, defined by the concept of *ionization potential*. The ionization potential corresponds to the maximum of the wavelengths (thus to the minimum of the energies) for which ionization occurs. Table 2.6 shows the ionization potential for a few species and atoms.

Ionization takes place in the upper atmosphere. Once the X-ray region is filtered, the remaining part of the spectrum is not energetic enough so that ionization is no longer possible. As expected, the electron density is an increasing function of

Table 2.5 Main absorption bands for infrared radiation. To be compared to Fig. 2.12. Source: [141]

Species	band center (μm)	band (μm)
CO_2	4.3	[4.1, 4.8]
	10.6	[8, 12]
	15	[12, 18]
O_3	9.6	[9, 10]
H_2O	6.2	[5.3, 6.9]
stratosphere	7.4	[6.9, 8]
	8.5	[8, 9]
H_2O	15	[12.5, 20]
troposphere	24	[20, 29]
	57	[29, 100]
CH_4	7.6	[6, 10]
N_2O	4.5	[4.4, 4.8]
	7.9	[7.4, 8.4]

altitude. For example, high values are responsible for the so-called *black out* that affects the communications of a space shuttle in the reentry phase (at an altitude of about 90–100 kilometers).

Ultraviolet and Visible Radiation The comparison between the spectrum of solar radiation at the top of the atmosphere and that at sea level is shown in Fig. 2.11.

For the ultraviolet solar radiation, the absorption is strong for molecular oxygen (O_2), ozone (O_3), water vapor (H_2O) and carbon dioxide (CO_2). These species, especially stratospheric ozone, filter the ultraviolet radiation, which anneals its adverse effects to health and vegetation. This motivates the focus on stratospheric ozone (Chap. 4).

Note the splitting between the ultraviolet and visible radiation:

- the shortwave solar radiations are absorbed in the ionosphere (X-ray region), in the mesosphere (Schumann-Runge continuum for O_2, $\lambda \in [150, 200]$ nm) and in the stratosphere (Hartley continuum for O_3, $\lambda \in [200, 300]$ nm);
- the atmosphere is transparent for the visible solar radiations: this property defines the so-called *atmospheric window*. This is a key point since it makes it possible to heat and to light the Earth's surface.

2.2.1.3 Absorption of the Terrestrial Radiation

The longwave infrared radiations, corresponding to terrestrial and atmospheric emissions, are absorbed by water vapor (H_2O), CH_4, CO_2 and O_3. These gases are characterized by their strong absorption of the infrared radiations (*greenhouse gases*).

Table 2.6 Ionization potential (wavelength in nm)

O_2	H_2O	O_3	H	O	CO_2	N	N_2	Ar
102.7	99	97	91.1	91	90	85	79.6	79

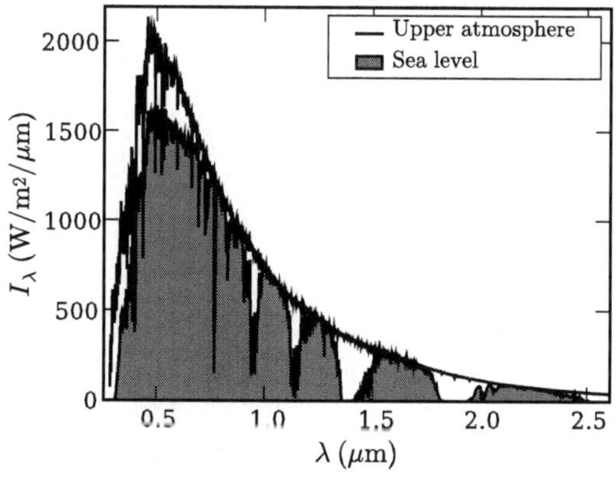

Fig. 2.11 Radiance spectrum at the top of the atmosphere and at sea level, respectively. The difference between the two curves corresponds to the absorption of solar radiation in the atmosphere

Figure 2.12 shows the radiance spectrum in the infrared region, as it would be measured by a sensor at the altitude of 70 kilometers, above a region with a temperature of 305 K. The sprectrum is computed by a numerical model that solves the radiative transfer equation (MODTRAN) for a standard atmosphere (USA 1976, clear sky). The Planck's distributions for the blackbody emissions are plotted for a few temperatures.

Note the main absorption bands related to the greenhouse gases. A simplified model for the greenhouse effect is investigated in Sect. 2.2.3. The altitude at which the absorbing (and then emitting) gases are located can be obtained by comparing the spectrum with the Planck distribution. The corresponding emission temperature results in an altitude (by using the vertical distribution of temperature). We refer to Exercise 2.3 for the study of this peaked altitude, with the concept of absorption layer.

Remark 2.2.1 (Passive Remote Sensing) *Passive remote sensing* (by satellital platforms) is based on the absorption of the infrared radiations by the atmosphere. The spectrum that is measured by the satellite gives a direct indication of the vertical distributions of the atmospheric trace species (typically water vapor) and of the temperature. This information is useful for numerical weather prediction.

We refer to Sect. 6.4 for a general introduction to the underlying methods (data assimilation and inverse modeling). The *forward model* is provided by the radiative

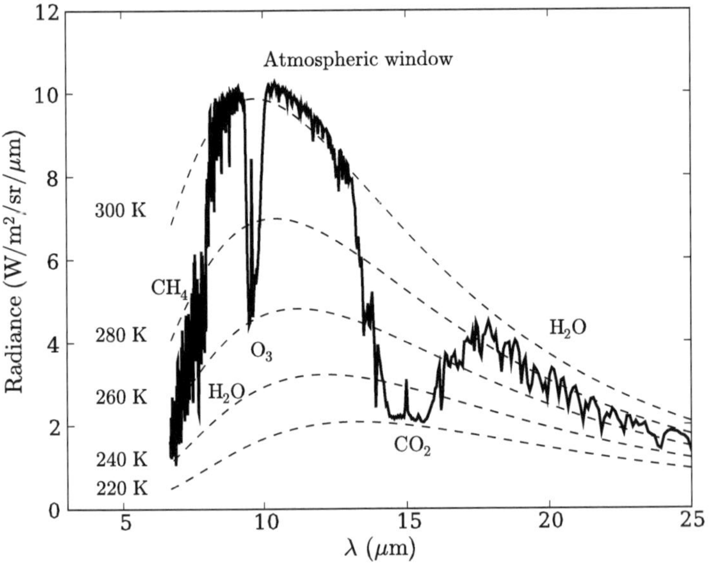

Fig. 2.12 Spectrum of the terrestrial infrared radiations, as measured by a sensor at the altitude of 70 km, above a surface with a temperature at 305 K. The sprectrum is a virtual spectrum computed by the numerical model MODTRAN for the standard atmosphere (USA 1976, clear sky). The Planck's distributions (blackbody emissions) are given at 220, 240, 260, 280 and 300 K. The corresponding greenhouse gases are indicated near the absorption peaks

transfer equation, for example in the form (2.32), where the temperature distribution, $T(\tau)$, is supposed to be known. Inverse modeling results in an estimation of this distribution on the basis of radiance observations.

Exercise 2.3 (Chapman's Theory of Absorption Layers) This exercise aims at introducing the concept of *absorption layer*. Consider a molecule with a known absorption spectrum for solar radiation. The maximum of the absorption is supposed to be peaked for a given wavelength. Emission and scattering by the atmosphere are neglected and we only take into account absorption in the following.

The vertical profile of concentration is given by $n(z) = n_0 \exp(-z/H)$, with H a scale height and z the vertical coordinate (increasing with increasing altitudes). Let θ be the angle of the incident radiation with respect to the vertical direction, and s be the abscissa along the radiation direction (decreasing with increasing altitude), respectively.

1. Calculate the distribution of the absorption rate $(-\mathrm{d}I/\mathrm{d}s)$.
2. Prove that there is a maximum at an altitude z_{max}. Comment.
3. Calculate z_{max} in the case of ozone (apply with $\theta = 0$).

Data for ozone:
– $\sigma_a = 4 \times 10^{-17}$ cm^2 molecule^{-1};
– $n(z) = n_0 \exp(-(z - z_0)/H)$ for $z \geq z_0 = 35$ km, $n_0 = 10^{12}$ molecule cm^{-3}, $H =$ 5 km.

Solution:

1. The evolution of the incident radiation is governed by the Beer-Lambert law, $\mathrm{d}I/\mathrm{d}s = -\sigma_a n(z(s))I$. At the top of the atmosphere, the boundary condition is $I(\infty)$. Since $\mathrm{d}z = -\cos\theta\, \mathrm{d}s$, this yields straightforward
$$I(z) = I(\infty) \exp\left(-\frac{H\sigma_a n_0}{\cos\theta} \exp(-z/H)\right).$$
The intensity (absorption, respectively) is an increasing (decreasing, respectively) function of the altitude, as expected. The absorption rate is given by
$$-\frac{\mathrm{d}I}{\mathrm{d}s} = \sigma_a n_0 I(\infty) \exp\left(-z/H - \frac{H\sigma_a n_0}{\cos\theta} \exp(-z/H)\right).$$
2. Setting to zero the second derivative of I gives the maximum of the absorption rate. The corresponding altitude is then
$$z_{max} = H \ln\left(\frac{H\sigma_a n_0}{\cos\theta}\right).$$
z_{max} depends on the incidence angle and on the properties of the absorbing medium (σ_a), but not on the incident radiation.
The absorption is a concave function of the altitude. Above z_{max}, the absorption is limited by the low value of the concentration; below z_{max}, the absorption is limited by the low value of the incident intensity since a large part of the intensity has been already absorbed.
3. For ozone, $z_{max} \simeq 50$ km (see Fig. 1.1, Chap. 1). This maximum of the absorption results in a maximum of an photolysis rate. We refer to Exercise 4.5 (Chap. 4) for an evaluation of the resulting increase in the atmospheric temperature.

2.2.2 *Radiative Budget for the Earth/Atmosphere System*

2.2.2.1 Solar Constant and Emission Effective Temperature

The solar constant of a given planet is defined by the solar radiation flux per unit area of the planet surface (Fig. 2.13, with the detail of the main notations). It can be calculated upon application of the Stefan-Boltzmann law to the Sun. The power emitted by the Sun (in W) is $4\pi R_s^2 \times \sigma T_s{}^4$ (with the Sun's radius $R_s = 6.96 \times 10^5$ km and the Sun's emission temperature $T_s = 5783$ K). At a distance r from the Sun, this generates a flux (expressed in $\mathrm{W\,m^{-2}}$)

$$S = \frac{R_s^2}{r^2} \sigma T_s{}^4. \tag{2.40}$$

With $r = 1.5 \times 10^8$ km (mean distance between the Sun and the Earth), we obtain the solar constant for the Earth, $S \simeq 1368\,\mathrm{W\,m^{-2}}$.

For a given planet, the *emission effective temperature* is defined as the emission temperature of a blackbody in radiative balance with the radiative fluxes received by the planet. The radiative budget for the Earth/atmosphere system is calculated as shown in Fig. 2.13: the solar radiation is intercepted by a surface πR_t^2 (with R_t the Earth's radius), the fraction A is reflected back to space (with A the global albedo of the Earth/atmosphere system, $A \simeq 0.3$). Finally, upon division by the Earth's surface $4\pi R_t^2$, this gives the following radiative budget (in $\mathrm{W\,m^{-2}}$),

$$\sigma T_e{}^4 = \frac{\pi R_t^2 S(1-A)}{4\pi R_t^2} = \frac{S(1-A)}{4}, \tag{2.41}$$

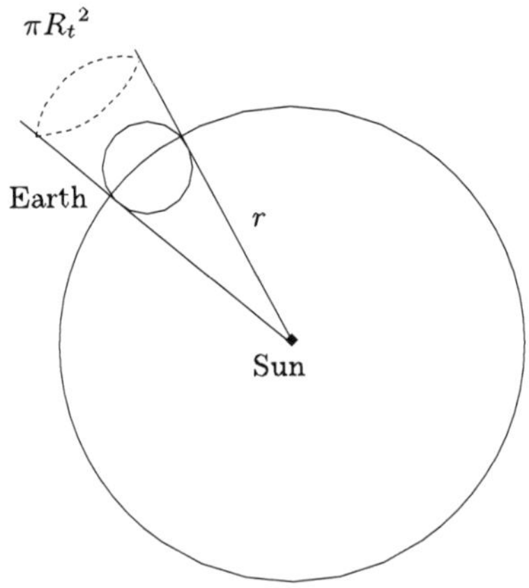

Fig. 2.13 Radiative flux received by the Earth (of radius R_t, at a distance r from the Sun). The solar flux per unit area of surface (expressed in $\mathrm{W\,m^{-2}}$) received by a sphere of radius r and centered at the Sun, is $S = 4\pi R_s^2 \sigma T_s{}^4/(4\pi r^2)$ with R_s and T_s the radius and the temperature of the Sun, respectively. The interception surface defined by the Earth is $\pi R_t{}^2$, resulting in a received flux $\pi R_t{}^2 \times S$ and in a flux per unit area of the Earth's surface $\pi R_t{}^2 \times S/(4\pi R_t{}^2) = S/4$

with T_e the emission effective temperature of the Earth/atmosphere system. Rearranging yields

$$T_e^4 = \frac{S(1-A)}{4\sigma}. \tag{2.42}$$

The emission effective temperature is therefore a function of the albedo and of the distance to the Sun (which defines the solar constant).

With $A \simeq 0.3$ we calculate $T_e \simeq 255$ K, to be compared with the mean temperature at the Earth's surface (about 288 K). The difference (33 K) corresponds to the greenhouse effect (which makes it possible to have a surface temperature greater than −18 °C!) and results from an energy redistribution from the atmosphere to the ground.

The available flux at the Earth's surface is $S/4 \simeq 342\ \mathrm{W\,m^{-2}}$. It is usually written as F_s.

Exercise 2.4 (Variation of the Solar Constant) Calculate the variation of the solar constant S during one year (the distance from the Sun to the Earth varies from 1.469 to 1.520×10^8 km)?

Solution:

We apply (2.40), which yields $S \in [1320, 1410]\ \mathrm{W\,m^{-2}}$. Thus, the variation is up to $90\ \mathrm{W\,m^{-2}}$.

2.2.2.2 Energy Budget for the Earth/Atmosphere System

The temperature of the Earth/atmosphere system is mainly fixed by the radiative properties of the Earth and of the atmospheric compounds (gases, clouds and aerosols).

Let us express the global energy budget for the Earth/atmosphere system (Fig. 2.14). We consider the received solar energy, the energy fluxes for the Earth and the energy fluxes for the atmosphere.

The resulting budget is a global budget and does not take into account the seasons, the diurnal cycles (day/night) and the spatial location. Except for the solar flux, the relative uncertainties are at least of 10%. Thus, the fluxes are only crude estimations.

Received Solar Energy The solar energy is the only energy source for the Earth/atmosphere system. The incident solar radiation represents $342\ \mathrm{W\,m^{-2}}$, to be split as follows.

- $77\ \mathrm{W\,m^{-2}}$ is reflected back to space by clouds and aerosols (namely 22%);
- $67\ \mathrm{W\,m^{-2}}$ is absorbed by gases and clouds (namely 20%);
- $168\ \mathrm{W\,m^{-2}}$ is scattered and then absorbed by the Earth (namely 49%);
- $30\ \mathrm{W\,m^{-2}}$ is reflected by the Earth and then scattered to space (namely 9%).

The planetary albedo is therefore about 0.31 (the reflected energy is $107\ \mathrm{W\,m^{-2}}$, to be compared with the received energy, $342\ \mathrm{W\,m^{-2}}$).

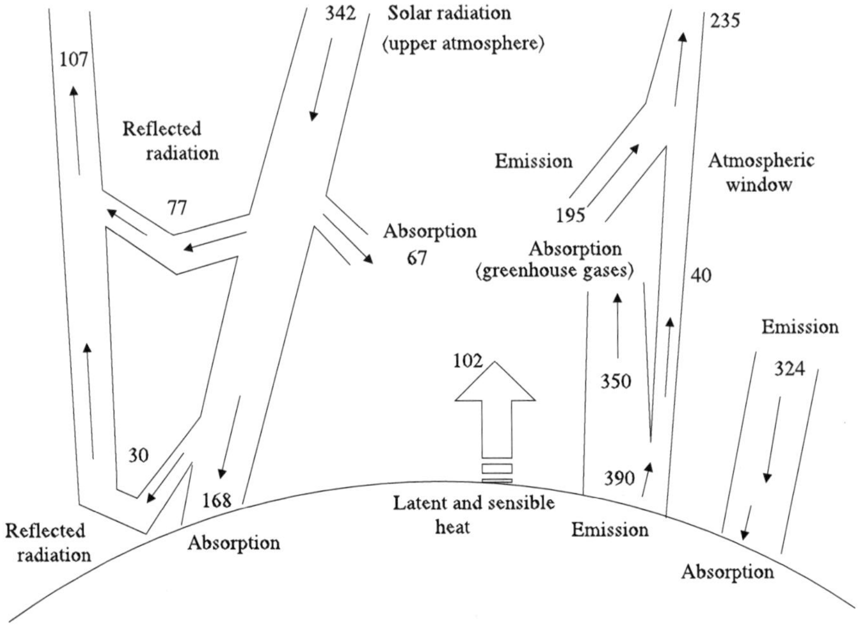

Fig. 2.14 Global energy budget for the Earth/atmosphere system. The fluxes are expressed in $\mathrm{W\,m^{-2}}$. The values are indicative. Source: [74]

Earth's Energy Budget The energy fluxes for the Earth are related to:

- radiative energy;
- *sensible* heat (Chap. 3), connected to the vertical turbulent motions;
- *latent* heat, produced by the water cycle (see Exercise 3.1);
- heat conduction in the soil (neglected in a first approximation).

With the same units as above, the Earth absorbs $168\,\mathrm{W\,m^{-2}}$ in the shortwave solar radiation. It emits $390\,\mathrm{W\,m^{-2}}$ in the longwave radiation into the atmosphere.

As seen before, the emitted radiation is strongly absorbed by the atmosphere: actually, up to $350\,\mathrm{W\,m^{-2}}$ is absorbed (namely 90%) and $40\,\mathrm{W\,m^{-2}}$ (10%) is transmitted to space through the *atmospheric window*.

Moreover, the Earth receives $324\,\mathrm{W\,m^{-2}}$ from the longwave radiation emitted by the atmosphere. The whole part is supposed to be absorbed at the Earth's surface (in a first approximation).

The radiative budget is then positive for the Earth: *there is a gain of radiative energy for the Earth* ($168 + 324 - 390 = 102\,\mathrm{W\,m^{-2}}$).

At equilibrium, the energy budget for the non-radiative part is therefore negative: the Earth has an energy loss ($-102\,\mathrm{W\,m^{-2}}$) by latent heat (water evaporation) and sensible heat (vertical turbulent motion).

Atmosphere's Energy Budget The atmosphere absorbs 67 W m^{-2} in the shortwave solar radiations and 350 W m^{-2} in the longwave terrestrial radiations. It emits 519 W m^{-2} in the longwave radiations (including 324 to the Earth and 195 to space).

The radiative budget for the atmosphere is therefore negative ($350+67-519=-102$ W m^{-2}): *the atmosphere has a loss of radiative energy*. At equilibrium, the energy gain is provided by latent and sensible heat fluxes (102 W m^{-2}, coming from the Earth).

The Earth/atmosphere system absorbs 235 W m^{-2} from the solar radiation ($168+67$), to be compared with the value given in Table 2.1.

2.2.3 Greenhouse Effect

2.2.3.1 A Toy Model for the Greenhouse Effect

The concept of greenhouse effect was introduced in order to justify the elevated temperature at Venus' surface (more than 700 K, see Exercise 2.5), as compared to the emission effective temperature (230 K): the hypothesis was formulated by Rupert Wildt in the 1930s and Carl Sagan in 1962, before the confirmation by the measurements of the CO_2 mixing ratio in the atmosphere of Venus.

Exercise 2.5 (Jupiter, Mars and Venus) The characteristics of Jupiter, Mars and Venus are given in Table 2.7. Assess the possibility of a greenhouse effect for these planets. What could be the other source of energy for Jupiter?

Solution:

We use (2.40) to get the general formula

$$T_e = T_s\sqrt{\frac{R_s}{2r}}(1-A)^{0.25}, \qquad (2.43)$$

with A the planet albedo and r the distance from the Sun. This gives $T_e = 88, 232$ and 216 K for Jupiter, Venus and Mars, respectively. Thus, there exists a strong greenhouse effect for Venus on the contrary to Mars. The composition of the Jovian atmosphere does not support the existence of a greenhouse effect. The high value of the surface temperature can be explained by internal energy sources. If ΔE_{int} stands for the internal energy source, the energy balance should read

$$\sigma T^4 = \sigma T_e{}^4 + \Delta E_{int},$$

Table 2.7 Characteristics of Jupiter, Venus and Mars

Planet	Distance (km) from the Sun	Albedo	C_{CO_2} (%)	T at ground (K)
Jupiter	7.8×10^8	0.73		130
Venus	1.08×10^8	0.75	0.96	700
Mars	2.28×10^8	0.15	0.95	220

namely $\Delta E_{int}/(\sigma T_e{}^4) = (T/T_e)^4 - 1 \simeq 4$. The internal energy source is up to four times as large as the solar energy flux. For the Earth, the contribution due to the internal energy (geothermy) can be neglected (Table 2.1).

Consider the atmosphere as a virtual layer at a given distance from the Earth's surface. Let us assume that nothing happens in the region between the surface and the layer (Fig. 2.15).

From the radiative properties of the atmosphere (strong absorption of the infrared radiations and atmospheric window for the visible radiations), we assume that:

- the layer and the Earth are at radiative equilibrium: in a first approximation, we neglect the non-radiative energy flux;
- the layer reflects a fraction A of the incident solar radiation F_s. It absorbs a fraction a_S and transmits to the Earth a fraction $(1 - a_S)$ of the remaining radiation $(1 - A)F_s$. The Earth is supposed to absorb the whole received radiation;
- the Earth emits longwave radiations U (*up*): a fraction (a_T) is absorbed by the layer while the remaining part $(1 - a_T)$ is transmitted to space;
- the layer is then heated and emits a radiation D (*down*): a fraction a is transmitted to the Earth and then absorbed.

Let us investigate the sensitivity of the Earth's temperature (T) with respect to the absorption coefficient of the layer for the terrestrial radiation (a_T).

Taking into account the previous data (Fig. 2.14) yields

$$a_T = \frac{350}{390}, \qquad a_S = \frac{198}{265}, \qquad a = \frac{324}{519}. \tag{2.44}$$

The budget energy for the Earth and the atmospheric layer reads

$$\begin{cases} (1 - a_S)(1 - A)F_s + aD = U = \sigma T^4 \\ a_S(1 - A)F_s + a_T U = D. \end{cases} \tag{2.45}$$

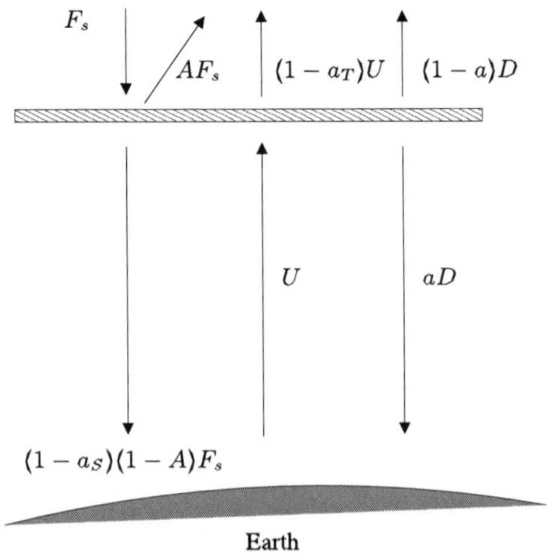

Fig. 2.15 A toy model for the greenhouse effect

Thus,

$$\sigma T^4 = U = \frac{(1-a_S)+a\,a_S}{1-a\,a_T}(1-A)F_s. \tag{2.46}$$

The function $T(a_T)$ is an *increasing* function with respect to the absorption coefficient of the layer (a_T). The more absorbing the layer is, the higher the Earth's temperature is: this is the *greenhouse effect*, to be measured by the difference between the emission effective temperature T_e given by (2.42) and the surface temperature T. This gives straightforward

$$T = \left(\frac{(1-a_S)+a\,a_S}{1-a\,a_T}\right)^{0.25} T_e. \tag{2.47}$$

As $T_e \simeq 255$ K, we get $T \simeq 288$ K (not far from the observed value).

2.2.3.2 Radiative Forcing, Feedbacks and Global Warming Potentials

The estimation of the impact on the temperature, and more generally on the climate, is an active resarch topic. It can also generate controversies since the scientific results play a leading role for the decision-making. We refer to the IPCC reports (*Intergovernmental Panel on Climate Change*, [106]). A key point is the attention paid to the "robustness" of the results. What are the levels of uncertainties? What is the level of scientific understanding (LOSU)?

In the following, we focus on a few key elements, namely the concepts of *radiative forcing*, *feedbacks* and *global warming potential*.

Radiative Forcing The *radiative forcing* is defined in the following way ([106]):

> The radiative forcing of the surface-troposphere system due to perturbation in or the introduction of an agent (say, a change in greenhouse gas concentrations) is the change in net (down minus up) irradiance (solar plus long-wave; in W m^{-2}) at the tropopause AFTER allowing the stratospheric temperatures to readjust to radiative equilibrium, but with surface and tropospheric temperatures and state held fixed at the unperturbed values.

For example, a modification of the albedo perturbs the balance (2.41). In coherence with the definition above (*down minus up*), the balance reads

$$F = F_s(1-A) - \sigma T_e{}^4 = 0. \tag{2.48}$$

Applying a perturbation ΔA for the albedo leads to a forcing $\Delta F = -F_s \Delta A$. An increase in the albedo results, as expected, in a negative radiative forcing (cooling effect).

Similarly, a modification ΔF_s of the received solar flux $F_s = S/4$ leads to a forcing $\Delta F = \Delta F_s(1-A)$. An increase in F_s results in a positive radiative forcing.

The impact on the effective temperature T_e (and then on the surface temperature T) can be estimated (see Exercise 2.7 for a more rigorous approach) by assuming that the radiative forcing is an energy flux to be added to the received solar energy. The resulting equilibrium is

$$F_s(1-A)+\Delta F-\sigma(T_e+\Delta T_e)^4=0. \tag{2.49}$$

The *climate sensitivity parameter* is often written as λ_0[4] and is defined by

$$\lambda_0=\frac{\Delta T_e}{\Delta F}. \tag{2.50}$$

It is expressed in $\mathrm{K\,(W\,m^{-2})^{-1}}$. Actually, the resulting value, about 0.3 (see Exercise 2.6), is an underestimation because couplings and *feedbacks* are not taken into account. We refer to Exercise 2.7 for the concept of feedbacks.

Exercise 2.6 (Estimation of the Climate Sensitivity Parameter) Calculate the sensitivity of the emission effective temperature with respect to a radiative forcing.
Solution:
Linearizing (2.49) yields $\Delta F\simeq 4\sigma T_e^3\Delta T_e$, namely

$$\lambda_0=\frac{\Delta T_e}{\Delta F}\simeq\frac{1}{4\sigma T_e^3}.$$

As $T_e=255\,\mathrm{K}$, we obtain $\lambda_0\simeq 0.27\,\mathrm{K\,(W\,m^{-2})^{-1}}$. This value does not take into account the feedbacks and is an underestimation.

This concept can be generalized to the radiative forcing related to a greenhouse gas, let us say X_i. Increasing the concentration of X_i results in decreasing the outgoing flux of the terrestrial radiations (we omit the possible feedbacks). Equivalently, this can be viewed as an increase in the incoming solar radiation. This calculation has to be carried out with the other parameters considered at constant values. A few values, taken from the IPCC report of 2001 ([106]), are shown in Table 2.8. Note the large uncertainties related to aerosols (see Sect. 2.2.4). For example, the sign of the radiative forcing related to mineral aerosols (e.g. dust) is not fixed (Problem 2.1).

Feedbacks The sensitivity, as presented above, does not take into account the resulting modifications of the other components of the Earth/atmosphere system, due to radiative forcing and temperature modification. The key feedbacks to describe are listed below.

- Water Vapor Feedback
  A warmer atmosphere is wetter, which results in an increasing greenhouse effect related to water vapor (positive feedback). Actually, the saturation vapor pressure of water vapor is an increasing function of temperature, which favors the gas-phase state of water (Sect. 5.2.2).
  Moreover, the increasing temperature can amplify the water evaporation from the oceans. An extreme case corresponds to the so-called *runaway greenhouse effect*.

[4]This is a standard notation (not to be mixed up with that used for wavelength).

The evaporation of water from oceans is then no longer compensated by cloud formations and subsequent precipitations, since the saturation vapor pressure is too high (due to the high values for the temperature). Water evaporates but cannot condense in the atmosphere (Fig. 2.16). This runaway is the hypothesis that is usually formulated for justifying the lack of water at Venus' current surface. Actually, the effective temperature of Venus was, initially, higher than that of the Earth.

- Albedo
  The resulting modification of the albedo, due to the modification of the Earth's surface (e.g. the decrease of the ice cover), is another positive feedback. A warmer atmosphere results in ice melting, which contributes to decrease the albedo and therefore to increase the absorption of solar radiation at the Earth's surface.
- Clouds and Aerosols
  Taking into account the impact of the cloud cover is more complicated. The clouds reflect the solar radiation but, meanwhile, they also absorb part of the infrared radiations. The impact of aerosols is a major uncertainty (Sect. 2.2.4).

The concept of feedback is formalized in Exercise 2.7.

Exercise 2.7 (Formalization of the Feedback Concept) A simple way for formalizing the concept of feedback is to consider that the temperature (T) is a function not only of the radiative flux F but also of the other variables, $\{x_i\}_i$ (e.g. albedo, cloud cover, concentrations of greenhouse gases, ...): $T = f(F, \{x_i\}_i)$. Assuming that the variables x_i also depend on T, calculate the sensitivity $\mathrm{d}T/\mathrm{d}F$.
Solution:
The sensitivity reads

$$\lambda = \frac{\mathrm{d}T}{\mathrm{d}F} = \underbrace{\frac{\partial T}{\partial F}}_{\lambda_0} + \sum_i \frac{\partial T}{\partial x_i}\frac{\partial x_i}{\partial F}.$$

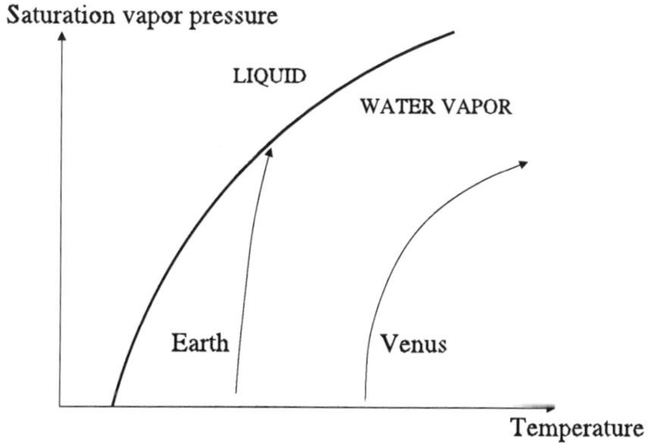

Fig. 2.16 Runaway greenhouse effect (schematic historical evolution of the atmospheres of Venus and of the Earth). Source: [58]

Taking into account the dependence upon T, $x_i(T)$, we obtain $\partial x_i/\partial F = \mathrm{d}x_i/\mathrm{d}T \times \mathrm{d}T/\mathrm{d}F$. Thus,

$$\lambda = \frac{\lambda_0}{1 - \sum_i f_i}, \qquad f_i = \frac{\partial T}{\partial x_i}\frac{\mathrm{d}x_i}{\mathrm{d}T}.$$

The variable f_i is the so-called *feedback factor* for the variable x_i (connected to a physical process). It is dimensionless. For the calculation of f_i, we have to distinguish $\partial T/\partial x_i$, corresponding to the temperature dependence on the variable (through the radiative balance), from $\mathrm{d}x_i/\mathrm{d}T$, corresponding to the fact that the variable is driven by the temperature (through the physical and chemical atmospheric processes). If both values are positive, (that is to say if an increase in the temperature results in an amplification of the process, leading to an increase in the temperature), we have a positive feedback.

From the IPCC works ([106], Table 2.8), the antropogenic radiative forcing between 1750 (preindustrial times) to 1998 can be estimated as 2.43 W m^{-2} (including 1.46 W m^{-2} for CO_2, 0.48 W m^{-2} for CH_4, 0.34 W m^{-2} for halogen compounds and 0.15 W m^{-2} for N_2O). For an increase in the temperature estimated as 0.6 K, the climate sensitivity parameter is therefore about 0.25.

Table 2.8 Mean yearly radiative forcing between 1750 and 2000, as estimated by the IPCC works in 2001. "LOSU" stands for *level of scientific understanding*. Source: [106]

Species X_i	ΔF_{X_i} (W m^{-2})	Uncertainties	LOSU
Greenhouse gases	2.43	10%	High
incl. CO_2	1.46	–	–
incl. CH_4	0.48	–	–
incl. N_2O	0.14	–	–
incl. halogens	0.34	–	–
Stratospheric O_3	−0.15	67%	Medium
Tropospheric O_3	0.35	43%	Medium
Sulfate aerosols (direct)	−0.4	[−0.8, −0.2]	Low
Biomass burning aerosols (direct)	−0.2	[−0.6, −0.07]	Very low
Soot (elemental carbon, direct)	0.1	[0.03, 0.3]	Very Low
Organic aerosols (direct)	−0.1	[−0.3, −0.03]	Very Low
Mineral aerosols	[−0.6, 0.4]	–	Very Low
Indirect effect of aerosols	[−2, 0]	–	Very Low
Condensation trails (aircrafts)	0.02	350%	Very Low
Cirrus formation (aircrafts)	[0, 0.04]	–	Very Low
Surface albedo (land use cover)	−0.2	100%	Very Low
Solar activity	0.3	67%	Very Low

Global Warming Potential A third key concept is the atmospheric residence time of the species. For a species X_i, the *global warming potential*, GWP_i, is defined by comparing, over a time interval $[0, t_f]$, the radiative forcing resulting from a 1 kg emission of X_i at $t = 0$, with that resulting from the same emission of a reference species (usually carbon monoxide CO_2). Thus,

$$GWP_i = \frac{\int_0^{t_f} \Delta F_{X_i}(t)\, dt}{\int_0^{t_f} \Delta F_{CO_2}(t)\, dt}. \tag{2.51}$$

We refer to Table 2.9 for a few values (note that GWP depends on t_f). For example, if the final time is 100 years, reducing the emissions of CFC-11 by 1 kg is as efficient as a 4600 kg emission reduction for CO_2.

Actually, this indicator only takes into account the so-called *direct* effects. It is sometimes referred to as *direct GWP*. It does not describe the indirect effects, resulting from the physical and chemical processes induced by the emission of a given species for other species. The *indirect* GWP is difficult to estimate. For example, the emission of methane (CH_4) has several indirect effects, among which:

- an increased chemical production of ozone (this indirect effect is estimated to be up to 25% of the direct effect);
- an increased production of water vapor in the stratosphere (5% of the direct effect).

2.2.4 Aerosols, Clouds and Greenhouse Effect

The uncertainties related to the radiative behavior of the aerosols and clouds are a challenging issue for providing an accurate estimation of the anthropogenic greenhouse effect.

Table 2.9 *Direct* global warming potential (normalized with respect to CO_2) for a few species, at 20, 100 and 500 years. Source: [106]

Species	Residence time (year)	GWP at 20 years	GWP at 100 years	GWP at 500 years
CO_2	100	1	1	1
CH_4	12	62	23	7
N_2O	114	275	296	156
CFC-11	45	6300	4600	1600
CFC-12	100	10200	10600	5200
CFC-13	640	10000	14000	16300

2.2.4.1 Direct Effect of Clouds

The clouds play a leading role in the radiative forcing with two opposite effects:

- a cooling effect with respect to the solar radiation.
They constitute a scattering medium for the solar radiation and contribute to the global albedo. The reflection of the incoming solar radiations back to space depends on the cloud type and on the microphysical properties: the cloud albedo (for solar radiation) varies from 0.20 (for thin stratus) to 0.90 (for cumulus), as shown in Table 2.4. It is usually estimated that the cloud contribution is up to 0.15 for the albedo of the Earth/atmosphere system, namely the half of the total albedo.
For the "current" atmosphere, the resulting cloud contribution to the radiative budget is estimated to be $-50\,\mathrm{W\,m^{-2}}$ (for the solar radiation).
- a greenhouse effect for the infrared radiation.
On the contrary, the clouds increase the scattering of the shortwave radiations to the Earth's surface. The key point is that they also increase the emissivity and the absorptivity of the atmosphere for longwave radiations ($\epsilon \simeq 0.97$ for clouds). Hence, the clouds act as strong "greenhouse gases". Note that they emit at temperatures lower than the temperature surface (or than the temperature for a clear sky). The resulting contribution to the radiative balance is estimated to be about $+25\,\mathrm{W\,m^{-2}}$.

The aggregated effect is then a cooling effect (with a contribution of about $-25\,\mathrm{W\,m^{-2}}$ for the radiative budget). A more accurate estimation of this impact is required: remember that a doubling of the CO_2 mixing ratio results in a perturbation of a few $\mathrm{W\,m^{-2}}$.

The impact on the clouds resulting from a modification in the atmospheric composition is difficult to estimate, which provides another illustration of the concept of feedback.

For the solar radiation, the increase of the water vapor mixing ratio results in an increase of the cloud cover, and then of the global albedo (cooling effect). Meanwhile, it can also result in an increase of precipitations, leading to a decreasing cloud lifetime and therefore to a decreasing albedo.

For the infrared radiation, the clouds take part in the greenhouse effect. For a finer estimation, the cloud altitude has to be taken into account: for example, clouds at high altitudes (e.g. cirrus) have a warming effect.

2.2.4.2 Direct Effects of Aerosols

Cooling Effect of Sulfate Aerosols Due to their radiative properties, the sulfate aerosols have a direct effect for the solar radiation. This is a cooling effect due to an increase in the planetary albedo. This is sometimes referred to as the *whitehouse effect* ([127]).

Decreasing temperatures have been measured after eruption of the Pinatubo Mount (1991). The eruption resulted in an increase in the stratospheric sulfate concentrations: the resulting radiative forcing was estimated to be up to $-4\,\mathrm{W\,m^{-2}}$ in 1992 with a fast decrease down to $-0.1\,\mathrm{W\,m^{-2}}$ in 1995. The cooling effect for the temperature was estimated to be of a few tenths of K.

The radiative forcing related to the increase in the albedo (to be applied for the effects above) is $\Delta F = -F_s \Delta A$. With Exercise 2.8, a coarse estimation for the aerosol albedo is about 2×10^{-2}. Considering that 25% of the aerosols are anthropogenic aerosols, we obtain an albedo connected to the anthropogenic aerosols of $A_1 = 5 \times 10^{-3}$. The perturbation for the global albedo cannot be calculated directly. The calculation of the albedo for a two-layer atmosphere is detailed in Exercise 2.2. The resulting perturbation for the global albedo is actually $\Delta A = A_1(1 - A_0)^2$, where A_0 is the albedo of the other atmospheric compounds (0.3 in a first approximation).

Exercise 2.8 (Estimation of the Aerosol Albedo) Estimate, for the aerosols, the contribution to global albedo with respect to the solar radiation.
Data:

- total optical depth associated to the aerosols: $\tau \simeq 0.12$ (mean value above the oceans, [57]);
- suppose that a fraction $\beta = 23\%$ of the scattered radiation is scattered back to space).

Solution:
The incident radiation, I, can be split in a transmitted radiation, $\exp(-\tau)I$, and in a scattered radiation, $(1 - \exp(-\tau))I$. A fraction β of the scattered radiation is reflected back to space, resulting in an albedo $A_1 = \beta(1 - \exp(-\tau)) \simeq \beta\tau$. Thus, $A_1 \simeq 0.026$.

Finally, we obtain $\Delta F \simeq -0.85\,\mathrm{W\,m^{-2}}$. Note that this value is similar to the sum of the direct radiative forcings related to anthropogenic aerosols in Table 2.8.

Hence, the particulate pollution (a more localized pollution) has reduced part of the anthropogenic greenhouse effect. The indirect effect of an improvement of the local air quality is an increase of the anthropogenic greenhouse effect: this is an example of an *atmospheric dilemma* (see Introduction and Exercise 2.9). Another application of the cooling effect related to aerosols is the so-called *nuclear winter* (the strong cooling induced by a nuclear war; Problem 2.4).

Exercise 2.9 (Climate Engineering and Greenhouse Effect) In order to counterbalance the reduction of the cooling effect related to sulfate aerosols, due to the improvement of the local air quality, P. J. Crutzen (who was awarded the Nobel Prize in 1995), suggests that we emit sulfate particles into the stratosphere, "[*as an*] *escape route against strongly increased temperature*" (see Introduction). The stratospheric emission is motivated by the higher residence times of the stratospheric particles (from 1 to 2 years, versus one week in the troposphere, Chap. 5). This exercise aims at giving the basis of Crutzen's arguments.

In 1991, the volcanic eruption of the Pinatubo Mount emitted about 10 Tg of sulfur into the stratosphere. A few months later, 6 Tg were still in the stratosphere.

This resulted to a radiative forcing of about $-4.5\,\mathrm{W\,m^{-2}}$ and to a diminution of the mean surface temperature of about 0.5 K in 1992.

Assume that the cost for sending 1 Tg S into the stratosphere is about 25 billions of dollars (on the basis of technologies supposed to be available). Calculate the project cost in order to compensate the radiative forcing due to the improvement of air quality (supposed to be about $+1.5\,\mathrm{W\,m^{-2}}$) and that due to a doubling of CO_2 concentration (supposed to be about $+4\,\mathrm{W\,m^{-2}}$). For indication, the magnitude of the current anthropogenic sulfur emissions is 55 Tg $\mathrm{year^{-1}}$.

Note that there are possible adverse effects of such emissions. For example, this could result in an increase of the stratospheric ozone destruction (as observed after the Pinatubo eruption), due to the role of the aerosol sulfates in heterogeneous processes (Sect. 4.2).

Solution:

Using the data for the Pinatubo eruption, with a linear assumption, the radiative forcing induced by a 1 Tg S emission is about $-0.75\,\mathrm{W\,m^{-2}}$ (4.5/6). As the lifetime varies from 1 to 2 years in the stratosphere, it is then required to inject from 1 to 2 Tg S per year into the stratosphere, so that the effect due to the improvement of the local air quality could be compensated. The resulting cost ranges from 25 to 50 billions of dollars. For compensating the doubling of CO_2 concentration, one should emit from 2.5 to 5 Tg S per year, with an annual cost ranging from 60 to 120 billions of dollars.

To know more ([26]):

P. CRUTZEN, *Albedo enhancement by stratospheric sulfur injections: a contribution to resolve a policy dilemma*? Climatic Change, **77** (2006), pp. 211–219

Sensitivity with Respect to the Aerosol Composition: Black Carbon and Liquid Water Content Actually, the radiative properties of aerosols depend on the size, on the chemical composition, on the mixing state (the way the components are mixed) and on the liquid water content. We refer to Problem 2.1 for the study of the sensitivity of the direct effect with respect to the aerosol chemical composition.

We investigate two characteristics in order to illustrate the related uncertainties: the first one is related to black carbon, the second one to the aerosol liquid water content.

Even if its mass contribution is rather low, *black carbon* has a major radiative impact that is difficult to assess. The resulting radiative forcing is positive since black carbon absorbs the infrared radiations. The amplitude is however highly uncertain: the aerosol mixing state strongly impacts the radiative forcing, with a multiplying factor up to 2 or 3 (Table 2.10). In case of *internal mixing*, the aerosol species are well mixed and are represented by one or a few families. In case of *external mixing*, the aerosols species are supposed not to be mixed (Sect. 5.1.1.4). Another key property is the aerosol geometry: in the so-called carbon core/shell model, the aerosol core is composed of black carbon while the surface shell is composed of organic and inorganic species (Fig. 2.17).

As shown in Table 2.10, the radiative forcing due to black carbon increases from the case of external mixing to the case of internal mixing. This results straightfor-

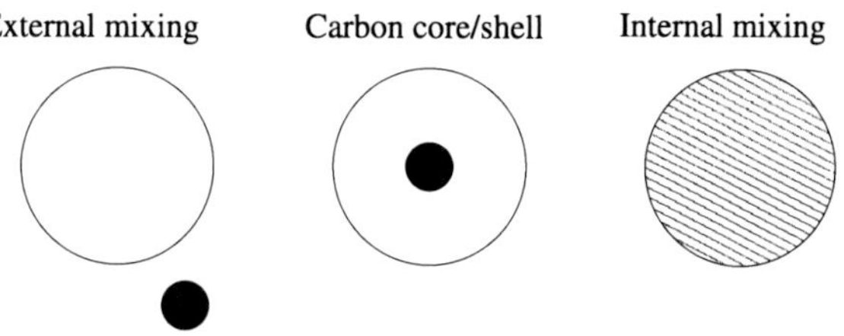

Fig. 2.17 Schematic representation of the aerosol mixing state: external mixing, carbon core/shell model, internal mixing. The black part stands for carbon (soot, elemental carbon or *black carbon*)

Table 2.10 Radiative forcing (in W m^{-2}) due to carbon aerosols (soot), as a function of the mixing state. Sources: [62, 63]

Mixing state	[62]	[63]
external mixing	0.27	0.31
carbon core/shell	0.54	0.55
internal mixing	0.78	0.62

ward from the ranking of the extinction coefficients. To date, many studies have been based on the assumption of external mixing (easier to implement for modelers). It may be therefore possible that the positive radiative forcing related to black carbon has been underestimated. The values obtained with the carbon core/shell model in the case of multiple internally-mixed families (due to hetero-coagulation of aerosols; probably the most realistic assumption), could rank black carbon as the second contributor to the greenhouse effect (Table 2.8). This could therefore motivate the reduction of black carbon emissions.

Assessing the impact of the aerosol liquid water content is another challenging issue. For aerosols containing hydrophilic compounds (that is, compounds that favor water conditions for appropriate values of the relative humidity), an increase in the humidity results in an increase in the aerosol size, which affects the radiative properties. A typical illustration is given in Fig. 2.18. An accurate estimation of this effect requires an accurate description of humidity and of the aerosol microphysical properties. One key property is the so-called deliquescence relative humidity, defined as the relative humidity above which water condensation takes place (Sect. 5.1, Chap. 5).

2.2.4.3 Indirect Effects of Aerosols

Part of the aerosol distribution provides the so-called *cloud condensation nuclei*, from which the cloud drops are produced (Sect. 5.2.3). Since the clouds affect the radiation, especially the solar radiations, this generates an *indirect effect* related to aerosols (Table 2.12). A key point is the decisive role of the cloud microphysical properties that are connected to aerosols: for example, the cloud albedo with respect to the visible radiations is a decreasing function of the drop size, or, equivalently, an increasing function of the drop number density (Twomey effect, Problem 2.2 and Fig. 2.19). These effects have been observed during the international field campaign INDOEX, devoted to the study of the Asian "brown cloud" (Exercise 2.10 and Table 2.11).

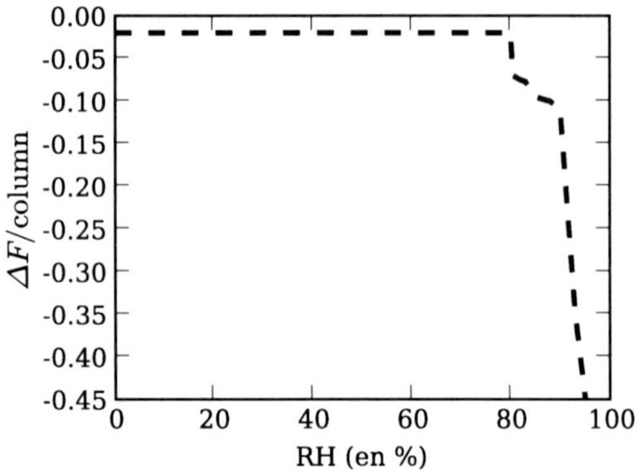

Fig. 2.18 Sensitivity of the radiative forcing (normalized with respect to the aerosol column; expressed in $\mathrm{W\,m^{-2}/(mg\,m^{-2})}$) with respect to the relative humidity. The size distribution is centered at a dry radius 0.1 µm; the aerosol is composed of 90% sulfate and 10% black carbon (carbon core). The water condensation occurs for a relative humidity above 80% (deliquescence relative humidity for ammonium sulfate, Table 5.6). Source: [85]

Table 2.11 Microphysical characteristics measured over the Indian Ocean: comparison between a *pristine* cloud and a *brown* cloud (Exercise 2.10) with the same liquid water content $L = 0.15\,\mathrm{g\,m^{-3}}$. The aerosols are taken into account for a diameter $d_p \geq 50\,\mathrm{nm}$. Source: [118]

Cloud	aerosol number density	cloud drop number density	effective radius (r_e)
pristine cloud	$500\,\mathrm{cm^{-3}}$	$90\,\mathrm{cm^{-3}}$	$r_e \geq 7.5\,\mu\mathrm{m}$
brown cloud	$1500\,\mathrm{cm^{-3}}$	$315\,\mathrm{cm^{-3}}$	$r_e \leq 6.5\,\mu\mathrm{m}$

Table 2.12 Indirect effects of aerosols. ΔF_0 stands for the radiative forcing at the Earth's surface and p_0 is the rain intensity (expressed in $\mathrm{mm\,hr^{-1}}$). Source: [90]

Effect	description	impact
Indirect effect of aerosols for clouds with a fixed liquid water content (cloud albedo, Twomey effect)	Increasing reflection of the solar radiation for small drops	$\Delta F_0 < 0$
Indirect effect of aerosols for clouds with a variable liquid water content (lifetime effect)	Decreasing precipitations and increasing cloud lifetime for small drops	$\Delta F_0 < 0$, $p_0 \downarrow$
Semi-direct effect	The absorption of solar radiation (soot) can increase the cloud drop evaporation	$\Delta F_0 < 0?$, $p_0 \downarrow$

Exercise 2.10 (Brown Clouds) Several studies, especially over the southern Asia and the Indian Ocean, have indicated the existence of extended persistent plumes of particulate matter, downwind urban polluted areas. These plumes are usually re-

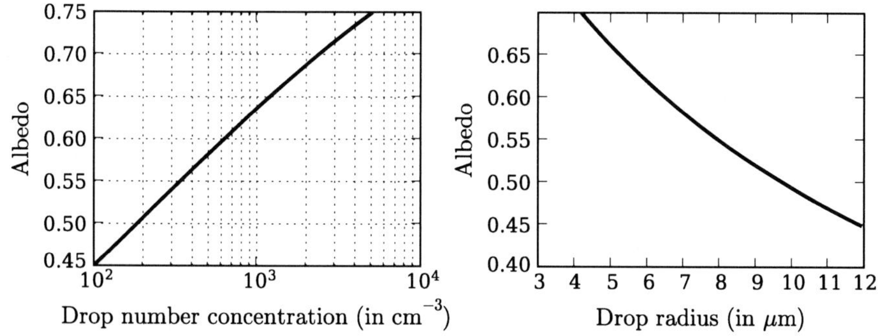

Fig. 2.19 Evolution of the cloud albedo as a function of the cloud drop number density and of the cloud drop effective radius. The cloud drop distribution is supposed to be monodispersed. See Problem 2.2

ferred to as *atmospheric brown clouds* (ABC). For example, the Asian brown cloud has a thickness up to 3 kilometers. Similar clouds can be measured over the North Atlantic Ocean (North Atlantic plume), over north-eastern Europe, over the Pacific Ocean (Chinese plume) and over the South Atlantic Ocean (biomass burning plume from the Amazonian forest).

There are many impacts. First, the resulting radiative forcing is negative ($-20 \pm 4\, W\, m^{-2}$ for the Asian brown cloud). Second, the water cycle is modified: the cloud drops are smaller and their evaporation may be also increased (according to a few measurements). Table 2.11 gives a few data for a pristine cloud and a brown cloud. The observational data indicate that the brown cloud is much stronger during the dry season and in the tropical zone. Why?

Solution:

Precipitation is weaker during the dry season, which results in a decrease of rain scavenging (Chap. 5). Moreover, there are always residual precipitations at mid-latitudes.

To know more ([118, 119]):

V. RAMANATHAN AND P. CRUTZEN, *Atmospheric Brown Clouds*, Atmos. Env., (2003), pp. 4033–4035

V. RAMANATHAN ET AL., *Indian Ocean Experiment (INDOEX): an integrated analysis of the climate forcing and effects of the great Indo-Asian haze*, J. Geophys. Res., **106** (2001), pp. 28371–28398

Contradictory effects resulting from the same cause may also occur. For example, an increase in the aerosol number results in an increase in the cloud drop number. This induces a reduction of the precipitation efficiency and, meanwhile, an increase in the cloud lifetime. A first impact is the decrease of snow falls (due to the reduced precipitation efficiency; microphysical effect). The second impact is the increase in reflection of the solar radiation, leading to a decrease in the temperature, and then in increasing snow falls (radiative effect). The global impact, as far as the snow falls are concerned, is then ambiguous ([90]).

2.2.4.4 Scientific Controversies

Other effects have been subject to scientific controversies during the last years. The typical example is the case of the *cosmic rays*. H. Svensmark ([140]) formulated the hypothesis that there was a correlation between the cosmic rays (depending on the solar activity) and the formation of cloud condensation nuclei, through ionization processes, especially for low clouds (those involved in the reflection of the solar radiation, namely implied in a cooling effect). Since the observational data are likely to show a decrease of 25% for the cosmic rays during the last century, this could explain the climate change. These works have generated scientific controversies. First, there is no well-defined microphysical processes that could support this hypothesis. Second, a few studies, using new data (after 1995) have shown that the correlation between the cosmic rays and the cloud cover may not be significant ([79]).

There are other scientific debates. For instance, a few studies investigate the possible underestimation of the solar radiation absorption by clouds (down to 40% of the usual value; [40] and [88], among many other references). New observational data (sometimes used by the same scientific teams) seem to indicate that this issue is not prevailing.

2.2.5 Atmospheric Pollution and Visibility

The reduction in visibility is one of the most spectacular impacts resulting from a pollution event in urban areas. This is usually defined with the concept of *visual contrast*. Consider a dark body in a clear medium. The visual contrast, C_v, is the relative difference between the intensity (radiance) of the body and that of the medium. It depends on the distance (x) between the observer and the body (located at $x = 0$). Let I_b be the intensity related to the medium, supposed to have a constant value, and I be that of the object, respectively. Thus, the visual contrast is

$$C_v(x) = \frac{I_b - I(x)}{I_b}. \tag{2.52}$$

Since the body is supposed to absorb all radiations (it does not emit nor reflect radiation), $I(0) = 0$ and the maximum of the visual contrast is met at the body location: $C_v(0) = 1$. For $x > 0$, there are two medium contributions to the evolution of I: first, there is a scattering of the ambient intensity due to gases and particles, and, second, there is an extinction of I (due to absorption and scattering). The first contribution is evaluated by $b^{ext} I_b$ while the second one is in the form $-b^{ext} I$ with b^{ext} the extinction coefficient (supposed to be the scattering coefficient), namely

$$\frac{dI}{dx} = b^{ext} I_b - b^{ext} I. \tag{2.53}$$

As I_b is constant (it does not depend on x), we obtain an equation similar to the Beer-Lambert law,

$$\frac{dC_v}{dx} = -b^{ext} C_v, \tag{2.54}$$

whose solution is $C_v(x) = \exp(-b^{ext} x)$.

The decrease in visibility is usually estimated by a distance, written as x_v, corresponding to the distance at which the reduction in the visual contrast is below a given threshold. The threshold is defined so that a "mean" observer would not see the contrast between the body and the medium. The threshold is typically 2%. As $\ln(50) = 3.912$, this yields the so-called Koschmieder equation (1922),

$$x_v = \frac{3.912}{b^{ext}}. \tag{2.55}$$

The contributions for the extinction include Rayleigh scattering (due to gases except NO_2), extinction due to NO_2 and that due to aerosols. Nitrogen dioxide colors the polluted plumes in red, brown or yellow (Fig. 2.22). The major contribution is provided by aerosols: from 50 to 95% for sulfate and nitrate aerosols, from 5 to 50% for organic aerosols and soot. A pollution event is characterized by a strong increase in the aerosol contribution to b^{ext} (Table 2.13 and Figs. 2.20 and 2.21), especially for scattering.

In the framework of the North-American reglementation (*Regional Haze Rule* of the US EPA, *Environmental Protection Agency*, 2003), specific attention has been

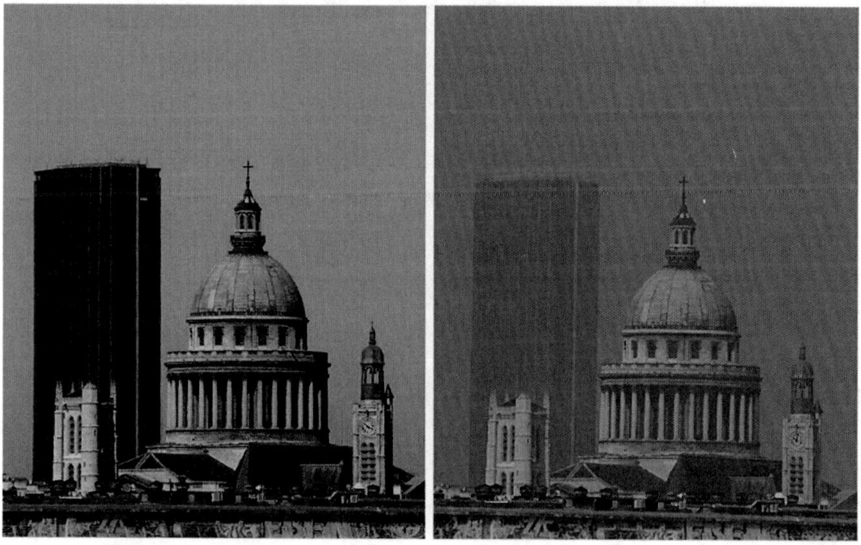

Fig. 2.20 Visibility reduction due to particulate matter. *Left*: Rayleigh sky (17 June 2004, $PM_{10} = 20\ \mu g\, m^{-3}$). *Right*: polluted event with high aerosol concentrations (9 June 2004, $PM_{10} = 80\ \mu g\, m^{-3}$). PM_{10} stands for the mass of particles whose radius (in a first approximation) is less than or equal to $10\ \mu m$. Credit: Airparif

Table 2.13 Comparison of the extinction coefficient for a "clean" day and a polluted day, in Los Angeles. The wavelength is in the visible region of the spectrum, $\lambda = 550\,\text{nm}$. For the polluted day, the visibility reduction is related to the aerosol scattering. Source: [82]

Extinction coefficient ($10^{-4}\,\text{m}^{-1}$)	"Clean" day (7 April 1983)	Polluted day (25 August 1983)
Aerosol scattering	0.26	4.08
Gas-phase scattering	0.11	0.11
NO_2 absorption	0.01	0.03
Soot absorption	0.09	0.78
Total	0.47	5

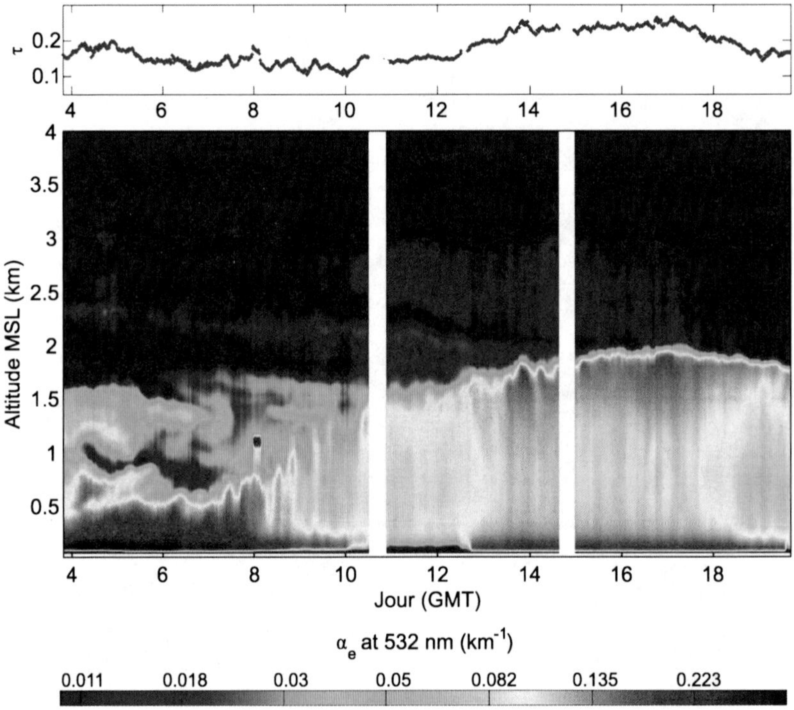

Fig. 2.21 *Upper panel*: evolution of the total optical depth. *Lower panel*: evolution of the vertical profile of the extinction coefficient, b_{ext} (here written as α_e; in km^{-1}). The wavelength is 532 nm (visible). The observational data were measured at Paris center ("place de l'Hotel-de-Ville", 4–18 May 2005, LISAIR campaign). Credit: Patrick Chazette, CEA

paid to visibility in the national parks. The visibility reduction is measured by the so-called *haze index*,

$$HI = 10 \ln \frac{b^{ext}}{10}, \tag{2.56}$$

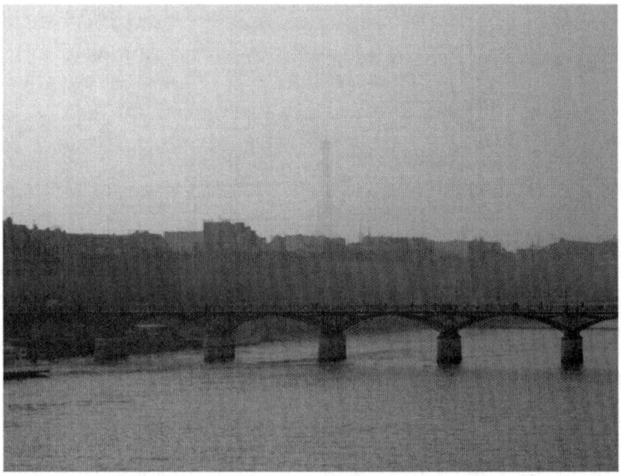

Fig. 2.22 Pollution event with high nitrogen dioxide concentrations over Paris, 1st February 2006. A maximum of about 350 $\mu g\,m^{-3}$ was measured. Credit: Airparif ([4])

with b^{ext} the extinction coefficient. When b^{ext} is expressed as $10^{-6}\,m^{-1}$ (Mm^{-1}), the haze index is expressed in *deciview* (dv). Note the connection to x_v. A "clean" reference atmosphere corresponds to $b^{ext} = 10\,Mm^{-1}$. The "Rayleigh atmosphere" (namely, without any particles) is sometimes defined by $b^{ext} = 13\,Mm^{-1}$.

A few parameterizations are available for calculating b^{ext} as a function of the aerosol chemical composition, for example ([108])

$$b^{ext} = 2f(RH) \times ([(NH_4)_2SO_4] + [NH_4NO_3]) + 1.4[OC] + 10[EC] + [mineral] + 0.6[coarse] + 10, \quad (2.57)$$

where $f(RH)$ is a function of the relative humidity RH (in order to describe the aerosol growth due to water condensation on nitrate and sulfate aerosols; typically, $f(.)$ ranges from 2 to 3), OC stands for the organic carbon, EC for the elemental carbon, "mineral" is related to the mineral part of $PM_{2.5}$[5] and "coarse" to the coarse part of the aerosol distribution (diameter above 2.5 μm). In the absence of aerosols, HI is equal to 0.

As expected, the visibility strongly depends on humidity.[6] The atmospheric dust burden is another key factor. For example, in southern Europe, the extinction coefficient may sometimes increase up to hundreds of Mm^{-1}, due to the transport of Saharian dust.

Problems Related to Chap. 2

Problem 2.1 (Radiative Forcing due to Aerosols and Direct Effect) The radiative behavior of aerosols depends not only on the size distribution but also on the chem-

[5] $PM_{2.5}$ stands for the mass of particles whose diameter is less than or equal to 2.5 μm.

[6] Think about the visibility in a fog!

ical composition (Table 2.14). This problem (taken from [58]) illustrates the sensitivity of the radiative behavior with respect to the aerosol type.

1. Estimate the *a priori* impact on the surface temperature,
   - first, of an increase in the scattering of solar radiation (sulfate aerosols),
   - and, second, of an increase in the absorption of infrared radiation (soot).

   What can conclude for the mineral aerosol?
2. We use a toy model for radiative transfer. The atmosphere is supposed to be a layer of temperature T_1. The layer does not absorb the solar radiation, absorbs the infrared radiation with an absorption coefficient f, and emits to the Earth and space a power per unit area of surface $2f\sigma {T_1}^4$ (half to the Earth, half to space). The Earth/atmosphere system has an albedo A with respect to the solar radiation. Motivate the expression used for the emission. Calculate the Earth's temperature as a function of F_s (solar flux), A and f. Give a coherent value for f so that $T_0 = 288\,\text{K}$.
   Data: $A = 0.3$; $\sigma = 5.67 \times 10^{-8}\,\text{W}\,\text{m}^{-2}\,\text{K}^{-4}$; $F_s = 342\,\text{W}\,\text{m}^{-2}$.
3. Consider the variations δT_0 due to the modifications of the radiative properties of the atmospheric layer ((δf and δA). To which aerosols are associated variations $\delta f > 0$ and $\delta A > 0$?
   Express δT_0 as a function of δf and δA (use the logarithmic derivative: $\mathrm{d}(\ln x) = \mathrm{d}x/x$). Prove that there exists a critical value of $\delta A/\delta f$.

Three-dimensional simulations illustrate the uncertainties of the radiative impact due to mineral aerosols. The negative radiative forcing in the shortwave radiations is likely to have the same magnitude as the positive radiative forcing in the longwave radiations. The resulting total contribution is estimated to be $[-0.6, +0.4]\,\text{W}\,\text{m}^{-2}$ ([106], see the reference below). This means that the sign of the radiative forcing is not known, even if it is likely to be negative.

See also Problem 2.3 for another wiewpoint.

Solution:

1. An increase in the scattering of the solar radiation leads to an increase in the albedo of the Earth/atmosphere system, namely to a cooling. An increase in the absorption of the infrared radiation is similar to an increase in the greenhouse effect and should result in an increasing temperature.
   Soot has an effect similar to a greenhouse gas, on the contrary to sulfate aerosols. We cannot conclude for mineral aerosols.
2. The emitted fraction corresponds to the absorbed fraction, which motivates the use of f for emission. At the radiative equilibrium for the Earth/atmosphere system,

$$(1-A)F_s = f\sigma {T_1}^4 + (1-f)\sigma {T_0}^4, \qquad 2f\sigma {T_1}^4 = f\sigma {T_0}^4,$$

Table 2.14 Radiative properties of aerosols with respect to the solar and infrared radiation

Aerosol type	Solar radiation	Infrared radiation
Sulfate	scattering	–
Soot	absorption	absorption
Mineral	scattering	absorption

thus

$$\left(1-\frac{f}{2}\right)\sigma T_0{}^4 = (1-A)F_s.$$

In order to have $T_0 = 288$ K (mean surface temperature), we take $f \simeq 0.78$.

3. $\delta f > 0$ corresponds to the case of soot; $\delta A > 0$ corresponds to the case of sulfate aerosols. The mineral aerosols are associated to both variations.
   Taking the logarithmic derivative of the previous equation yields

$$4\frac{\delta T_0}{T_0} = \frac{\delta f}{2(1-\frac{f}{2})} - \frac{\delta A}{1-A}.$$

   In order to assess the sign of δT_0, there exists a critical value

$$\frac{\delta f}{\delta A} = \frac{2(1-\frac{f}{2})}{1-A}.$$

To know more ([106]):
I.P. ON CLIMATE CHANGE, *Climate Change 2001. IPCC Third Assessment Report. The Scientific Basis*, 2001. WMO and UNEP (Chap. 6, pp. 372–373)

Problem 2.2 (Cloud Albedo and Twomey Effect) This problem aims at calculating the cloud albedo in the visible region of the electromagnetic spectrum.

Consider a cloud with thickness h and with a liquid water content L (defined as the mass of liquid water per air volume). The cloud is composed of water drops with a number density n, supposed to have the same *effective* radius (r_e). The effective radius is computed as a mean radius weighted by the drop cross sections (that play a leading role for the radiative properties).

1. Calculate the cloud optical depth τ_c in the visible region. Give two formulations: the first one as a function of the drop number density n, the second one as a function of the effective radius r_e. Let ρ_w be the water density.
2. The observational data indicate that the cloud effective radius ranges from 5 to 8 µm over urban (polluted) areas, from 8 to 10 µm over remote continental regions, and from 10 to 15 µm over oceans. Investigate the impact of a division by two on the effective radius for the drop number density, the drop surface and the optical depth.
3. The cloud albedo can be parameterized as a function of the optical depth, as

$$A_c = \frac{\tau_c}{\tau_c + 7.7}.$$

   Plot the evolution of A_c as a function of the cloud microphysical properties. Compare the cloud albedo in a marine environment ($n \simeq 100\,\mathrm{cm}^{-3}$) and in a polluted area ($n \simeq 1000\,\mathrm{cm}^{-3}$).
   Data: $L - 0.5\,\mathrm{g\,m}^{-3}$ and $h - 100\,\mathrm{m}$.

Solution:

1. Assuming that the cloud is homogeneous, $\tau_C = b^{ext} \times h$ (we omit the dependence with respect to the wavelength in the notations). We use $b^{ext} = \pi r^2 Q^{ext} n$, with $Q^{ext} \simeq 2$ (Sect. 2.1.7). Thus
$$\tau_C = 2\pi r_e^2 nh.$$
The use of the liquid water content, L, leads to the elimination of n or of r_e since
$$L = \frac{4}{3}\pi {r_e}^3 \times n \times \rho_w.$$
We obtain the possible expressions, as a function of the radius,
$$\tau_C = \frac{3}{2}\frac{Lh}{r_e \rho_w},$$
or as a function of the drop number density,
$$\tau_C = h\left(\frac{9}{2}\frac{n\pi L^2}{\rho_w^2}\right)^{1/3}.$$
The optical depth is therefore an increasing function of the drop number density and a decreasing function of the drop radius.
2. Dividing by 2 the effective radius implies a multiplication by 8 of n (number density). The total surface and the optical depth are proportional to nr^2 and, hence, are multiplied by 2.
3. The cloud albedo is an increasing function of the optical depth. The sensitivity with respect to the microphysical properties is then similar to that of the optical depth.
We refer to Fig. 2.19. The albedo for a marine cloud is about 0.45; it is about 0.63 for a cloud in an urban area.

Problem 2.3 (Albedo of an Aerosol Layer) This problem, partially taken from [130], is the follower of Problem 2.1. We want to investigate the impact of the size distribution.

Consider an aerosol layer defined by its optical depth τ and its scattering albedo ω_d. Let β be the fraction of the scattered solar radiation that is reflected back to space. Moreover, the radiation transmitted by the layer is supposed to reach directly the Earth's surface, on which it is reflected with an albedo A_s.

1. Evaluate the fraction r of the incident radiation that is directly reflected back to space. Let t be the transmitted fraction. We do not take into account the contributions due to the reflection at ground.
2. Calculate ΔA the modification, due to the aerosol layer, of the albedo for the Earth/atmosphere system. Hint: take into account the multiple reflections between the Earth's surface and the aerosol layer.
3. The impact of the aerosol layer on the global albedo depends on the competition between scattering and absorption (both define extinction). Show that there exists a critical value ω_d that determines the transition from a cooling effect to a warming effect. Use asymptotic expansions with respect to τ (its value is about 0.1).

4. Show that a non-absorbing aerosol layer has always, as expected, a cooling effect. This is typically the case of sulfate aerosols.
5. We consider the visible radiation ($\lambda = 550\,\text{nm}$). For a mineral aerosol, ω_d is a decreasing function of the diameter (d_p). Its value is about 0.96 for $d_p = 0.2\,\mu\text{m}$ and 0.72 for $d_p = 8\,\mu\text{m}$. Estimate the resulting impact.

Data: surface albedo $A_s = 0.15$, $\beta \simeq 0.5$ for fine aerosols ($d_p \simeq 100$ nm), $\beta \simeq 0.2$ for coarse aerosols ($d_p > 1\,\mu\text{m}$).

Solution:

1. The incident radiation has a *direct* transmitted fraction ($\exp(-\tau)$) and a fraction subject to extinction ($1 - \exp(-\tau)$).
   For the extinction, a fraction ω_d is scattered while a fraction $(1 - \omega_d)$ is absorbed.
   A fraction β of the scattered fraction is reflected back to space. A fraction $(1 - \beta)$ is transmitted to the Earth's surface upon scattering, and has to be added to the direct transmitted fraction.
   Finally, we obtain
   $$r = \beta\omega_d(1 - \exp(-\tau)), \qquad t = \exp(-\tau) + (1 - \beta)\omega_d(1 - \exp(-\tau)).$$
2. Let t_n be the fraction of the radiation that is transmitted to the Earth's surface after n "interactions" with the aerosol layer. After a reflection on the Earth's surface, the fraction becomes $A_s \times t_n$. After a new interaction with the aerosol layer, a fraction t is transmitted to space (to be taken into account in the global albedo) while a fraction r is reflected to the ground. Therefore, the iteration reads
   $$t_{n+1} = r \times A_s \times t_n, \qquad r_{n+1} = t \times A_s \times t_n,$$
   and finally $t_n = (rA_s)^{n-1}t$, $r_n = tA_s(rA_s)^{n-1}t$. The total reflected fraction is
   $$A_a = r + \sum_{n=1}^{\infty} r_n = r + \frac{t^2 A_s}{1 - rA_s}.$$
   The albedo variation is $\Delta A = A_a - A_s$.
3. The aerosol albedo has a cooling effect or a warming effect, depending on the sign of ΔA: for example, if $\Delta A > 0$, there is an increase in the global albedo, which results in a cooling.
   Using the asymptotic expansions $r \simeq \beta\omega_d\tau$ and $t \simeq 1 - \tau + (1 - \beta)\omega_d\tau$, up to second order in τ, we get
   $$\Delta A = [\omega_d(\beta(1 - A_s)^2 + 2A_s) - 2A_s]\tau.$$
   The critical value is then
   $$\omega_d^{\star} = \frac{2A_s}{\beta(1 - A_s)^2 + 2A_s}.$$
   If $\omega_d > \omega_d^{\star}$, $\Delta A > 0$ (cooling effect). Note that a cooling effect is expected for high values of ω_d (scattering is dominant with respect to absorption). Moreover, the critical value is a decreasing function of β (increasing the fraction scattered back to space favors the cooling effect).
   With the numerical data, we calculate for coarse aerosols ($\beta \simeq 0.2$), $\omega_d^{\star} \simeq 0.67$, and for fine aerosols ($\beta \simeq 0.5$), $\omega_d^{\star} \simeq 0.45$.

4. For a non-absorbing layer (the extinction is only composed of scattering: $\omega_d = 1$), we obtain

$$\Delta A = \beta(1 - A_s)^2 \tau > 0.$$

The effect is a cooling effect, as expected.

5. The values of ω_d are much greater than the critical value. The impact is a cooling effect in the visible region and a warming effect in the infrared region.

To know more ([24, 99]):

S. NEMESURE, R. WAGENER, AND S. SCHWARTZ, *Direct shortwave forcing of climate by the anthropogenic sulfate aerosol: sensitivity to particle size, composition, and relative humidity*, J. Geophys. Res., **100** (1995), pp. 26105–26116 (for the sulfate aerosols)

R. CHARLSON, J. LANGNER, H. RODHE, C. LEOVY, AND S. WARREN, *Perturbation of the northern hemisphere radiative balance by backscattering from anthropogenic sulfate aerosols*, Tellus, **43AB** (1991), pp. 152–163

Problem 2.4 (Nuclear Winter) In 1982, Crutzen and Birks ([27]) described the impact of a nuclear war: it would lead to a strong global cooling, which was referred to as *nuclear winter*. The context was the "Cold War" between the USA and the USSR, with the peak of the so-called "Euromissile crisis" (early 1980s). In the 1980s, many studies, based on the use of climate models, have confirmed this preliminary study.

More attention has been recently paid to this subject due to the possibility of a *regional* nuclear conflict.

Describe the processes that could explain the cooling effect. Comment on the long-term impact (more than ten years), as compared to the impact of volcanic eruptions (a few years, Exercise 2.9).

Data ([121]):

– emissions of smoke and ashes after the explosions: 150 Tg;
– stratospheric residence time of smoke and ashes: 5 years;
– radiative forcing for the solar radiation averaged over one decade (after ten years): $-100\,\mathrm{W\,m^{-2}}$ ($-20\,\mathrm{W\,m^{-2}}$);
–ΔT at the Earth's surface during one decade (after ten years): -7 K (-4 K);
– precipitations: -50%.

Solution:

A huge amount of smoke and ashes (particles) would be injected into the atmosphere, just after the explosion. The negative radiative forcing in the shortwave radiation is similar to that of the sulfate aerosols.

The high temperatures following the explosions would favor direct injection into the upper stratosphere (*lofting*), while the sulfate aerosols are rather near the tropopause. This results in a higher residence time and then a longer impact (more than one decade).

To know more ([27, 121]):

P. CRUTZEN AND J. BIRKS, *The atmosphere after a nuclear war. Twilight at noon*, Ambio, **11** (1982), pp. 114–125

A. ROBOCK, L. OMAN, AND G. STENCHIKOV, *Nuclear winter revisited with a modern climate model and current nuclear arsenals: still catastrophic consequences*, J. Geophys. Res., **112** (2007), p. 13107

Chapter 3
Atmospheric Boundary Layer

The intensity of pollution events is mainly governed by meteorological conditions in the lower part of the troposphere. The dispersion of emitted pollutants is driven by two distinct processes:

- *horizontal* transport by the wind field, which explains the long-range transport of long-lived pollutants;
- *vertical* mixing due to atmospheric turbulence, which is directly induced by boundary layer effects.

Atmospheric turbulence has two sources. *Mechanical* turbulence is induced by vertical wind shear: the wind is constrained at the Earth's surface. The *thermal* (or *convective*) turbulence is induced by the vertical distribution of temperature: the Earth's surface is heated by solar radiation during daytime, cooled during nighttime.

In simple words, the *atmospheric boundary layer* (ABL hereafter) is defined as the part of the atmosphere that is sensitive, in short-term (namely in a few hours), to the varying conditions at the Earth's surface. A typical example is the day/night cycle.

Investigating the ABL is required for the study of urban pollution. A key parameter is the so-called *mixing height*, that is to say the height of the volume in which the emitted pollutants are mixed. This defines the *dilution* volume for the pollutants. Consider a pollutant emitted with a mass m in a "pristine" volume with a fixed area of surface S (given by the urban area) and a *variable* height H (Fig. 3.1). Assuming that the pollutant is well mixed, the pollutant concentration is then $c = m/(S \times H)$. As a result, the concentration is governed not only by the emissions but also by the dilution height, varying as a function of the meteorological conditions.

The dilution state of the ABL is usually classified by distinguishing:

- the *neutral* ABL;
- the *unstable* ABL, corresponding to high dilution, namely to strong turbulent mixing;
- the *stable* ABL, for which the pollutants accumulate at the ground level (weak dilution), typically at night.

B. Sportisse, *Fundamentals in Air Pollution*,

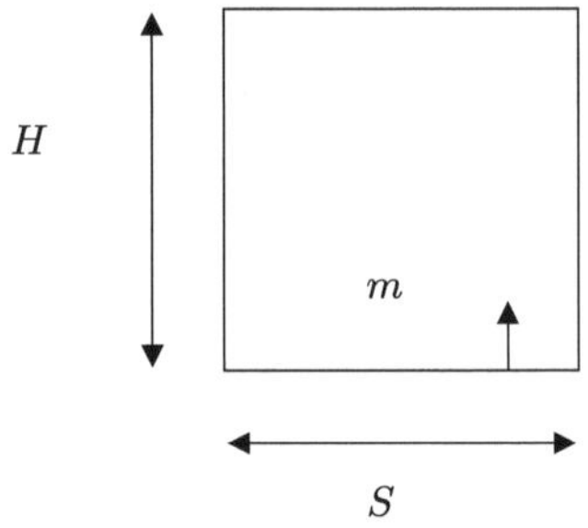

Fig. 3.1 Emission of a pollutant mass m in a volume defined by the area of surface S and a dilution height H. The resulting concentration is $c = m/(S \times H)$ and depends on the dilution height. For weak dilution (e.g., $H = 400\,\text{m}$), the concentration is three times as great as that corresponding to strong dilution (e.g., $H = 1200\,\text{m}$)

In fluid mechanics, this classification is classical for any flow presenting a temperature stratification (a vertical gradient of the temperature field). The underlying concept is the so-called *buoyancy* (resulting from the Archimedes' principle).

Even though this classification is an idealized description of real-life conditions (that appear to be much more complicated), it makes it possible to describe the leading processes in a simple way.

This chapter aims at introducing the behavior of the ABL and is organized as follows.

The main meteorological scales are introduced in Sect. 3.1. The phenomenology of the ABL is briefly presented in Sect. 3.2. A stability analysis is carried out in Sect. 3.3 in order to give a rigorous basis for the above classification. Another viewpoint is provided by investigating atmospheric turbulence in Sect. 3.4. The fundamentals of ABL dynamics are given in Sect. 3.5 with the so-called Boussinesq approximation. To conclude, Sect. 3.6 is devoted to the specific characteristics of the "urban climate".

3.1 Meteorological Scales

Atmospheric circulation is characterized by a wide range of scales .

The horizontal dimension, L, is the leading parameter for classifying the spatial scales. We can distinguish (the bounds have illustrative values):

- the synoptic or planetary scale ($L > 1000\,\text{km}$);
- the regional or meso scale ($10\,\text{km} < L < 1000\,\text{km}$);
- the local or urban scale ($L < 10\,\text{km}$).

Actually, these scales can be subdivided into many scales, due to the leading processes (Table 3.1).

The different scales can also be defined by considering the flow characteristic timescales (or frequencies). The fundamental frequencies of the atmospheric flows are usually defined as follows:

Table 3.1 A *possible* classification of meteorological scales as a function of the horizontal dimension L. Source: [150] (Chap. 9)

L	scale
10 000 km	planetary
1000 km	synoptic
100 km	meso-α
10 km	meso-β
1 km	meso-γ/submeso
100 m	micro-α (boundary layer)
10 m	micro-β (surface layer)
1 m	micro-γ
1 mm	micro-δ
0.1 nm	viscous dissipation (molecular)

- the *Brunt-Vaisala frequency* is associated to the vertical oscillations in a stratified atmosphere,

$$f_b = \sqrt{\frac{g}{\theta}\frac{\partial \theta}{\partial z}} \simeq 10^{-2}\,\mathrm{s}^{-1}, \tag{3.1}$$

  where z stands for the vertical coordinate, g is the acceleration of gravity and θ is the potential temperature (see Sect. 3.3);
- the *inertial frequency* is associated to the Earth's rotation and to the resulting Coriolis force,

$$f = 2\Omega \sin\phi \simeq 10^{-4}\,\mathrm{s}^{-1}, \tag{3.2}$$

  where Ω is the angular velocity vector of the Earth and ϕ is the latitude;
- the *planetary frequency* is associated to the so-called β effect, corresponding to the variation of the Coriolis force with respect to latitude,

$$f_p = \sqrt{U\beta} \simeq 10^{-6}\,\mathrm{s}^{-1}, \tag{3.3}$$

  with U the modulus of the mean wind velocity and β the variation rate of the Coriolis force with respect to latitude.

These three frequencies define timescales of about one minute, one hour and one day, respectively. The first two frequencies are investigated in the following (in Sect. 3.3.2.2 for f_b and in Exercise 3.4 for f).

The meteorological scales can be defined from the characteristic timescales of the response to a given forcing. If F stands for the corresponding frequency, the scale is said to be

- local for $F > f_b$;
- regional (meso) for $f < F < f_b$;
- synoptic for $f_p < F < f$;
- planetary for $F < f_p$.

The most relevant scales for urban air quality are the local and regional scales.

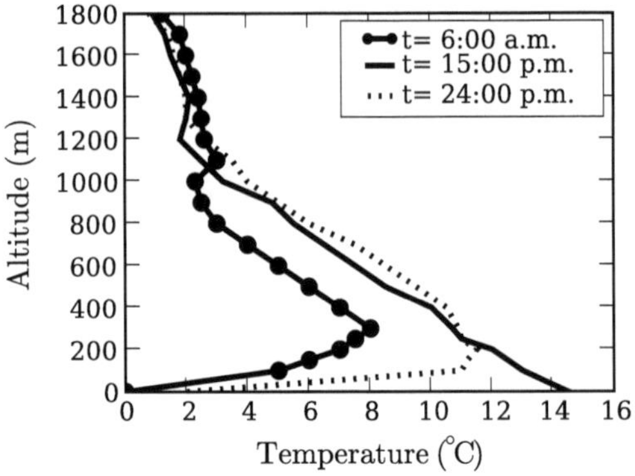

Fig. 3.2 Schematic daily evolution of the vertical temperature profile

3.2 Atmospheric Boundary Layer

3.2.1 Background

The ABL may be defined as the part of the atmosphere subject to surface effects. The timescale of the atmospheric response to a given forcing at the surface (heat transfer, evaporation, friction effect—effect of roughness on the wind field—, etc.) is about one hour ([139]). This timescale is much greater above the ABL, in the so-called *free atmosphere* (FA hereafter; "free" with respect to the influence of the Earth's surface).

For example, the temperature evolution in the ABL is governed by the diurnal cycle, contrary to that in the FA (in a first approximation). This evolution is not connected with the solar radiation absorption in the ABL since the major part of solar radiation is transmitted to the Earth's surface (Chap. 2), but with the variation of the vertical turbulent heat flux:

- during the day, the surface is heated by solar radiation, and turbulent heat fluxes are generated in the ABL;
- during the night, the surface is cooled, which anneals the heat flux.

As shown in Fig. 3.2, the temperature does not undergo a strong time variation above the altitude of 1500 meters, which gives an estimation of the ABL height.

The ABL is a complicated environment due to the surface effects (orography, radiative properties of the surface, etc.). Turbulence plays a leading role in the ABL, which favors vertical motions and the dilution of pollutants far from the emission sources (otherwise, the concentrations would be too high). This *permanent* influence of turbulence is another specificity of the ABL. Note that there is only an intermittent influence in the FA through strong convective processes.

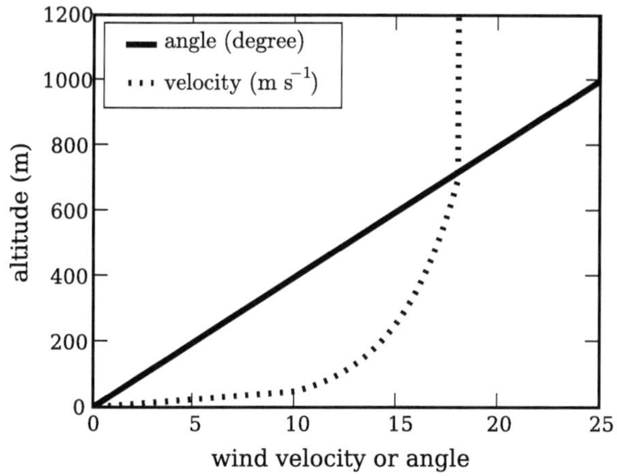

Fig. 3.3 Neutral ABL

3.2.2 Classification

The ABL is systematically composed of a thin *laminar layer* (with a thickness of a few centimeters), just above the ground. The molecular diffusivity dominates turbulence in this layer. At the top of the ABL, the boundary conditions define the FA (*geostrophic wind* given by the equality between the Coriolis force and the pressure-gradient force, Sect. 3.5).

Considering the turbulence intensity classifies the ABL as follows:

- **Neutral ABL** (Fig. 3.3)
  In the *surface boundary layer* (SBL, Sect. 3.5), with a thickness of a few tens of meters, the Coriolis and pressure-gradient forces can be neglected. In a first approximation, the wind velocity is horizontal and has a constant direction. The vertical profile of its modulus is given by a logarithmic distribution, as a function of the altitude z ($u(z) \propto \ln z$).
  Above the surface layer, the wind direction tends toward that of the geostrophic wind (upper boundary condition). The resulting layer is the so-called *Ekman layer*, in which the horizontal wind velocity follows a spiral (Sect. 3.5).
  To a large extent, this state is an academic viewpoint. It can be met above the sea or when the cloud cover is very strong.
- **Unstable ABL** (Fig. 3.4)
  Just above the ground, the vertical profile of potential temperature is decreasing with respect to altitude. The thermal production of turbulence results in an upward-oriented turbulent heat flux.
  Above, the *mixing layer* (also referred to as the *convective layer*) is characterized by strong turbulent motions, contributing to the homogenization of vertical profiles (of potential temperature and of horizontal wind velocity).
  The upper part of the ABL is more stable due to the inversion of the temperature profile, which defines the so-called *inversion layer*.

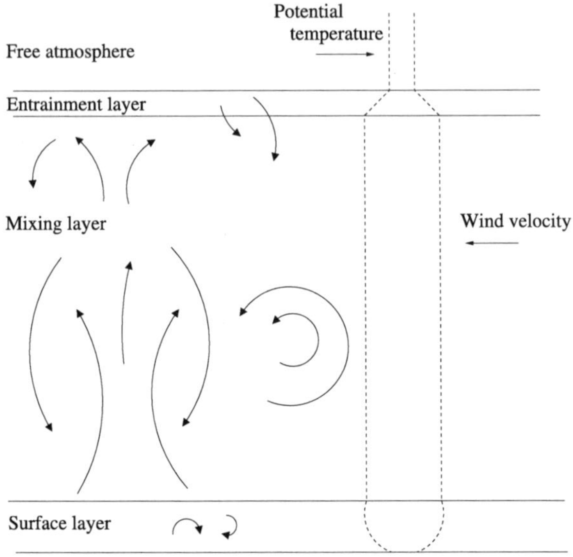

Fig. 3.4 Unstable ABL. Source: [30]

The unstable ABL is a common state during daytime since the ground is heated by solar radiation.

- **Stable ABL** (Fig. 3.5)
  The surface temperature is lower than the air temperature in the layer just above the ground: this results in a downward-oriented vertical sensible heat flux (turbulent heat flux), which can be viewed as a destruction of the turbulence induced by the wind shear.
  The resulting turbulence is weak, which leads to the accumulation of emitted pollutants near the surface.
  Above, the *inversion layer* is characterized by an increase in the modulus of the horizontal wind velocity. This modulus can even be greater than that of the geostrophic wind in the upper part of the ABL. This defines the so-called *low-level jet streams* (Exercise 3.4).
  This stable state is often met during nighttime.

The above classification is academic and the real-life situations are more complicated. For example, the effects of orography have to be taken into account. In practice, the observed vertical profiles are much more "flacky".

3.3 Thermal Stratification and Stability

This section aims at giving a rigorous basis for the classification of the ABL state. We introduce the concept of buoyancy (given by the Archimedes' principle) and investigate the stability of vertical motions.

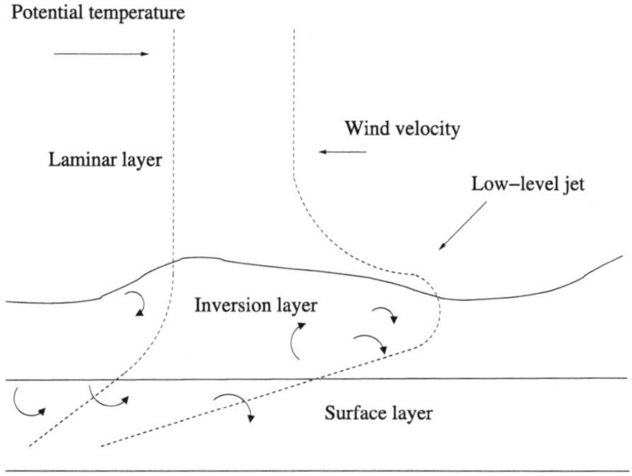

Fig. 3.5 Stable ABL. Source: [30]

For the sake of clarity, the influence of clouds and humidity are not taken into account. We first investigate the case of *dry air*. Therefore, the convective motions are related to the vertical temperature distribution (and not to that of humidity).

3.3.1 A Few Useful Concepts

3.3.1.1 Potential Temperature

Consider an infinitesimal transformation of an air parcel (to be defined below). The first law of thermodynamics gives the change in internal energy, U,

$$\mathrm{d}U = -P\mathrm{d}V + \delta Q, \tag{3.4}$$

where P is the pressure, V the volume and δQ is the heat flux between the parcel and the environment. Defining the enthalpy as $H = U + PV$ yields

$$\mathrm{d}H = V\mathrm{d}P + \delta Q. \tag{3.5}$$

The specific heat at constant pressure, c_p (in $\mathrm{J\,kg^{-1}\,K^{-1}}$), is the energy required to increase the temperature of 1 kg of air at constant pressure. The change in enthalpy for a parcel of mass m is also given as a function of the change in temperature (T),

$$\mathrm{d}H = mc_p\mathrm{d}T. \tag{3.6}$$

An *adiabatic process* is defined as a process for which no heat is transferred between the parcel and the environment, namely $\delta Q = 0$. The first law of thermodynamics becomes

$$mc_p\mathrm{d}T = V\mathrm{d}P. \tag{3.7}$$

We rewrite the ideal gas law as $P = \rho r_{air} T$ where $r_{air} = R/M_{air}$ (R universal gas constant and M_{air} molecular weight of dry air) and $\rho = m/V$ (air density). Thus,

$$\mathrm{d}\left[\ln(T\,P^{-r_{air}/c_p})\right] = 0. \tag{3.8}$$

The *potential temperature*, θ, is then defined as the temperature that the air parcel should have at ground, after an *adiabatic* motion from the thermodynamic state (T, P) to the ground. Let P_s be the ground pressure (or a reference pressure, namely 1000 hPa). We obtain the so-called Poisson equation,

$$\theta(P,T) = T\left(\frac{P_s}{P}\right)^{r_{air}/c_p}. \tag{3.9}$$

For the case of dry air, we calculate $r_{air}/c_p \simeq 0.286$ since $c_p = 1005\,\mathrm{J\,kg^{-1}\,K^{-1}}$.

3.3.1.2 Dry Adiabatic Lapse Rate

Note that, by definition, the potential temperature is conserved during an adiabatic process. The potential temperature has therefore a constant vertical profile if the atmosphere is supposed to be adiabatic. Taking the logarithmic derivative of θ with respect to the altitude z yields

$$\frac{1}{\theta}\frac{\partial\theta}{\partial z} = \frac{1}{T}\frac{\partial T}{\partial z} - \frac{r_{air}}{c_p}\frac{1}{P}\frac{\partial P}{\partial z}. \tag{3.10}$$

The adiabatic lapse rate, defined by $\partial\theta/\partial z = 0$, is then

$$\left(\frac{\partial T}{\partial z}\right)_{ad} = \frac{r_{air}}{c_p}\frac{T}{P}\frac{\partial P}{\partial z}. \tag{3.11}$$

Finally, for any vertical distribution,

$$\frac{\partial\theta}{\partial z} = \frac{\theta}{T}\left[\frac{\partial T}{\partial z} - \left(\frac{\partial T}{\partial z}\right)_{ad}\right]. \tag{3.12}$$

The potential temperature profile is then related to the deviation from the adiabatic case, which motivates its common use.

Using the hydrostatic assumption (Sect. 1.2.2), the vertical pressure distribution is

$$\frac{\partial P}{\partial z} = -\rho g. \tag{3.13}$$

For an adiabatic atmosphere, we obtain from (3.11),

$$\left(\frac{\partial T}{\partial z}\right)_{ad} = -\frac{\rho r_{air} T}{P}\frac{g}{c_p} = -\frac{g}{c_p}. \tag{3.14}$$

This defines the *dry adiabatic lapse rate*, usually written as $-\Gamma_d$ (*d* for *dry*), with $\Gamma_d \simeq 9.8\,\mathrm{K\,km^{-1}}$. The deviation from this profile will determine the stability of the ABL.

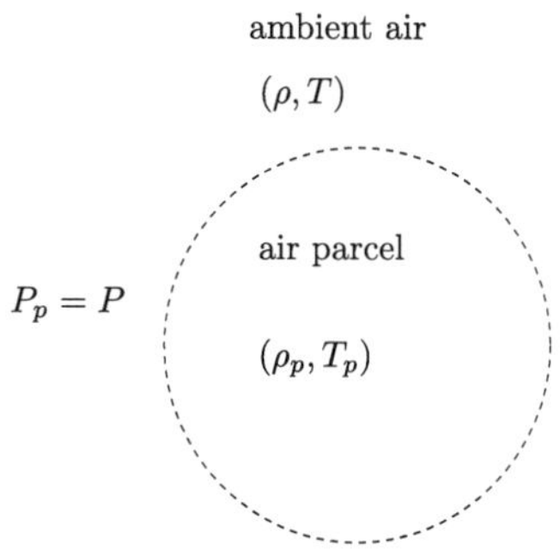

Fig. 3.6 Air parcel viewed as a virtual atmospheric balloon

3.3.2 Stability

3.3.2.1 Air Parcel

It is common in meteorology to use the idealized concept of *air parcel*. An air parcel is defined as follows:

- the parcel motion is *adiabatic* since the heat fluxes with the environment have no time to take place;
- the air parcel has its own temperature, that may be different from that of the ambient air;
- there is a balance between the pressure of the air parcel and that of the ambient air at the interface.

An air parcel can be viewed as an atmospheric balloon without any material membrane (Fig. 3.6).

3.3.2.2 Buoyancy Effects

Let us consider the vertical adiabatic motion of an air parcel with a volume of 1 m^3. Let (T_p, ρ_p) and (T, ρ) be the thermodynamic states of the parcel and of the ambient air, respectively.

The parcel is subject to gravity and to Archimedes' principle. The resulting force is $\mathbf{F} = (\rho - \rho_p)\mathbf{g}$ where $\mathbf{g}$ is downward oriented. Upon application of Newton's law of motion, the parcel acceleration, $\mathbf{a}$, satisfies $\rho_p \mathbf{a} = \mathbf{F}$. Thus,

$$\mathbf{a} = \left(\frac{\rho}{\rho_p} - 1\right)\mathbf{g} = \left(\frac{T_p}{T} - 1\right)\mathbf{g}, \tag{3.15}$$

using the balance of pressures, namely $\rho T = \rho_p T_p$. As expected, the acceleration $\mathbf{a}$ is upward-oriented if and only if $T_p > T$. This is the basis for the motion of atmospheric balloons: warm air is lighter and is subject to upward-oriented motions.

Consider an infinitesimal displacement δz from an equilibrium state at the altitude z_0. We assume that, at z_0, the air parcel is well mixed in the ambient air, namely

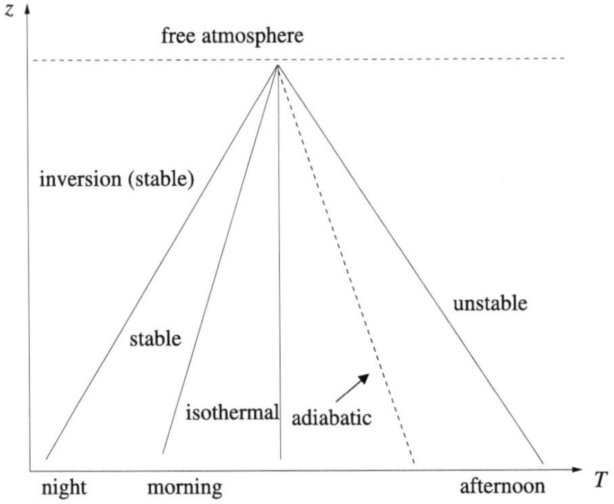

Fig. 3.7 Sketch of the daily variation of the vertical temperature profile in the ABL, and resulting stability. The ground temperature increases during daytime and decreases during nighttime. The comparison with respect to the adiabatic profile determines the stability (*left*: stable ABL; *right*: unstable ABL)

that $T_p(z_0) = T(z_0)$. We want to investigate the impact of the buoyancy effect on the vertical motion.

The air parcel undergoes an *adiabatic* motion, given by the lapse rate $(\partial T/\partial z)_{ad}$, while the atmosphere is characterized by the gradient $(\partial T/\partial z)$. Linearizing at the altitude $z = z_0 + \delta z$ yields

$$T_p(z) \simeq T(z_0) + \left(\frac{\partial T}{\partial z}\right)_{ad} \delta z, \qquad T(z) \simeq T(z_0) + \frac{\partial T}{\partial z}\delta z, \tag{3.16}$$

and finally

$$\mathbf{a} = -\frac{\frac{\partial T}{\partial z} - (\frac{\partial T}{\partial z})_{ad}}{T} \times \delta z \mathbf{g} = -\frac{\partial \theta}{\partial z}\frac{\delta z}{\theta}\mathbf{g}, \tag{3.17}$$

since the expression of the potential temperature gradient is given by (3.12).

Studying the buoyancy effect is then connected to the sign of the buoyant force: the motion is stable when the sign of the vertical component of **a** has the opposite sign of δz. The resulting classification is then governed by the sign of the potential temperature gradient (Fig. 3.7):

- if $\partial\theta/\partial z > 0$, the ABL is said to be *stable*.
  This corresponds to $\partial T/\partial z > (\partial T/\partial z)_{ad}$.
- if $\partial\theta/\partial z < 0$, the ABL is said to be *unstable*.
  This corresponds to $\partial T/\partial z < (\partial T/\partial z)_{ad}$.
- if $\partial\theta/\partial z = 0$, the ABL is said to be *neutral*.
  This corresponds to the adiabatic case, $\partial T/\partial z = (\partial T/\partial z)_{ad}$.

Remark 3.3.1 (Brunt-Vaisala Frequency) Note that we have obtained $a = d^2(\delta z)/dz^2 = -{f_b}^2\delta z$, where f_b is the Brunt-Vaisala frequency (Sect. 3.1). Depending on the case, ${f_b}^2$ may have a negative value.

This is a simple illustration of linearization approaches in meteorology, with the study of the resulting *waves* (here the *gravity waves* associated to a temperature gradient).

3.3.2.3 Temperature Inversion

The adiabatic lapse rate is negative. An isothermal profile ($\partial T/\partial z = 0$) is then a specific case of a stable ABL. When the temperature profile is characterized by an inversion (namely $\partial T/\partial z > 0$), there is therefore a strong stability.

We can distinguish two typical cases for temperature inversion:

- the case of *surface inversion* corresponds to the cooling of the Earth's surface (e.g. at night) or to a heating of the upper ABL;
- the case of *subsidence* corresponds to anticyclonic meteorological conditions ("nice weather").

In the case of subsidence, downwelling air masses tend to increase the pressure of the lower layers, which results in an increasing temperature: the clouds evaporate and the air becomes drier. Subsidence often corresponds to the strongest pollution events (due to the stability of the ABL).

3.3.3 Moist Air

In the case of moist air (w stands for *wet* in the following), the specific heat at constant pressure is given by

$$c_{p,w} = (1 - q_s)c_{p,d} + q_s c_{p,v}, \tag{3.18}$$

with $c_{p,d}$ for dry air (1005 J kg^{-1} K^{-1} at 273 K) and $c_{p,v}$ for water vapor (1952 J kg^{-1} K^{-1} at 273 K). The specific humidity q_s is the ratio of the water vapor mass to the moist air mass (Chap. 1). The previous formula can be obtained from the variations in specific enthalpy,

$$(m_d + m_v)dh = m_d \times c_{p,d}dT + m_v \times c_{p,v}dT, \tag{3.19}$$

and from the definition $dh = c_{p,w}dT$. Similarly to the case of dry air,

$$\left(\frac{\partial T}{\partial z}\right)_{w,ad} = -\frac{g}{c_{p,w}} = -\Gamma_w. \tag{3.20}$$

As $c_{p,d} < c_{p,w}$ (it is more difficult to heat a gas in a moist environment), the specific heat at constant pressure is greater in the case of moist air. As a result, $\Gamma_w < \Gamma_d$. A dry atmosphere is therefore more stable (see the classification in Fig. 3.7).

Actually, the impact of the discrepancy between the specific heats is not the leading point. The possible large discrepancy between the dry and moist cases (typically $\Gamma_w \sim \Gamma_d/3$ in the tropical regions) is related to the *latent* heat flux, resulting from water condensation (Exercise 3.1). Indeed, water condensation is an exothermic process. The resulting heat transfer is given by

$$\delta Q = -m \times L_v \,\mathrm{d}q_s, \tag{3.21}$$

with m the mass of moist air, L_v the so-called *specific latent heat of evaporation* (it is defined for the evaporation of 1 kg of water vapor; $L_v = 2.5 \times 10^6\,\mathrm{J\,kg^{-1}}$ at 273 K). Note that condensation ($\mathrm{d}q_s < 0$) is an exothermic process ($\delta Q > 0$).

We refer to Exercise 3.1 for the calculation of the *moist adiabatic saturated lapse rate*, Γ_w. The stability criteria of the moist atmosphere are then given for the virtual temperature (Chap. 1) as follows:

- if $\partial T_v/\partial z < -\Gamma_d$: absolutely unstable ABL;
- if $\partial T_v/\partial z = -\Gamma_d$: unsaturated neutral ABL;
- if $\partial T_v/\partial z \in]-\Gamma_d; -\Gamma_w[$: conditionally unstable ABL (stable if the parcel is not saturated or dry; unstable if the parcel is saturated, thus subject to condensation);
- if $\partial T_v/\partial z = -\Gamma_w$: saturated neutral ABL;
- if $\partial T_v/\partial z > -\Gamma_w$: absolutely stable ABL.

The concept of *conditional stability* is connected with humidity. Let us consider a virtual temperature profile $\partial T_v/\partial z \in]-\Gamma_d, -\Gamma_w[$:

- the stability of a parcel that is dry or not saturated (the parcel is moist but not subject to condensation) is determined by the dry adiabatic lapse rate: the parcel is stable since $\partial T_v/\partial z > -\Gamma_d$.
- the stability of a saturated parcel (thus subject to condensation) is determined by the moist saturated adiabatic lapse rate: the parcel is unstable since $\partial T_v/\partial z < -\Gamma_w$.

The unstability of the moist saturated parcel results from water condensation, which heats the parcel. For example, this generates the so-called *convective clouds*.

The mean adiabatic lapse rate for the moist real atmosphere is $-6.5\,\mathrm{K\,km^{-1}}$ (see the application to altimetry in Exercise 3.2).

Exercise 3.1 (Moist Saturated Adiabatic Lapse Rate) Consider a moist upwelling air parcel. During its rise, the parcel is cooling and the pressure is decreasing down to the saturation vapor pressure of water. Then, water condensation can occur. If the motion is an adiabatic process, this defines thereafter the so-called *moist saturated adiabatic lapse rate*.

Calculate the moist saturated adiabatic lapse rate. Compare with the dry adiabatic lapse rate.

Solution:

Due to preexisting particles, the so-called condensation nuclei (Sect. 5.2.3), the water vapor can condense. This results in a heat flux transferred from the parcel to the ambient air (exothermic process), given by (3.21). The first law of thermodynamics (3.7) becomes

$$mc_p\mathrm{d}T = V\mathrm{d}P - mL_v\,\mathrm{d}q_s,$$

with V the air volume. Under the hydrostatic assumption, $\mathrm{d}P/\mathrm{d}z = -\rho g$, we obtain

$$-\frac{\mathrm{d}T}{\mathrm{d}z} = \underbrace{\frac{g}{c_p} + \frac{L_v}{c_p}\frac{\mathrm{d}q_s}{\mathrm{d}z}}_{\Gamma_w}.$$

As $\mathrm{d}q_s/\mathrm{d}z < 0$ (during the condensation process, the fraction of water vapor is decreasing), the resulting moist saturated adiabatic lapse rate is lower than the dry adiabatic lapse rate. Thus, a moist saturated air mass is less stable than a dry air mass (see the stability criteria).

Following the calculation and using the saturation condition yields an expression for Γ_w as a function of temperature and of the fraction of water vapor (derivation beyond the scope of this exercise).

Exercise 3.2 (Altimetry) An altimeter is an airborne instrument used for measuring altitude. An *aneroid barometer* is designed to register changes in atmospheric pressure accompanying changes in altitude. State the relation between the altitude and the pressure assuming that the vertical temperature distribution is given by $-\Gamma_w$ (in practice, $\Gamma_w \simeq 6.5\,\mathrm{K\,km^{-1}}$).

Solution:

With $\mathrm{d}T/\mathrm{d}z = -\Gamma_w$, $\mathrm{d}P/\mathrm{d}z = -\rho g$ (hydrostatic equation) and $P = \rho r T$ (ideal gas law), it is straightforward to obtain

$$\frac{\mathrm{d}P}{P} = \frac{g}{r\Gamma_w}\frac{\mathrm{d}T}{T},$$

and then

$$\ln\frac{P(z)}{P_0} = \ln\left(\frac{T(z)}{T_0}\right)^{\frac{g}{r\Gamma_w}},$$

where P_0 and T_0 are the temperature and pressure at ground, respectively. Using $T(z) = T_0 - \Gamma_w z$ yields

$$z = \frac{T_0}{\Gamma_w}\left(1 - \left(\frac{P}{P_0}\right)^{\frac{r\Gamma_w}{g}}\right).$$

3.3.4 Daily Variation of the ABL Stability

Two properties govern the evolution of the ABL stability during the diurnal cycle (Fig. 3.8):

- in a first approximation, the solar radiation is absorbed at the Earth's surface rather than in the lower layers of the atmosphere;
- the soil has a better thermal conductivity than the atmosphere.

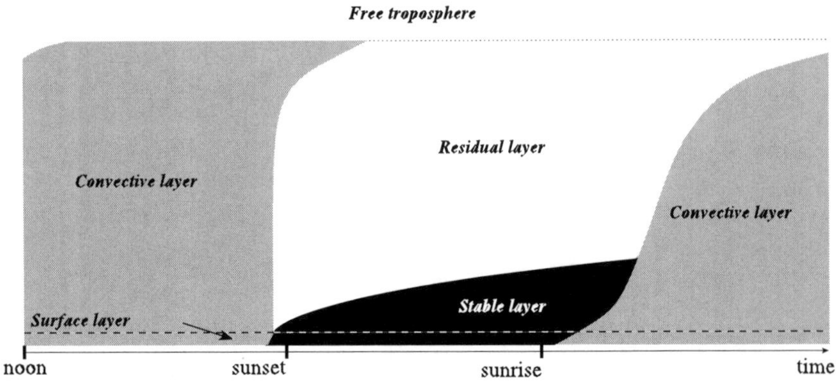

Fig. 3.8 Daily evolution of the ABL stability. Source: [139] (credit: Vivien Mallet, CEREA)

During the Day After sunrise, the Earth's surface is heated by solar radiation. Uprising air masses, the so-called *thermals* (also referred to as *updrafts*) take part in the homogenization of the ABL. A large part of the ABL is in an adiabatic state (Fig. 3.4): this defines the so-called *mixing layer* (also referred to as the *convective layer*).

Just above the ground, in the *surface boundary layer*, the vertical profile of θ is decreasing and turbulence motions are generated. At the top of the mixing layer, an inversion of the vertical profile defines the so-called *entrainment layer* (at the interface between the free atmosphere and the mixing layer: Fig. 3.4, Exercise 3.3). Actually, the thermals coming from the mixing layer can enter into the free atmosphere before "diving" back into the mixing layer: this may cause the entrainment of air masses from the free atmosphere into the ABL.

During the day, the convective layer is increasing (Fig. 3.8) due to the entrainment effect, which results in the destruction of the residual nocturnal boundary layer (see below).

During the Night A temperature inversion occurs at the ground, due to the cooling of the soil: the resulting stable layer is referred to as the *nocturnal boundary layer* (NBL). The concentrations of pollutants are increasing. Above this layer, the mixing layer of the previous day becomes a neutral *residual layer*, containing the pollutants that have been mixed the day before.

3.4 ABL Turbulence

Another approach to describe the stability of the ABL is based on turbulence. In this section, we recall the background of turbulent flows and then introduce the so-called *thermal* and *mechanical* turbulence.

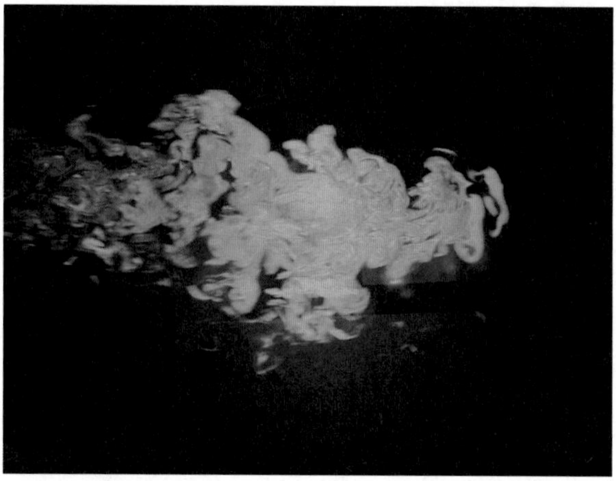

Fig. 3.9 Visualization of a turbulent flow in the wind tunnel of Centrale Lyon (reproduction of the Bugey nuclear power plant). Credit: Centrale Lyon and French Institute of Nuclear Safety and Radiological Protection (Olivier Isnard)

3.4.1 Background

A turbulent flow is characterized by "whirls" (see Fig. 3.9) covering a wide range of time and space scales. Kinetic energy is transferred from large scales to small scales, at which the energy is dissipated into heat through molecular diffusion. This results in an increasing mixing within the flow.

The turbulent nature of the ABL is directly related to boundary layer effects:

- the *mechanical* boundary layer is defined by the wind shear and the friction effect due to wind velocity boundary condition at the Earth's surface ($V = V_{ground}$);
- the *thermal* boundary layer is defined by the temperature shear and is governed by the diurnal cycle affecting the surface temperature.

3.4.1.1 Mechanical Boundary Layer

The mechanical boundary layer is defined by the wind shear above the ground, due to boundary conditions at the Earth's surface. Above this boundary layer, the variation of the horizontal wind velocity is much weaker. The wind velocity is then governed by the gradient-pressure force (*geostrophic wind*, Sect. 3.5).

The turbulence strength can be evaluated by calculating the ratio of the inertial forces (convection and pressure-gradient terms), tending to render the flow unstable, to the viscous diffusion force, tending to make the flow stable. Considering the evolution equation for the horizontal component of wind velocity, let say u (Sect. 3.5), written as $\partial u/\partial t + u \cdot \nabla u = \nu \Delta u + \cdots$ (ν is the kinematic viscosity), this defines

Table 3.2 Atmospheric turbulence

	Mechanical turbulence	Thermal turbulence
Free atmosphere	intermittent	intermittent
ABL	always	depends on stability

the Reynolds number

$$R_e = \frac{(u.\nabla)u}{\nu \Delta u} \simeq \frac{U\,L}{\nu}, \tag{3.22}$$

where U is a velocity scale and L a length scale. Typical values in the ABL are $U \simeq 10\,\mathrm{m\,s^{-1}}$, $L \simeq 1000\,\mathrm{m}$ and $\nu \simeq 10^{-5}\,\mathrm{m^2\,s^{-1}}$. Thus, $R_e \simeq 10^9$, which indicates a turbulent environment.

3.4.1.2 Thermal Boundary Layer

The usual experiment is the so-called Rayleigh-Benard experiment. Consider two parallel horizontal plates. Let L be the thickness of the layer between both plates and T be the plate temperature. The experiment consists in heating the lower plate to $T + \Delta T$. If $\Delta T > 0$ (heating), the air between the plates becomes unstable, which generates upwelling motions, referred to as *convective motions*.

Thermal turbulence can be measured by the Rayleigh number defined as

$$R_a = \frac{g}{T}\frac{L^3}{\nu}\frac{\Delta T}{\nu_\theta}, \tag{3.23}$$

where ν_θ is the thermal viscosity. The flow is said to be turbulent for strongly negative values of R_a (e.g. less than -50000).

An application to the ABL with $T \simeq 300\,\mathrm{K}$ and $\nu_\theta \simeq 2 \times 10^{-5}\,\mathrm{m^2\,s^{-1}}$ yields a critical value of ΔT of $10^{-12}\,\mathrm{K}$. This implies that once there is a decreasing temperature profile,[1] convective turbulence can be generated.

In the unstable case, both effects (mechanical and thermal effects) generate turbulence (Table 3.2). In the stable case, there is a mechanical production of turbulence that may even be canceled by the thermal effects due to buoyancy (Sect. 3.4.3).

3.4.2 Scale Range and Averaging

Turbulent flows are characterized by a wide range of scales (of "whirl sizes"). Let L be the length scale for the large-scale motions and l be the length scale for the small-

[1] Actually, the appropriate variable is the *potential* temperature.

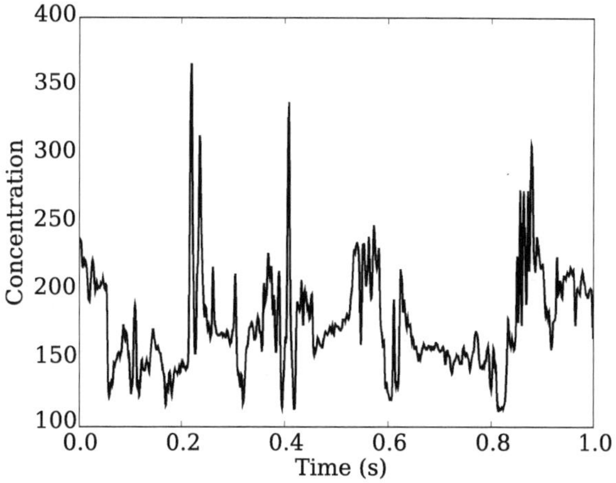

Fig. 3.10 Time evolution of a trace species concentration in a turbulent flow. The concentration is measured at a given point of the flow presented in Fig. 3.9. Credit: Centrale Lyon and French Institute of Nuclear Safety and Radiological Protection

scale motions. In the case of mechanical turbulence, the so-called Kolmogorov analysis provides an estimation of the ratio of the characteristic spatial scales, namely

$$\frac{L}{l} \sim Re^{3/4}. \tag{3.24}$$

In the ABL, the large-scale motions are constrained by the ABL height (let us say 1000 meters) while the small-scale motions can occur at scales of a few millimeters. As expected, the Reynolds number, Re, is very large.

It is therefore not possible to *solve* (to describe) all the scales. An *averaging* approach has to be used. Let Ψ be a spatial flow field (temperature, velocity, density, etc.). It may be split into an average $\langle\Psi\rangle$ (the rigorous definition is not detailed here) and a fluctuation Ψ' (Fig. 3.10), so that

$$\Psi = \langle\Psi\rangle + \Psi'. \tag{3.25}$$

The Navier-Stokes equations that govern the time evolution of Ψ are then rewritten with this decomposition and are averaged. The averaging procedure is required to satisfy a few properties: it has to commute with the derivatives, $\langle\Psi'\rangle = 0$, $\langle\lambda\Psi\rangle = \lambda\langle\Psi\rangle$ if λ is constant, etc. The resulting equations give the evolution of the average fields $\langle\Psi\rangle$. They are the basis for the so-called RANS model (*Reynolds Averaged Navier Stokes*).

Such models require a *closure scheme*. Actually, the nonlinear terms of the initial model, let us say $\Psi_1\Psi_2$ for a quadratic nonlinearity, generate *correlations* between fluctuations. From

$$\Psi_1\Psi_2 = \left(\langle\Psi_1\rangle + \Psi_1'\right) \times \left(\langle\Psi_2\rangle + \Psi_2'\right), \tag{3.26}$$

averaging yields

$$\langle\Psi_1\Psi_2\rangle = \langle\Psi_1\rangle\langle\Psi_2\rangle + \left\langle\Psi_1'\Psi_2'\right\rangle, \tag{3.27}$$

due to the properties of the averaging procedure. The correlation term $\langle \Psi_1' \Psi_2' \rangle$ has to be expressed as a function of the average fields in order to "close" the equations. For the vertical motion of pollutants in the ABL, a challenging issue is the closure scheme for the vertical turbulent flux (Sect. 3.5.2.3).

3.4.3 *Turbulent Kinetic Energy*

Turbulence increases the flow mixing properties. The turbulence intensity is usually assessed by the *specific turbulent kinetic energy*,

$$TKE_s = \frac{1}{2}\left(\langle u'^2 \rangle + \langle v'^2 \rangle + \langle w'^2 \rangle\right), \tag{3.28}$$

where u', v', and w' stand for the fluctuations of the wind velocity components, $V = (u, v, w)$.

3.4.3.1 Competition Between Mechanical and Thermal Effects

The time evolution of TKE_s is governed by

$$\frac{\partial TKE_s}{\partial t} + \langle \mathbf{V} \rangle \cdot \nabla(TKE_s) = P_m + P_t + T_r - \varepsilon, \tag{3.29}$$

where P_m and P_t stand for the turbulence production terms related to the mechanical and thermal effects, respectively,[2] T_r is a redistribution term due to turbulent motions and ε is the viscous dissipation term at small scales, due to molecular diffusion ($\varepsilon > 0$).

The expression of the production terms can be obtained from the flow equations.

Thermal Production For thermal production, we obtain

$$P_t = \langle \theta' w' \rangle \frac{g}{\theta}, \tag{3.30}$$

where w is the vertical component of the wind velocity. The sign of P_t is therefore that of $\langle \theta' w' \rangle$, to be interpreted as a vertical turbulent flux of potential temperature.

This sign is deeply related to the stability of the ABL. Consider an increasing vertical distribution of potential temperature (stable case). A phenomenological argument yields $\langle \theta' w' \rangle > 0$ (Fig. 3.11).

Actually, a fluctuation of the vertical component of the wind velocity has a negative correlation with a potential temperature fluctuation: a rising air mass ($w' > 0$) tends to decrease the ambient temperature after mixing ($\theta' < 0$). On the contrary, a negative fluctuation of w is associated with a positive temperature fluctuation.

[2]Note that *production* does not imply that the source term is positive.

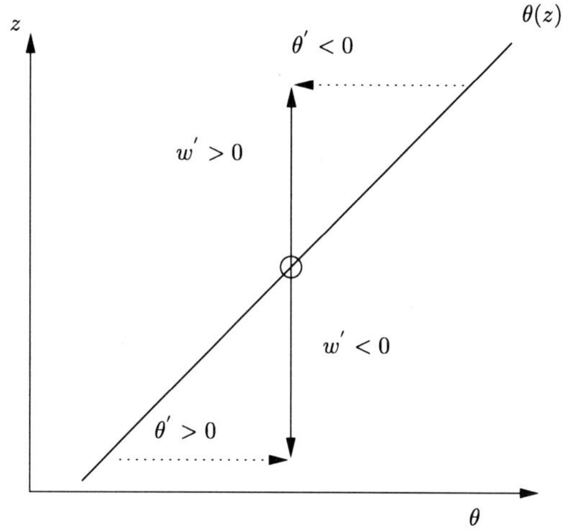

Fig. 3.11 Sign of the vertical turbulent flux for θ, as a function of the atmospheric stability (here in the case of a positive vertical gradient of $\theta(z)$)

The sign of $\langle\theta' w'\rangle$ is therefore the opposite sign to that of $\partial\langle\theta\rangle/\partial z$. This is coherent with the stability criteria for the ABL:

- if buoyancy amplifies the vertical motions, the turbulence is increasing and the ABL is unstable: $P_t > 0$;
- if buoyancy dampens the vertical motions, the (mechanical) turbulence is decreasing and the ABL is stable: $P_t \langle 0$;
- if buoyancy has no effect, the ABL is in the neutral case and the only source of turbulence is the mechanical production: $P_t = 0$

Mechanical Production For mechanical production, we obtain

$$P_m = -\left\langle u' w' \right\rangle \frac{\partial \langle u \rangle}{\partial z} - \left\langle v' w' \right\rangle \frac{\partial \langle v \rangle}{\partial z}. \tag{3.31}$$

Using a similar argument to that used for thermal production, it can be shown that the sign of the vertical turbulent flux of a given variable has the opposite sign to that of the vertical gradient of the average variable: for u, $\langle u' w'\rangle \partial\langle u\rangle/\partial z\langle 0$. The mechanical production term is then *always* positive, which corresponds to turbulence production by wind shear.

3.4.4 Mixing Height and Turbulence Indicators

The *mixing height* is usually defined as the height at which TKE_s can be neglected. This defines the height of the volume in which the pollutants are mixed. The mixing height varies during the day (see Fig. 3.12).

In the stable case, (destruction of turbulence by the thermal effects), the turbulence intensity (equivalently, the mixing intensity) can be evaluated by the *flux*

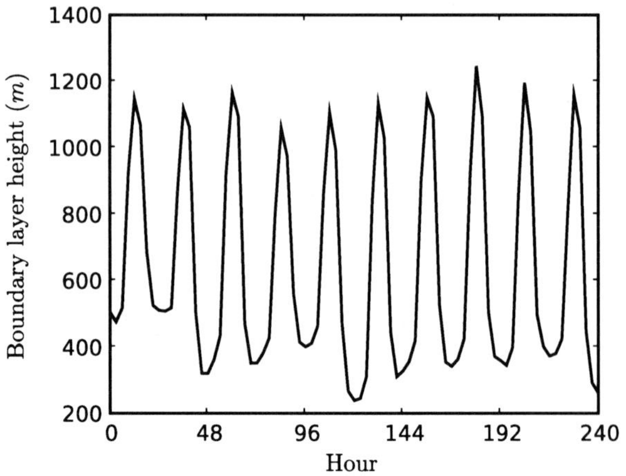

Fig. 3.12 Evolution of the mixing height computed with the POLYPHEMUS system from meteorological fields of ECMWF (*European Centre for Medium-Range Weather Forecasts*). Average over Europe from 22 April 2001. Credit: Vivien Mallet, CEREA

Richardson number,

$$Ri_f = -\frac{P_t}{P_m}. \tag{3.32}$$

This number evaluates the intensity of the turbulence destruction due to the thermal effects. In the stable case, $Ri_f \geq 0$. It is usually admitted that the flow becomes turbulent once $Ri_f \leq 0.21$ ("academic" value).

Using the closure schemes proposed in Sect. 3.5.2.3, the production terms can be parameterized as functions of the vertical distributions of horizontal wind velocity and potential temperature. Namely,

$$Ri_f \simeq -\frac{\frac{g}{\langle\theta\rangle}\frac{\partial\langle\theta\rangle}{\partial z}}{\left(\frac{\partial\langle u\rangle}{\partial z}\right)^2 + \left(\frac{\partial\langle v\rangle}{\partial z}\right)^2}. \tag{3.33}$$

Using the Richardson number and the turbulent kinetic energy (TKE_s) provides estimations of the mixing height, a key parameter for describing the dispersion of pollutants in the boundary layer.

Exercise 3.3 (Entrainment and Residual Layers) The so-called two-box dispersion models describe the mixing layer and the residual layer (Fig. 3.8 and Fig. 3.13).

Let $H(t)$ be the mixing height, and $c_1(z,t)$ and $c_2(z,t)$ be the concentrations in the mixing and residual layers, respectively. Both layers are supposed to be homogeneous in the horizontal direction. Formulate a model for the time evolution of the vertical average of c_1.

Solution:

We write $\overline{c}_1(t) = \int_0^{H(t)} c_1(z,t)\mathrm{d}z / H(t)$. Thus, $\frac{\mathrm{d}(H\overline{c}_1)}{\mathrm{d}t} = \overline{c}_1\frac{\mathrm{d}H}{\mathrm{d}t} + H\frac{\mathrm{d}\overline{c}_1}{\mathrm{d}t}$. This can be also computed by

$$\frac{\mathrm{d}}{\mathrm{d}t}\left(\int_0^{H(t)} c_1(z,t)\mathrm{d}z\right) = \int_0^{H(t)} \frac{\partial c_1(z,t)}{\partial t}\mathrm{d}z + \frac{\mathrm{d}H}{\mathrm{d}t}c_1(H(t),t),$$

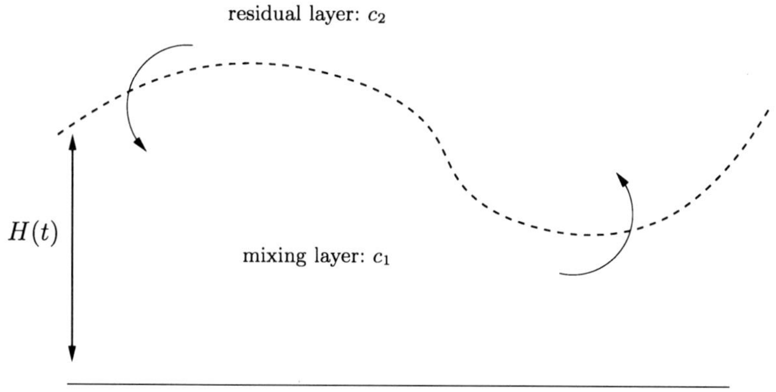

Fig. 3.13 Entrainment of the residual layer (Exercise 3.3)

and hence,

$$\frac{d\bar{c}_1}{dt} = \frac{1}{H(t)} \int_0^{H(t)} \frac{\partial c_1(z,t)}{\partial t} dz + \frac{1}{H} \left[\frac{dH}{dt} (c_1(H(t),t) - \bar{c}_1) \right].$$

The first term can be approached by $\partial \bar{c}_1 / \partial t$ (given by the chemical reactions and the ground boundary conditions). The second term represents the entrainment flux from the residual layer, with the so-called *entrainment velocity* dH/dt (sign to be commented on).

3.5 Fundamentals of Atmospheric Dynamics

This section introduces a few fundamentals for the calculation of the meteorological fields in the atmospheric boundary layer and in the free troposphere. It is recommended to readers not familiar with fluid mechanics to omit this section.

The leading processes depend on the altitude:

- in the free troposphere, there is a balance between the pressure-gradient force and the Coriolis force, which defines the so-called *geostrophic wind*;
- in the atmospheric boundary layer, the surface effects play a key role and a simplified version of the fluid mechanics equations, the *Boussinesq approximation*, can be used.

3.5.1 Primer for Fluid Mechanics

In order to alleviate the notations, the partial derivatives are written in the following form: $\partial_t \Psi = \partial \Psi / \partial t$, $\partial_x \Psi = \partial \Psi / \partial x$, etc.

For any vector $\mathbf{A} = (A_x, A_y, A_z)$, the *divergence* is the real number defined by $\text{div}\,\mathbf{A} = \partial_x A_x + \partial_y A_y + \partial_z A_z$. For any scalar field Ψ, the gradient is the vector $\nabla \Psi = (\partial_x \Psi, \partial_y \Psi, \partial_z \Psi)$.

3.5.1.1 Navier-Stokes Equations

Air is described by the following fields: wind velocity (**V**), density (ρ), energy or temperature (T) and specific humidity (q_s). We assume that air is an ideal gas: as a result, pressure can be computed from density and temperature ($P = \rho r_{air} T$).

The time evolution of the atmospheric fields is governed by equations expressing the conservation laws for mass, impulsion, energy and, for moist air, specific humidity.

Mass Conservation Mass conservation is expressed by the so-called *continuity equation*

$$\partial_t \rho + \text{div}(\rho \mathbf{V}) = 0. \tag{3.34}$$

Impulsion In a reference frame related to the rotating Earth, the forces exerted on air are the pressure-gradient force, the gravity force, the Coriolis force and the air viscosity (molecular diffusion).

Moreover, the particulate derivative, defined by $\text{D}_t \mathbf{V} = \partial_t \mathbf{V} + \mathbf{V} \cdot \nabla \mathbf{V}$, can be used in a Lagrangian approach, where the observer follows the air mass motion. Thus, Newton's law reads

$$\rho \text{D}_t \mathbf{V} = -\nabla P - \rho \mathbf{g} + \rho \mathbf{F}_c + \mu \Delta \mathbf{V}, \tag{3.35}$$

with μ the molecular diffusion coefficient (the kinematic viscosity is $\nu = \mu/\rho$ and is expressed in $\text{m}^2\,\text{s}^{-1}$). The vector **g** is upward-oriented (positive with increasing altitudes). The Coriolis force is $\rho \mathbf{F_c} = -2\rho \Omega \wedge \mathbf{V}$ where Ω is the angular velocity vector of Earth's rotation.

Energy Similarly, a continuity equation can be derived for temperature. The heat fluxes include the latent heat flux (phase transition for water), the radiative heat flux and the diffusion flux. The continuity equation reads

$$\rho c_p \text{D}_t T = Q + \mu_\theta \Delta T, \tag{3.36}$$

where μ_θ is the thermal conductivity and Q stands for the other heat fluxes (latent heat and radiative flux).

Specific Humidity The continuity equation for specific humidity is

$$\text{D}_t q_s = S_{q_s} + \nu_w \Delta q_s, \tag{3.37}$$

with ν_w the molecular diffusion coefficient for water vapor and S_{q_s} the water phase transition rate.

Equations (3.34)–(3.37) constitute the Navier-Stokes equations.

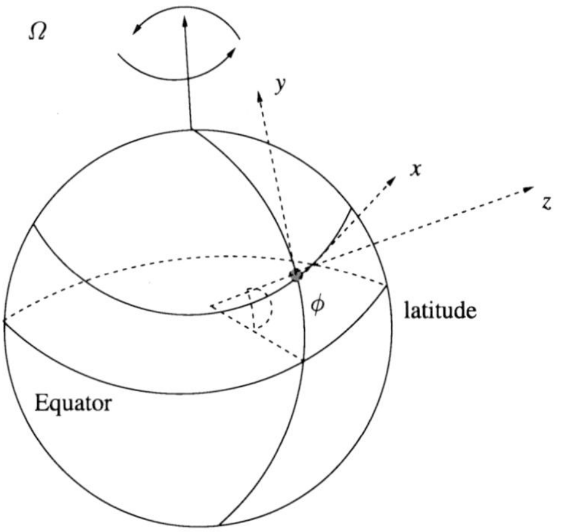

Fig. 3.14 Latitude-longitude-altitude referential

3.5.1.2 Projection in a Latitude-Longitude-Altitude System

Consider a frame of reference related to the rotating Earth (Fig. 3.14): the x axis is westward oriented (longitude), the y axis is northward oriented (latitude), and the z axis is upward-oriented (altitude). Note that such a referential is not strictly Cartesian because a curvature term should be included so that the Earth's rotation is taken into account. The resulting terms are omitted in this simplified presentation.

We focus on the wind velocity and on the air density (pressure is given by the hydrostatic approximation, temperature by the ideal gas law). Writing $\mathbf{V} = (u, v, w)$, the evolution equations are in the corresponding coordinate system

$$\begin{cases} \partial_t \rho + \partial_x(\rho u) + \partial_y(\rho v) + \partial_z(\rho w) = 0 \\ \partial_t u + u\partial_x u + v\partial_y u + w\partial_z u = -\frac{1}{\rho}\partial_x P + (F_c)_x + \nu\Delta u \\ \partial_t v + u\partial_x v + v\partial_y v + w\partial_z v = -\frac{1}{\rho}\partial_y P + (F_c)_y + \nu\Delta v \\ \partial_t w + u\partial_x w + v\partial_y w + w\partial_z w = -\frac{1}{\rho}\partial_z P + (F_c)_z + \nu\Delta w - g. \end{cases} \tag{3.38}$$

Let ϕ be the latitude, defined as the angle between the Equatorial plane and the vertical direction. As $\mathbf{\Omega} = (0, \Omega\cos\phi, \Omega\sin\phi)$, the Coriolis force is

$$\begin{cases} (F_c)_x = -2\Omega(w\cos\phi - v\sin\phi) \\ (F_c)_y = -2\Omega u\sin\phi \\ (F_c)_z = 2\Omega u\cos\phi. \end{cases} \tag{3.39}$$

In a first approximation, the vertical component w can be neglected with respect to the horizontal components, namely

$$(F_c)_x \simeq fv, \qquad (F_c)_y = -fu, \qquad (F_c)_z = 2\Omega u\cos\phi, \tag{3.40}$$

where $f = 2\Omega\sin\phi$ is the so-called *Coriolis parameter* (its unit is that of a frequency). At mid-latitudes ($\phi \simeq 45°$), f is about $10^{-4}\ \mathrm{s}^{-1}$.

Table 3.3 Magnitude of the horizontal components

Term	Magnitude	Estimation ($\mathrm{m\,s^{-2}}$)
Time derivative $\mathrm{D}u/\mathrm{D}t$	U^2/L	10^{-4}
Coriolis force	fU	10^{-3}
Pressure	$\delta P/(\rho L)$	10^{-3}
Molecular diffusion	$\nu U/H^2$	10^{-12}

Table 3.4 Magnitude of the vertical component

Time	Magnitude	Estimation ($\mathrm{m\,s^{-2}}$)
Time derivative $\mathrm{D}w/\mathrm{D}t$	UW/L	10^{-7}
Coriolis force	fU	10^{-3}
Pressure	$P_0/(\rho H)$	10
Gravity	g	10
Molecular diffusion	$\nu W/H^2$	10^{-15}

In the Northern Hemisphere, $\phi > 0$, and the eastward deviation is justified by $(F_c)_x > 0$. Note that the reverse case applies for the Southern Hemisphere.

3.5.1.3 Magnitude Analysis

At the synoptic scale, it is common to carry out a magnitude analysis ([54]). The characteristic scales are $U \simeq 10\,\mathrm{m\,s^{-1}}$ for the horizontal wind velocity and $W \simeq 1\,\mathrm{cm\,s^{-1}}$ for the vertical wind velocity.

Let us distinguish the horizontal components whose length scale is $L \simeq 1000\,\mathrm{km}$, from the vertical components (above the boundary layer) with the length scale $H \simeq 10\,\mathrm{km}$. The corresponding timescale is $T = L/U = 10^5\,\mathrm{s}$. For the horizontal component, the pressure-gradient term is of magnitude $\delta P/(\rho L) \simeq 10^{-3}\,\mathrm{m\,s^{-2}}$ (which corresponds, with $\rho \simeq 1\,\mathrm{kg\,m^{-3}}$ at ground, to a change of 1 hPa for 100 km). The vertical term is adimensionalized with a ground pressure of $P_0 \simeq 1000\,\mathrm{hPa}$. Moreover, for the diffusion terms, the vertical components dominate ($H \ll L$), which motivates the use of the length scale H.

The magnitude of the different terms arising in the evolution equation is shown in Tables 3.3 and 3.4, for the horizontal and vertical components, respectively.

This magnitude analysis motivates the approximation of *geostrophic winds*, valid at the synoptic scales and above the atmospheric boundary layer:

$$u \simeq u_G = -\frac{1}{\rho f}\partial_y P, \qquad v \simeq v_G = \frac{1}{\rho f}\partial_x P. \tag{3.41}$$

The geostrophic wind is therefore horizontal and tangent to the isobars (the surface with a constant pressure), with a clockwise rotation around a *high* (anticyclonic region) in the Northern Hemisphere (Fig. 3.15).

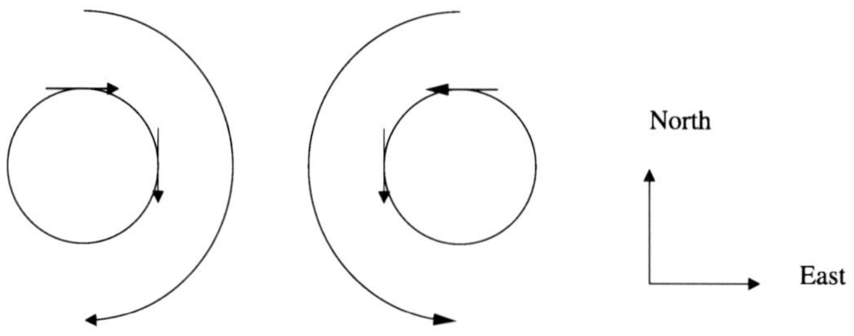

Fig. 3.15 Geostrophic wind in the Northern Hermisphere ($f > 0$). *Left*: rotation around a *high* (anticyclone); *right*: rotation around a *low* (depression)

Note that the vertical component satisfies the *hydrostatic approximation*, namely $\partial_z P \simeq -\rho g$.

3.5.2 ABL Flow

We now focus on the ABL. The evolution equations can be simplified by using the so-called *Boussinesq approximation*. Then, we investigate three applications: the *Ekman spiral*, the *low-level nocturnal jets* and the surface boundary layer.

3.5.2.1 Boussinesq Approximation

In the ABL, the air density does not deviate too much from the density provided by the hydrostatic approximation. Moreover, the density is nearly constant. It is however not possible to use a constant density because the density changes play a key role for the buoyancy effects (as illustrated by the stability analysis of the ABL). Therefore, the Boussinesq approximation assumes that the density is a constant field, except for the buoyant term.

It is common to carry out a *perturbation analysis*, on the basis of a model or scale hierarchy. Let us assume that the state, Ψ, is near a *reference* state, Ψ_r, namely that $\Psi_1 = \Psi - \Psi_r$ is "small". Using the decomposition $\Psi = \Psi_r + \Psi_1$, the resulting system is linearized by removing all the quadratic terms in the form $\Psi_1 \Psi_1$ and by keeping the linear terms in the form $\Psi_1 \Psi_r$. Note that the approach is different from that used for the averaging of turbulent flows (RANS), for which the caveat is the closure of the nonlinear terms.

The reference state for the Boussinesq approximation is

- at rest ($\mathbf{V_r} = 0$);
- hydrostatic ($\partial_z P_r = -\rho_r g$);
- adiabatic ($\partial_z T_r = -g/c_p$);

- air is supposed to be an ideal gas ($P_r = \rho_r r T_r$).

The thermodynamic variables (ρ_r, T_r, P_r) are then defined for any altitude z. Moreover, it is observed that the deviation to the average value over the ABL (of height H_{ABL}),

$$\Psi_0 = \frac{1}{H_{ABL}} \int_0^{H_{ABL}} \Psi_r(z)\mathrm{d}z, \tag{3.42}$$

is small:

$$\Psi_r = \Psi_0 + \Delta\Psi_r, \qquad \frac{\Delta\Psi_r}{\Psi_0} \ll 1. \tag{3.43}$$

Finally,

$$\Psi = \Psi_r + \Psi_1 = \Psi_0 + \Delta\Psi_r + \Psi_1, \tag{3.44}$$

where the discrepancies, first between the reference state and the average state, and second between the reference state and the "true" state, are small:

$$\frac{\Delta\Psi_r}{\Psi_0} \ll 1, \qquad \frac{\Psi_1}{\Psi_0} \ll 1. \tag{3.45}$$

After a few algebraic manipulations (omitted for the sake of clarity and for convenience), we obtain

$$\begin{cases} \partial_x u + \partial_y v + \partial_z w = 0 \\ \partial_t u + u\partial_x u + v\partial_y u + w\partial_z u = \frac{\mu}{\rho_0}\Delta u - \frac{1}{\rho_0}\partial_x P_1 + (F_c)_x \\ \partial_t v + u\partial_x v + v\partial_y v + w\partial_z v = \frac{\mu}{\rho_0}\Delta v - \frac{1}{\rho_0}\partial_y P_1 + (F_c)_y \\ \partial_t w + u\partial_x w + v\partial_y w + w\partial_z w = \frac{\mu}{\rho_0}\Delta w - \frac{1}{\rho_0}\partial_z P_1 + \frac{T_1}{T_0} g + (F_c)_z \\ \rho_0 c_p(\partial_t T_1 + u\partial_x T_1 + v\partial_y T_1 + w\partial_z T_1) = Q + \mu_\theta \Delta T_1. \end{cases} \tag{3.46}$$

Q stands for the possible heat sources, μ_θ for the thermal conductivity and μ for the dynamic viscosity. The resulting system comprises five equations for five unknown variables (u, v, w, P_1, T_1). It is a linear system with respect to the thermodynamic variables.

Remark 3.5.1 (Potential Temperature) An equation similar to that obtained for temperature T_1 can be derived for the potential temperature $\theta_1 = \theta - \theta_r$:

$$\rho_0 c_p(\partial_t \theta_1 + u\partial_x \theta_1 + v\partial_y \theta_1 + w\partial_z \theta_1) = Q + \mu_\theta \Delta\theta_1. \tag{3.47}$$

A key remark is that the potential temperature indicates a deviation to the adiabatic state (as mentioned before): $\theta_1 \simeq T - T_0$.

In a first approximation, the term $(T_1/T_0)g$, in the equation for the vertical velocity component w, can be replaced by $(\theta_1/\theta_0)g$.

3.5.2.2 Reynolds Averaging for the Turbulent ABL

Turbulence plays a leading role in the ABL, as illustrated by the magnitude of the Reynolds and Rayleigh numbers. This is associated to a wide range of scales and, as a result, it is not possible to solve all the scales. An averaging approach is then required. It is based on the partitioning of any field Ψ in an average $\langle\Psi\rangle$ and in a fluctuation Ψ': $\Psi = \langle\Psi\rangle + \Psi'$.

Upon averaging (3.46), we obtain

$$\begin{cases} \partial_x\langle u\rangle + \partial_y\langle v\rangle + \partial_z\langle w\rangle = 0 \\ \frac{\mathrm{D}\langle u\rangle}{\mathrm{D}t} = \frac{\mu}{\rho_0}\Delta\langle u\rangle - \frac{1}{\rho_0}\partial_x\langle P_1\rangle + \langle (F_c)_x\rangle - \mathrm{div}(\langle u'\mathbf{V}'\rangle) \\ \frac{\mathrm{D}\langle v\rangle}{\mathrm{D}t} = \frac{\mu}{\rho_0}\Delta\langle v\rangle - \frac{1}{\rho_0}\partial_y\langle P_1\rangle + \langle (F_c)_y\rangle - \mathrm{div}(\langle v'\mathbf{V}'\rangle) \\ \frac{\mathrm{D}\langle w\rangle}{\mathrm{D}t} = \frac{\mu}{\rho_0}\Delta\langle w\rangle - \frac{1}{\rho_0}\partial_z\langle P_1\rangle + \langle (F_c)_z\rangle + \frac{\langle\theta_1\rangle}{\theta_0}g - \mathrm{div}(\langle w'\mathbf{V}'\rangle) \\ \rho_0 c_p \frac{\mathrm{D}\langle\theta_1\rangle}{\mathrm{D}t} = Q + \mu_\theta\Delta\langle\theta_1\rangle - \mathrm{div}(\langle\theta_1{}'\mathbf{V}'\rangle), \end{cases} \tag{3.48}$$

where the particulate derivative of an average field $\langle\Psi\rangle$ is defined by

$$\frac{\mathrm{D}\langle\Psi\rangle}{\mathrm{D}t} = \partial_t\langle\Psi\rangle + \langle u\rangle\,\partial_x\langle\Psi\rangle + \langle v\rangle\,\partial_y\langle\Psi\rangle + \langle w\rangle\,\partial_z\langle\Psi\rangle, \tag{3.49}$$

and the divergence of the correlation with the velocity fluctuation $\mathbf{V}'$ is

$$\mathrm{div}\big(\langle\Psi'\mathbf{V}'\rangle\big) = \partial_x\langle\Psi'u'\rangle + \partial_y\langle\Psi'v'\rangle + \partial_z\langle\Psi'w'\rangle. \tag{3.50}$$

The turbulent fluxes $\langle\Psi'\mathbf{V}'\rangle$ have to be parameterized: this is the *closure problem* of turbulent models (here with nine correlations).

3.5.2.3 Closure Schemes

The closure schemes can be classified according to their *order*. A first order closure scheme is expressed as a function of the average fields (see below). A second order closure scheme is based on the average correlations. The so-called $k-\varepsilon$ model is usually referred to as a scheme of order 1.5. The viscous dissipation rate, ε (see (3.29)), is calculated by

$$\varepsilon = \frac{(TKE_s)^{1.5}}{L_\varepsilon}, \tag{3.51}$$

where TKE_s is the specific turbulent kinetic energy and L_ε a characteristic length scale for viscous dissipation.

The most popular approach is still the use of first-order schemes, especially with the *K theory* ([80]). By analogy with the molecular diffusion flux, the turbulent fluxes of Ψ are supposed to be proportional to the gradient of the average field $\langle\Psi\rangle$ for a given direction. For example, the vertical fluxes are given by

$$\langle\Psi'w'\rangle \simeq -K_{\Psi,z}\partial_z\langle\Psi\rangle, \tag{3.52}$$

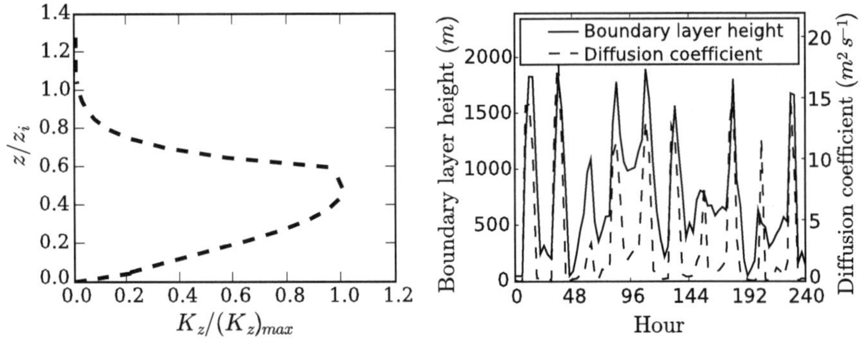

Fig. 3.16 *Left panel*: vertical profile of K_z (normalized) in the unstable case (z_i corresponds to the mixing height). Source: [130]. *Right panel*: evolution of the mixing height and of K_z at an altitude of 50 m above Paris, from 22 April 2001 (data simulated as in Fig. 3.12). Credit: Vivien Mallet, CEREA

where the eddy diffusion coefficient, $K_{\Psi,z}$, has to be parameterized. Note that this is coherent with the phenomenological argument of Sect. 3.4.3 when the coefficient is positive.

Actually, only four correlations are usually parameterized:

$$\begin{cases} \langle \mathbf{V}'w' \rangle = -K_m \partial_z \langle \mathbf{V} \rangle \\ \langle {\theta_1}'w' \rangle = -K_h \partial_z \langle \theta_1 \rangle, \end{cases} \tag{3.53}$$

where K_m is the eddy dynamic viscosity (supposed to be equal for the three directions) and K_h is the heat eddy viscosity. The coefficients K_m and K_h are usually positive, except in the so-called *counter-gradient models* where $\langle {\theta_1}'w' \rangle = -K_h(\partial_z \langle \theta_1 \rangle - \gamma)$ with $\gamma > 0$. These coefficients are calculated as functions of altitude and time, so that the time evolution of the turbulent ABL is described. For the pollutant dispersion, the key parameter is the vertical eddy coefficient $K_z(z,t)$ (Fig. 3.16). The precise form of the parameterizations is not detailed here (out of the scope of this presentation).

3.5.2.4 A Few Examples

Ekman Layer Consider a horizontal wind velocity, supposed to be homogeneous in the horizontal directions. The ABL is supposed to be adiabatic (neutral case). Mathematically, $\langle w \rangle = 0$, $\langle u \rangle(z)$, $\langle v \rangle(z)$, and $\langle \theta_1 \rangle = 0$.

Using the K theory as a closure scheme and neglecting the molecular diffusion with respect to the eddy diffusion, the equations become

$$\begin{cases} \partial_t \langle u \rangle = -\frac{1}{\rho_0} \partial_x \langle P_1 \rangle + f \langle v \rangle + K_m \partial^2_{zz} \langle u \rangle \\ \partial_t \langle v \rangle = -\frac{1}{\rho_0} \partial_y \langle P_1 \rangle - f \langle u \rangle + K_m \partial^2_{zz} \langle v \rangle. \end{cases} \tag{3.54}$$

At the top of the ABL, the daily variations are low ($\partial_t \simeq 0$), and the surface effects can be neglected (weak turbulence: $K_m \simeq 0$). These conditions define the geostrophic wind as

$$\langle u \rangle_g = -\frac{1}{\rho_0 f}\partial_y \langle P_1 \rangle, \qquad \langle v \rangle_g = \frac{1}{\rho_0 f}\partial_x \langle P_1 \rangle. \tag{3.55}$$

Assuming that the horizontal pressure gradient is constant, we obtain

$$\begin{cases} \partial_t \langle u \rangle = -f(\langle v \rangle_g - \langle v \rangle) + K_m \partial^2_{zz} \langle u \rangle \\ \partial_t \langle v \rangle = f(\langle u \rangle_g - \langle u \rangle) + K_m \partial^2_{zz} \langle v \rangle. \end{cases} \tag{3.56}$$

For convenience, the x axis is supposed to be aligned with the geostrophic wind, namely $\langle v \rangle_g = 0$. The boundary conditions are $\langle u \rangle(0) = \langle v \rangle(0) = 0$ at ground. The stationary solutions (given by $\partial_t = 0$) are then easily calculated by using complex variables $\langle u \rangle + j\langle v \rangle$ (with $j^2 = -1$),

$$\begin{cases} \langle u \rangle (z) = \langle u \rangle_g (1 - \cos(az)e^{-az}) \\ \langle v \rangle (z) = \langle u \rangle_g \sin(az)e^{-az}, \end{cases} \tag{3.57}$$

where $a = \sqrt{f/2K_m}$. The ABL height (H_{ABL}) is defined by the first altitude at which the velocity field is aligned with the direction of the geostrophic wind (x axis). It is given by $aH_{ABL} = \pi$, namely $H_{ABL} = \pi\sqrt{2K_m/f}$.

The stronger the surface (friction) effect is (large values of K_m), the larger the ABL height is. With $f \simeq 10^{-4}\,\mathrm{s}^{-1}$ and $K_m \simeq 10\,\mathrm{m}^2\,\mathrm{s}^{-1}$, we obtain a realistic value $H_{ABL} \simeq 1500\,\mathrm{m}$.

At this altitude, the horizontal component is nearly equal to that of the geostrophic wind, since $\langle u \rangle(H_{ABL}) = \langle u \rangle_g(1 + e^{-\pi}) \simeq \langle u \rangle_g$.

At the altitude z, the angle of the wind velocity with the direction of the geostrophic wind is exactly the angle with the x axis, $\alpha(z)$, namely

$$\tan\alpha(z) = \frac{\langle v \rangle}{\langle u \rangle}. \tag{3.58}$$

The Ekman spiral represents the angle evolution as a function of altitude (Fig. 3.17). Using asymptotic expansions in the vicinity of 0, $\sin\theta \simeq \theta$ and $1 - \cos\theta \simeq \theta^2/2$, we calculate an angle $\alpha(0) = 45$ degrees at ground. Actually, the real-life angle varies from 25 to 30 degrees, as shown in Fig. 3.3. The limitations of this simple model can be overcame by defining the surface layer (see below).

The surface friction implies that the wind velocities are oriented inward the spiral: this is sometimes referred to as *Ekman pumping*.

Exercise 3.4 (Low-Level Nocturnal Jets) This exercise aims at illustrating the low-level nocturnal jets. This phenomenon is related with the *inertial oscillations* of frequency f.

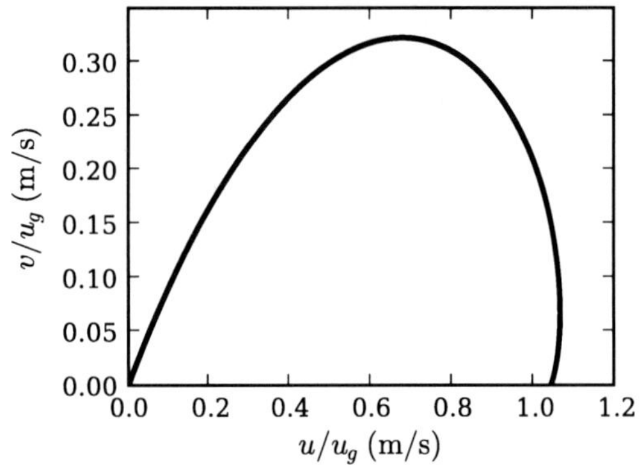

Fig. 3.17 Ekman spiral of wind velocity for $v_g = 0$

Consider the model (3.56). The geostrophic wind is supposed to be aligned with the x axis ($\langle v\rangle_g = 0$). Assume that a stationary state $\langle u\rangle_j(z)$, $\langle v\rangle_j(z)$ has been met during the day: it is therefore computed by the Ekman solution (3.57).

Calculate the evolution of the wind velocity during nighttime (omit turbulence in a first approximation).

Solution:

The subscript n refers to the nocturnal case. Equation (3.56) becomes

$$\partial_t \langle u\rangle_n = f\,\langle v\rangle_n\,, \qquad \partial_t \langle v\rangle_n = f(\langle u\rangle_g - \langle u\rangle_n),$$

with the initial conditions (at $t = 0$) $\langle u\rangle_n(0) = \langle u\rangle_j$ and $\langle v\rangle_n(0) = \langle v\rangle_j$.

It is easy to get

$$\begin{cases} \langle u\rangle_n(t) = \langle u\rangle_j \cos(ft) + \langle v\rangle_j \sin(ft) + \langle u\rangle_g (1 - \cos(ft)) \\ \langle v\rangle_n(t) = \langle v\rangle_j \cos(ft) + (\langle u\rangle_g - \langle u\rangle_j) \sin(ft). \end{cases}$$

The oscillation period, defined by the Coriolis parameter f ($f \simeq 10^{-4}\,\text{s}^{-1}$), is $2\pi/f \simeq 17$ hours. Note that the nocturnal wind velocity can be greater than the geostrophic wind (development of a horizontal jet).

Surface Boundary Layer (SBL) The Ekman layer does not correspond to the real-life ABL, especially due to the stationary assumption. A strong limitation is related to the so-called *surface boundary layer* (SBL), defined as a layer with *constant fluxes*, just above the ground. The SBL is the layer in which dry deposition occurs (Problem 3.1). The profiles of the main meteorological fields in the SBL can be computed with analytical formula.

The *friction velocity*, $u_\star$, is computed from the surface turbulent flux (supposed to be constant in the SBL),

$$(u_\star)^2 = \left|\langle u'w'\rangle\right|_{z=0}. \tag{3.59}$$

The terminology is quite logical: the rougher the ground is, the stronger the friction force is, which generates turbulence. The value of $u_\star$ is of a few tens of centimeter per second.

Table 3.5 A few typical values for the roughness height (z_0). Source: [155]

Environment	z_0
Sea	0.1 mm
Grass	1 cm
Culture	10 cm
Forest	1 m
Urban	≥2 m

Neutral Case In the *neutral* case, a first approach to compute the wind velocity gradient is to carry out a dimensional analysis. Inside the SBL, the two physical appropriate variables are $u_\star$ and z (the altitude can be viewed as the distance from the ground). The gradient is then in the form

$$\frac{\mathrm{d}\langle u\rangle}{\mathrm{d}z} = \frac{u_\star}{\kappa\, z}, \tag{3.60}$$

with κ a dimensionless constant. We have kept the usual notation: κ is the von Karman constant, ranging from 0.35 to 0.4.

Upon integration, we obtain a logarithmic profile

$$\langle u\rangle (z) = \frac{u_\star}{\kappa} \ln\left(\frac{z}{z_0}\right), \tag{3.61}$$

with z_0 the *roughness height* defined by $u(z_0) = 0$. Typically, z_0 is about one tenth of the obstacles, namely a few centimeters in a natural environment and a few meters in an urban area (Table 3.5).

Exercise 3.5 (Turbulent Diffusion in the SBL) Calculate the eddy diffusion coefficient (K_m) in the neutral case of the SBL.

Solution:

By definition, $\langle u'w'\rangle = -K_m \partial\langle u\rangle/\partial z$. At ground, this flux is also equal to $-u_\star^2$ (the sign is, as expected, negative). Moreover, with (3.60), one gets $\mathrm{d}\langle u\rangle/\mathrm{d}z = u_\star/(\kappa z)$. This yields

$$K_m = \kappa u_\star z.$$

The result is logical: turbulence increases with increasing friction and wind velocity.

A second approach is based on the so-called *mixing length theory* (also referred to as Prandtl theory). Writing ξ' the length of the vertical motion during which an air parcel is not mixed with the environment (the sign of ξ' is fixed by the direction), the motion generates a fluctuation of the horizontal wind velocity,

$$u' \simeq -\xi' \frac{\mathrm{d}\langle u\rangle}{\mathrm{d}z}. \tag{3.62}$$

Taking the opposite sign to that of the vertical gradient is motivated by the fact that, in the vicinity of the ground, $\mathrm{d}\langle u\rangle/\mathrm{d}z > 0$: an uprising parcel ($\xi' > 0$) generates a negative fluctuation of u.

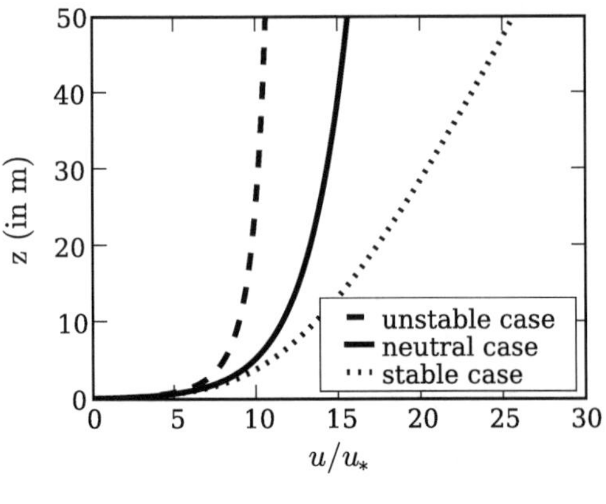

Fig. 3.18 Dimensionless profile of the horizontal wind velocity in the SBL (rural environment: $z_0 = 0.1$ m)

For the vertical velocity fluctuation, we can assume that the horizontal and vertical scales are similar. Thus,

$$w' = \xi' \frac{\mathrm{d}\langle u \rangle}{\mathrm{d}z}, \tag{3.63}$$

since an uprising motion ($\xi' > 0$) is associated to a positive vertical fluctuation ($w' > 0$). This gives

$$\langle u'w' \rangle = -\langle \xi'^2 \rangle \left(\frac{\mathrm{d}\langle u \rangle}{\mathrm{d}z} \right)^2. \tag{3.64}$$

Using (3.59), we obtain

$$u_\star = \sqrt{\langle \xi'^2 \rangle} \frac{\mathrm{d}\langle u \rangle}{\mathrm{d}z}. \tag{3.65}$$

A "logical" choice for the mixing length $\sqrt{\langle \xi'^2 \rangle}$ is $\kappa\, z$ (since the "whirl" size is given by the distance from the ground). We conclude by getting (3.61).

General Case The dimension analysis can be extended to stable and unstable cases. Other physical variables have to be taken into account, especially for describing buoyancy effects. Let us define a mixing length, the so-called Monin-Obukhov length (L_{MO}),

$$L_{MO} = -\frac{\theta_0}{\kappa\, g} \frac{u_\star^3}{Q_0}, \tag{3.66}$$

with Q_0 the heat flux at ground. The vertical velocity gradient is therefore

$$\frac{\mathrm{d}\langle u \rangle}{\mathrm{d}z} = \frac{u_\star}{\kappa z} \phi \left(\frac{z}{L_{MO}} \right), \tag{3.67}$$

where ϕ stands for given analytical functions (e.g. the Businger functions).

For example, in the neutral case, $\phi = 1$. In the stable case, ϕ is given by $\phi(z) = 1 + 8.1 z/L_{MO}$ and the integration is straightforward. The unstable case leads to a

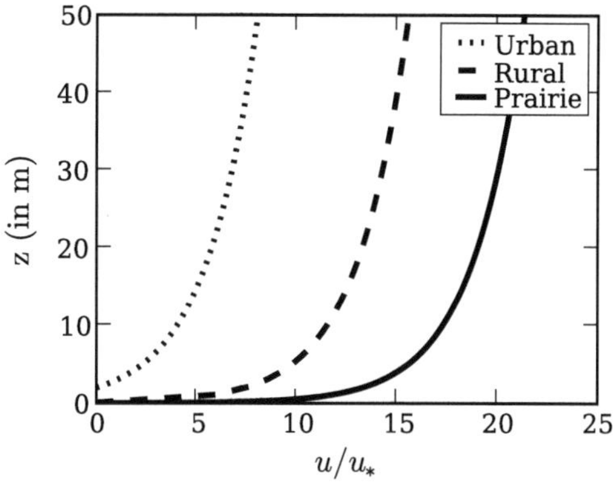

Fig. 3.19 Dimensionless profile of the horizontal wind velocity in the neutral case for a few roughness heights

more complicated expression (omitted here). The wind velocity profiles obtained for a rural case are shown in Fig. 3.18.

The change in the roughness induced by the land use cover may lead to strong differences in the wind profile (Fig. 3.19 for the neutral case).

3.6 A Few Facts for the Urban Climate

The effects of an urban area on the atmospheric boundary layer (and, as a result, on pollutant dispersion) have different sources:

- the geometrical factors (e.g. the building height) have many consequences: radiative effects, increasing roughness, increasing turbulence, etc.
- the surface factors (an urban surface is drastically different from a natural surface) induce flow changes due to the varying surface albedo, to an increasing thermal conductivity and to a decreasing evaporation (related to the lack of vegetation);
- the increasing emissions of aerosols and gases may modify the radiative budgets and the precipitation regime (e.g. with an increasing number of cloud condensation nuclei for cloud formation, Sect. 5.2.3).

In the following, a few key facts are given for understanding the urban climate. The attention is paid on four aspects: the urban breeze (similar to the sea breeze), the urban energy budget, the urban heat island and the formation of the internal urban boundary layer.

3.6.1 Thermal Forcing and Urban Breeze

Temperatures are higher in an urban area than in its vicinity. This thermal forcing generates the appropriate conditions for the development of the so-called *urban breeze*.

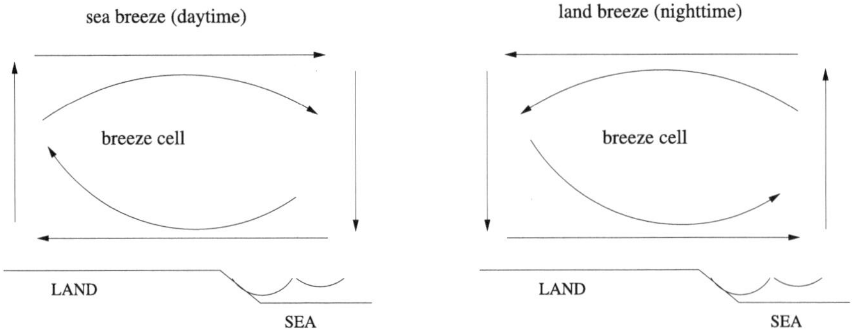

Fig. 3.20 Sketch of the sea breeze. The arrows indicate the circulation of the air masses

This phenomenon is usually compared to the sea breeze (Fig. 3.20): a continental surface has a lower thermal inertia than the sea (the temperature changes at the sea level do not exceed a few degrees during the day).

At day, uprising warm air masses are formed above the continental area. This generates pressure gradients: the highs are located above the continental area while the lows are above the sea. Indeed, the scale height of pressure is an increasing function of temperature (see (1.11), Chap. 1). At elevated altitudes, a sea breeze circulation takes place from the high pressures (above the continental area) to the low pressures (above the sea). As a result, the air column above the continental area is "lighter", which implies a decreasing pressure gradient at ground, from the continental area to the sea. This generates a ground circulation of air masses from the sea to the continental area (the so-called sea breeze).

At night, the cooling of the continental area is stronger than that of the sea. The circulation is reversed.

The analogy with the urban area is straightforward. Actually, the temperature of an urban area is higher than that of its vicinity, especially at night (see below the concept of urban heat island). This results in a circulation similar to the sea breeze, which can be viewed as a ventilation effect for the atmospheric pollutants.

3.6.2 Energy Budget

The energy budget over an urban area reads

$$Q_{rad} + Q_{source} = Q_{sensible} + Q_{latent} + Q_{stock} + Q_{advection}, \tag{3.68}$$

with Q_{rad} the net radiative flux, Q_{source} the anthropogenic heat sources, $Q_{sensible}$ the sensible heat flux (turbulent flux), Q_{latent} the latent heat flux (water condensation), Q_{stock} the flux related to the heat trapped inside the city and $Q_{advection}$ the advected flux. These fluxes are expressed in $\mathrm{W\,m^{-2}}$.

An urban area is characterized by a large source term: Q_{source} may vary from 50 to 100 $\mathrm{W\,m^{-2}}$ on average, with lower values in summertime and higher values

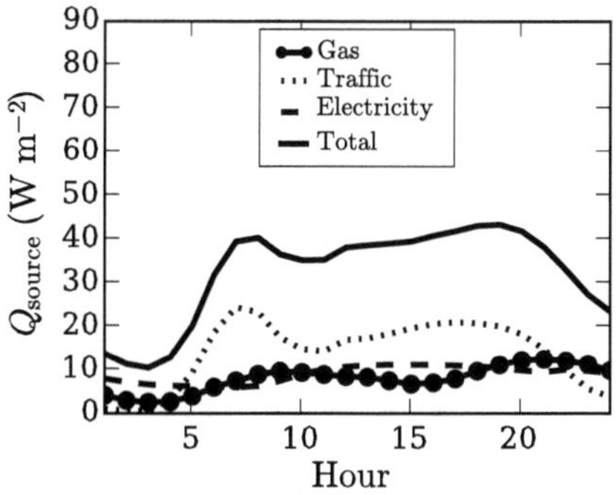

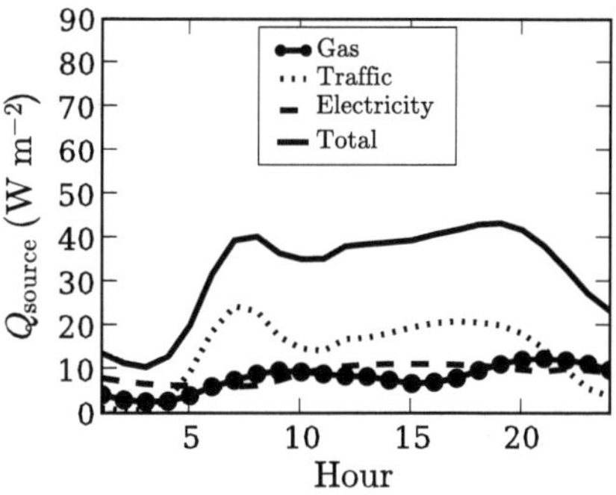

Fig. 3.21 Daily variation of Q_{source} for Tehran. *Left*: summertime; *right*: wintertime. Credit: Hossein Malakooti, CEREA

Table 3.6 Albedo of urban surfaces with respect to the solar radiation

Surface	Albedo
asphalt	$\simeq 0.1$
concrete	$\simeq 0.2$
tree	$\simeq 0.15$
roof	[0.1, 0.9] (depending on color)

in wintertime (up to hundreds of W m^{-2}). Figure 3.21 shows the daily variations estimated for Tehran (Iran).

A city has also a specific radiative behavior. The net radiative flux is given by $Q_{rad} = (1-A)F_{incident} + F_{\downarrow} - F_{\uparrow}$ with A the surface albedo, $F_{incident}$ the shortwave incident radiative flux, $F_{\downarrow}$ the longwave radiative flux emitted or scattered by the atmosphere, and $F_{\uparrow}$ the longwave radiative flux emitted by the urban surface.

The surface albedo is low for an urban area (typically $A \simeq 0.15$, Table 3.6 for the "urban" albedos; to be compared to Table 2.4 for natural surfaces). Moreover, the high particulate concentrations induce an increase in the shortwave scattering ($F_{incident}$). As a result, the energy absorbed at ground is higher. For longwave radiation, the scattering is also increasing ($F_{\downarrow}$), but the overall effect is complicated because the increasing temperature results in an increasing emission ($F_{\uparrow}$).

The sensible heat flux is increasing (up to 50%) due to the increasing turbulence.

Finally, all these contributions lead to an increasing temperature in an urban area (of a few degrees). The maximal increase defines the so-called *urban heat island*.

3.6.3 Urban Heat Island

The urban heat island is the most noticeable illustration of the urban climate. It can be particularly marked at nighttime, with a discrepancy between the "urban temperature" and the "rural temperature" of a few degrees (ΔT in Table 3.7). There

Table 3.7 Measured values of the urban heat island for a few cities during the last three decades. Source: [114]

City	ΔT	City	ΔT
Barcelona	8.2 K	Calgary	8.1 K
Mexico	9.4 K	Montreal	10 K
Moscow	9.8 K	New York	11.6 K
Tokyo	8.1 K	Vancouver	11.6 K

are old observations since the XVIIIth century: $\Delta T \simeq 6\,\text{K}$ for Mannheim in 1784 (Deurer), $\Delta T \simeq 4\,\text{K}$ for London in 1818 (Howard), etc.

The infrared measurements provide a powerful tool to evaluate the urban heat island ([35]). Figure 3.22 shows the assessment of the average ground temperature during the heat wave of summer 2003 over Greater Paris (Île-de-France region; period from 4 to 13 august 2003), respectively at nighttime between 20:00 and 23:00 (TU) and at daytime between 12:00 and 14:00 (TU). We can refer to [36].

The nocturnal case corresponds to the classical case: the urban heat island is located in the city center and can be up to 8 K. At daytime, the situation is quite different with a complicated map in the suburbs, especially in industrial areas. Due to the thermal inertia, these areas have a faster response time than the city center (due to the *trapping effect* in urban canopies).

Empirical laws have been proposed on the basis of observations in order to connect ΔT to the number of inhabitants (P) and to the meteorological conditions (especially for the wind velocity V, usually at a height of 10 meters, in a rural area). For example, the Oke's law ([104]) reads

$$\Delta T \simeq \frac{1}{4}\frac{P^{1/4}}{\sqrt{V}}. \tag{3.69}$$

In the case of a weak wind velocity, the measurements indicate a logarithmic function $\Delta T \propto \ln P$.

In case of a strong wind, the temperature difference is reduced and Oke's law is only valid for a value of V less than a critical value (from a few $\text{m}\,\text{s}^{-1}$, for small cities, up to $9\,\text{m}\,\text{s}^{-1}$, for megapoles; [104]). Such a law does not take into account the geometrical effects (a key factor for the radiative fluxes) or the relative humidity. Other laws depend on the city planning (strongly different between an European city and a North-American city),

$$\Delta T \simeq 7.54 + 3.97 \ln\left(\frac{H_{street}}{L_{street}}\right), \tag{3.70}$$

where H_{street}/L_{street} is a geometrical factor computed as the ratio of the street height to the street width in the city center. The temperature difference is then marked for the so-called *canyon streets* (due to heat trapping).

A fine understanding of the urban heat island is required for many applications, such as the forecast of the future urban climate (in the global context of the climate change), the optimization of city planning (role of vegetation, of building surface,

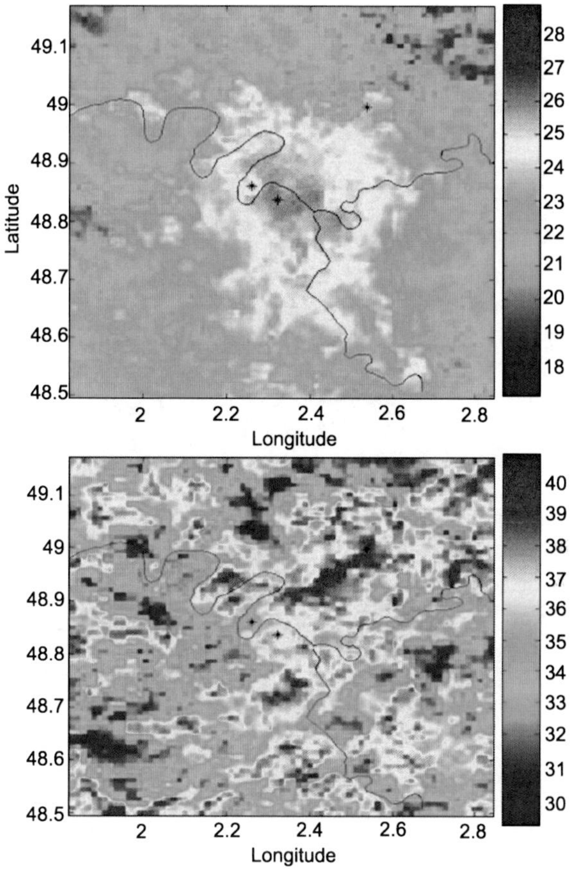

Fig. 3.22 Infrared measurements (NOAA-AVHRR sensor) of the urban heat island over Île-de-France, from 4 to 13 August 2003. Average brilliance temperature (in Celsius degrees). *Upper figure*: average over 7 images from 20:00 to 23:00 (TU). *Lower figure*: average over 10 images from 12:00 to 14:00 (TU). The data are not modified for the atmospheric absorption and emission, and for the surface emission: the values indicate the discrepancy between the urban and rural temperature. The stars (*) correspond to an industrial area (North), to Paris (center) and to the Boulogne forest (West). Credit: Benedicte Dousset (Gomer Laboratory and University of Hawaii). Source: [36]

etc.), the electricity demand forecast or the investigation of the urban artefacts of meteorological measurements.

3.6.4 Urban Boundary Layer

The urban boundary layer is also characterized by a decrease in the wind velocity, which is easy to check in the neutral case of stability. For an urban area, (3.61)

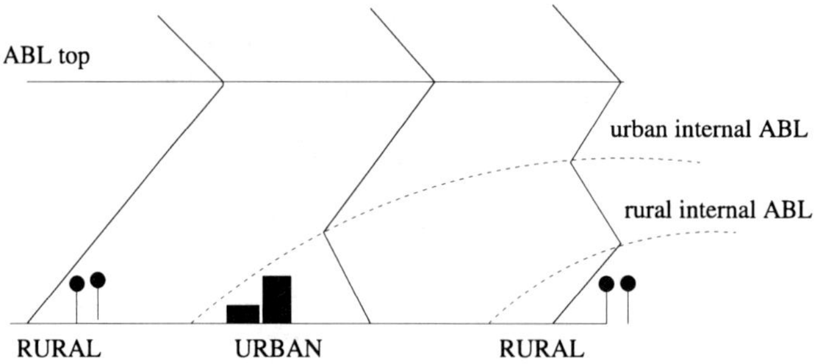

Fig. 3.23 Development of the internal urban ABL as a function of the temperature profile (discontinuous lines). Source: [105]

Fig. 3.24 Evolution of the dry deposition velocity as function of the particle size for several wind velocities (1, 5 and 10 m s^{-1}). Parameterization from [146]. Credit: Yelva Roustan, CEREA

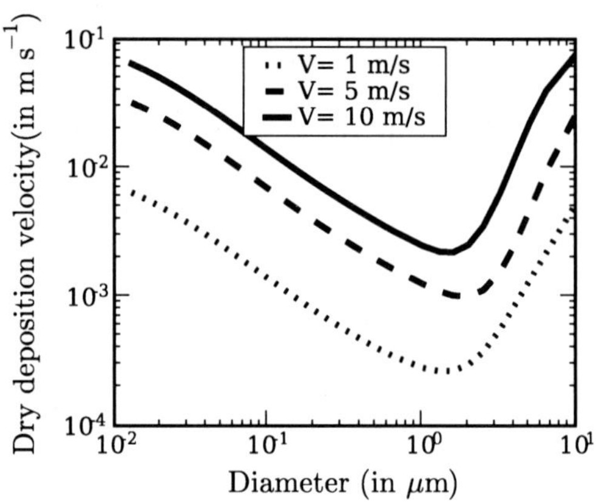

becomes

$$\langle u \rangle (z) = \frac{u_\star}{\kappa} \ln \left(\frac{z-d}{z_0} \right) \tag{3.71}$$

with d the displacement height due to obstacles (from a few meters to the building height). The increasing roughness height leads to a decrease in the wind velocity (for the same altitude).

Another urban phenomenon is the development of an *internal boundary layer*, due to the high surface temperature (Fig. 3.23).

Problems Related to Chap. 3

Problem 3.1 (Dry Deposition Velocity for Aerosols) The vertical distribution of a gas or particle concentration just above the Earth's surface is usually given by a layer with a constant diffusion flux, namely

$$K(z)\frac{\mathrm{d}C}{\mathrm{d}z} = F,$$

with $C(z)$ the concentration, $K(z)$ the vertical eddy diffusion coefficient and F the surface flux, supposed to have a constant value.

1. Upon integration, show that the flux can be written as $F = v_{dep}C$ where the dry deposition velocity v_{dep} can be calculated from $K(z)$. Assume that $C = 0$ at $z = 0$ and introduce the so-called *resistance*

$$R(z) = \int_0^z \frac{\mathrm{d}z'}{K(z')}.$$

2. Assume that there are two possible pathways for the transport in the surface layer, corresponding to two fluxes $F_1 = \alpha_1(z)\frac{\mathrm{d}C}{\mathrm{d}z}$ and $F_2 = \alpha_2(z)\frac{\mathrm{d}C}{\mathrm{d}z}$, where α_i stands for a transport coefficient.
   The total flux is then $F = F_1 + F_2$. Calculate the total dry deposition velocity as a function of the resistances for pathways 1 and 2.
3. In the case of a particle, the flux related to the gravitational settling has to be taken into account. Let v_s be the settling velocity (Exercise 5.1). The equation defining the surface layer becomes

$$K(z)\frac{\mathrm{d}C}{\mathrm{d}z} + v_s C = F.$$

   Calculate the dry deposition velocity as a function of the resistance R and of v_s.

Solution:

1. Upon integration of the gradient of C between $z = 0$ and z, $C(z) = RF$, namely $v_{dep} = 1/R$.
2. Integrating both flux equations yields $C(z) = R_1F_1$ and $C(z) = R_2F_2$. As

$$F = F_1 + F_2 = C(z) \times (1/R_1 + 1/R_2),$$

   this gives $v_{dep} = 1/R_1 + 1/R_2$. The rule is then similar to that used for resistances in parallel electric circuits.
3. Let us search for a solution in the form $C(z) = \alpha + \beta \exp(-R(z)v_s)$, where $\mathrm{d}R(z)/\mathrm{d}z = 1/K(z)$. Due to the boundary condition, $\alpha = -\beta$. With the flux equation,

$$\frac{\mathrm{d}C}{\mathrm{d}z} = -\frac{\beta v_s}{K}\exp(-R(z)v_s) = -\frac{v_s}{K}(C(z) - \alpha).$$

   Upon identification, $\alpha = F/v_s$ and

$$C(z) = \frac{F}{v_s}(1 - \exp(-Rv_s)),$$

   yielding

$$v_{dep} = \frac{v_s}{1 - \exp(-Rv_s)}.$$

This formulation is not the standard formulation. However, it is more rigorously founded than the previous formulation.

Figure 3.24 shows an illustration of this parameterization. The dry deposition increases with the wind velocity. It is also a nonlinear function of the particle size, with a minimal value for submicronic particles.

To know more ([146]):

A. VENKATRAM AND J. PLEIM, *The electrical analogy does not apply to modeling dry deposition of particles*, Atmos. Env., **33** (1999), pp. 3075–3076.

Chapter 4
Gas-Phase Atmospheric Chemistry

The chemical composition of the atmosphere governs its radiative behavior (Chap. 2):

- stratospheric ozone acts as a filter for solar ultraviolet radiation;
- greenhouse gases, such as carbon dioxide, methane or ozone, absorb the terrestrial infrared radiation, and contribute to an increase in temperature.

Whereas this interaction between the radiative properties and the chemical composition has been well known since the XIXth century,[1] the understanding of the *reactive* nature of the atmosphere is much more recent (Table 4.1).

A key point is the *oxidizing capacity* of the atmosphere, mainly governed by the amount of the hydroxyl radical OH. Oxidation of compounds emitted by anthropogenic activities (e.g. hydrocarbons) is a decisive process since it *scavenges* the atmosphere by transforming toxic volatile species to semi-volatile species which can condense onto aerosols and are then removed from the atmosphere by wet scavenging (Chap. 5).

Ozone[2] was identified in the middle of the XIXth century. It is one of the key atmospheric species, due to its radiative properties, the links to the oxidizing power of the atmosphere (it is the precursor of OH) and its role in photochemical air pollution. These properties motivate the focus put on ozone in the following.

This chapter aims at giving the key facts for understanding gas-phase atmospheric chemistry. The description of multiphase processes is detailed in Chap. 5.

The first section introduces the basic elements for chemical kinetics and a few specific characteristics of atmospheric chemistry:

- the role of oxidation reactions, with the catalysis by a few radicals (such as the hydrogen oxides, OH and HO_2, and the nitrogen oxides, NO and NO_2);

[1] John Tyndall measured the absorption of terrestrial radiation by greenhouse gases (water vapor, CO_2, CH_4) in 1859. Svante Arrhenius estimated in 1896 ([9]) that the mean temperature would increase by 4 degrees if CO_2 concentration doubles.

[2] The name is derived from the Greek word *ozein* (to smell): actually, above high concentrations, ozone has a specific odor.

B. Sportisse, *Fundamentals in Air Pollution*,

Table 4.1 The main steps in understanding the atmospheric composition. Source: [18]

−350	Aristotle: water is present in the air.
XV–XVIIth	A component of air favors fire (L. de Vinci and J. Mayow: *fire-air*).
XVIIIth	This component is isolated and named *oxygen* by A. L. Lavoisier.
	Identification of CO_2 in 1750 (J. Black) and N_2 (D. Rutherford).
XIXth	Identification of ozone (C. Schönbein, 1840).
	Identification of methane in the air (J. B. Boussingault, 1862).
late XIXth	Identification of argon (Lord Rayleigh, W. Ramsay) and other noble gases.
XXth	Identification of H_2 (1900–1920: J. Dewar, G. Claude and P. Schuftan).
	Identification of N_2O (1939, G. Adel).
	Detection of CO (1949, M. Migeotte), HNO_3 (1968, D. Murcray), etc.
	Detection of CFCs in the atmosphere (1971, J. Lovelock).

- the role of *photolytic reactions*, whose kinetic rates depend on solar radiation. This explains the relations between atmospheric chemistry and solar radiation, and as a result, the distinction between day-time chemistry and night-time chemistry, tropospheric chemistry and stratospheric chemistry, etc.;
- the concepts of *chemical lifetime* and of *reservoir* species.

The second section investigates the processes governing ozone in the stratosphere. The focus is put on the heterogeneous processes leading to the formation of the "ozone hole" under specific conditions. This example illustrates the successive steps of the understanding of atmospheric chemistry.

The third section is devoted to the tropospheric chemistry of ozone with the study of chemical reactions amongst nitrogen oxides and volatile organic compounds (VOCs) in the "polluted" troposphere. The objective is to understand the development of photochemical smogs (*Los Angeles smog*), one of the most spectacular manifestations of urban pollution. This section is a good illustration of the complexity of atmospheric chemistry with the existence of different *chemical regimes*. A good understanding of these regimes is required to evaluate accurately the impact of emission reductions or of changes in emissions (e.g. new fuels or new engines). The sensitivity of concentrations to emissions can actually strongly differ from one chemical regime to another.

Finally, a few topics related to indoor air quality are briefly introduced.

4.1 Primer for Atmospheric Chemistry

4.1.1 Background for Chemical Kinetics

Let us consider a set of gas-phase chemical species. A *chemical mechanism* describes a set of reactions among these chemical species. Let n_s be the number of species (i labels the species of chemical symbol X_i) and n_r be the number of reactions (labelled by r).

Reaction Order The reaction order is defined by the number of reacting species (*reactants*):

- *Monomolecular Reactions*
  Following the absorption of solar radiation, one molecule is dissociated,
  $$X + h\nu \longrightarrow X^*, \tag{R 2}$$
  with $h\nu$ the absorbed radiation in a given range of wavelengths and X^* the resulting excited state. The reaction is referred to as a *photolytic dissociation* (also *photodissociation* or *photolysis*) (Sect. 4.1.2).
- *Bimolecular Reactions*
  They imply two molecules, for example,
  $$X_1 + X_2 \longrightarrow X_3. \tag{R 3}$$
- *Trimolecular Reactions*
  The third molecule is referred to as a *third body* and is required to be an abundant species, namely N_2 or O_2. The third body is written as M and the reaction reads for example
  $$X_1 + X_2 + M \longrightarrow X_3 + M. \tag{R 4}$$

Higher order reactions are not taken into account because they are not significant (the probability of occurrence is much too low).

Stoichiometry The time evolution of reactants and products is governed by the so-called *stoichiometry* of reactions. The reaction r, written in the general form

$$\sum_{i=1}^{n_s} s_{ir}^- X_i \longrightarrow \sum_{i=1}^{n_s} s_{ir}^+ X_i, \tag{R 5}$$

induces the following time evolution of the concentration (molar or molecular) of X_i, written as $[X_i]$,

$$\left(\frac{d[X_i]}{dt}\right)_r = (s_{ir}^+ - s_{ir}^-)\omega_r, \tag{4.1}$$

where ω_r is the *reaction rate* of the reaction r and $(s_{ir}^{\pm})$ are the *stoichiometric coefficients* of X_i in reaction r.

The contributions of all the chemical reactions are added and the time evolution of species X_i for the whole mechanism is

$$\frac{d[X_i]}{dt} = \sum_{r=1}^{n_r} \left(\frac{d[X_i]}{dt}\right)_r = (S\omega)_i, \tag{4.2}$$

with S the *stoichiometric matrix* $(n_s \times n_r)$ and ω the vector of reaction rates (in $\mathbf{R}^{n_r}$).

Mass Action Law The reaction rate ω_r is given by the *mass action law* as a function of the reactant concentrations.

- In the monomolecular case:

$$\omega_r = J_r\,[\mathrm{X}], \tag{4.3}$$

  with J_r the photolytic rate;
- In the bimolecular and trimolecular cases:

$$\omega_r = k_r[\mathrm{X}_1][\mathrm{X}_2], \qquad \omega_r = k_r[\mathrm{X}_1][\mathrm{X}_2][\mathrm{M}], \tag{4.4}$$

  with k_r the kinetic rate.

Kinetic Rate For a *thermal* bimolecular reaction, the kinetic rate is usually parameterized as a function of temperature T by the Arrhenius' law,

$$k_r(T) = A_r\,T^{B_r} \exp\left(-\frac{E_{ar}}{RT}\right), \tag{4.5}$$

with R the universal gas constant, A_r the pre-exponential factor, B_r the exponential factor and E_{ar} the activation energy. The set of parameters (A_r, B_r, E_{ar}) is specific to the reaction r. The activation energy may be viewed as the amount of energy required for the transition from the reactants to the products (Fig. 4.1).

For a trimolecular reaction, the reaction rate may depend on pressure (through M) and the general form can be written as (Exercise 4.1)

$$k_r(T, P) = \frac{k_0}{1 + \frac{k_0}{k_\infty}[\mathrm{M}]}, \tag{4.6}$$

where k_0 and k_∞ correspond to the low-pressure and high-pressure cases, respectively.

Elementary reactions, defined as reactions that actually take place, are distinguished from *global reactions*, defined by aggregation of elementary reactions. Ac-

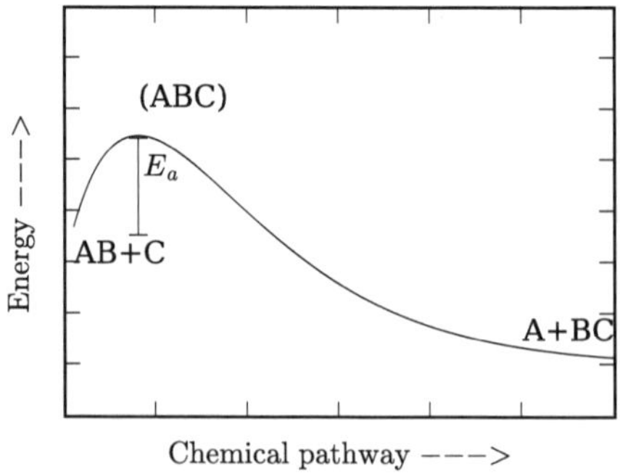

Fig. 4.1 Interpretation of the activation energy for the bimolecular reaction $AB + C \rightarrow A + BC$. The reaction path has an intermediate state (ABC)

cording to thermodynamic principles, an elementary reaction is reversible:

$$\sum_{i=1}^{n_s} s_{ir}^{-} X_i \rightleftharpoons \sum_{i=1}^{n_s} s_{ir}^{+} X_i. \qquad \text{(R 6)}$$

With obvious notations, the Van't Hoff law gives the ratio of the forward kinetic rate to the backward kinetic rate as

$$\frac{k_r^+(T)}{k_r^-(T)} = K_r^{eq}(T), \qquad (4.7)$$

where $K_r^{eq}(T)$ is the thermodynamic equilibrium constant. In practice, for many reactions, the forward reaction may be neglected.

Exercise 4.1 (Fall-Off Reaction) The reaction $A + B + M \rightarrow C + M$ is not an elementary reaction. It is a global reaction obtained from the mechanism

$$A + B \overset{1}{\rightleftharpoons} (AB)^\star,$$

$$(AB)^\star + M \overset{2}{\rightarrow} C + M,$$

where the intermediate (excited) state $(AB)^\star$ is stabilized after a collision with the third body M. Calculate the kinetic rate for the global reaction.
Hint: assume that the intermediate state is at equilibrium.
Solution:
The *effective* kinetic rate, k_{eff}, is defined by $d[C]/dt = k_{eff}[A][B][M]$. The time evolution of $[(AB)^\star]$ is

$$\frac{d[(AB)^\star]}{dt} = k_1^+[A][B] - (k_1^- + k_2[M])[(AB)^\star].$$

$(AB)^\star$ is a short-lived species and can be assumed to be at equilibrium, $d[(AB)^\star]/dt \simeq 0$, which provides an estimation of $[(AB)^\star]$. Using this expression in the evolution equation for C, $d[C]/dt = k_2[(AB)^\star][M]$, we obtain an expression for the effective kinetic rate

$$k_{eff} = \frac{k_1^+ k_2}{k_1^- + k_2[M]}$$

which depends on pressure through $[M] = P/(k_B T)$. Two limit cases can be distinguished: the "high-pressure" case ($k_2[M] \gg k_1^-$) for which the global reaction is similar to a bimolecular reaction ($\omega \simeq k_1^+[A][B]$), and the "low-pressure" case ($k_2[M] \ll k_1^-$) for which it is similar to a trimolecular reaction ($\omega \simeq (k_1^+ k_2/k_1^-)[A][B][M]$). In the general case, k_{eff} is rewritten as a function of the kinetic rates of the limit cases (k_0 for the low-pressure case and k_∞ for the high-pressure case, respectively),

$$k_{eff} = \frac{k_0}{1 + \frac{k_0}{k_\infty}[M]},$$

with $k_0 = k_1^+ k_2/k_1^-$ and $k_\infty = k_1^+$.

4.1.2 Photochemical Reactions

Atmospheric chemistry is a "cold" chemistry, the opposite of combustion chemistry. The temperature effects are not significant except for two cases:

- in the troposphere, the chemical lifetime of a few species strongly depends on temperature and therefore on altitude (e.g. peroxyacetylnitrate, PAN, Sect. 4.1.4.5);
- in the stratosphere, the low temperatures play a major role in the formation of polar clouds, which catalyze the heterogeneous consumption reactions of ozone (Sect. 4.2).

The energy source required by some reactions, especially for photodissociation, is provided by solar radiation: the absorption of the most energetic part of the electromagnetic spectrum (ultraviolet region) plays a decisive role for the dissociation of molecules (Exercise 4.2).

4.1.2.1 Photon Absorption

The energy that is required to initiate some reaction chains is not provided by an increase in temperature but by the absorption of photons by molecules. These chemical reactions are said to be *photochemical reactions*.

Consider a molecule AB (A and B stand for atoms or molecules). The dissociation has an intermediate state AB^*:

$$AB + h\nu(\lambda) \longrightarrow AB^* \qquad \text{(R 7)}$$

where λ is the wavelength of the absorbed radiation.

The resulting excited state AB^* is unstable and takes part in many possible reactions thereafter:

1. dissociation reaction:

$$AB^* \longrightarrow A + B^*. \qquad \text{(R 8)}$$

2. ionization reaction (see Sect. 2.2.1.2, Chap. 2):

$$AB^* \longrightarrow AB^+ + e^-. \qquad \text{(R 9)}$$

3. direct chemical reaction with a species C to lead to the formation of new products:

$$AB^* + C \longrightarrow D + E. \qquad \text{(R 10)}$$

4. fluorescence reaction with photon emission:

$$AB^* \longrightarrow AB + h\nu. \qquad \text{(R 11)}$$

5. stabilization reaction by collision (*quenching*) with a third body:

$$AB^* + M \longrightarrow AB + M. \qquad \text{(R 12)}$$

6. intermolecular (with another molecule) or intramolecular energy transfer.

In the atmosphere, the photodissociation is the most significant reaction at the chemical level.

Remark 4.1.1 (Oxygen Atom) The dissociation of molecular oxygen O_2 and ozone O_3 lead to different states of the oxygen atom, respectively the *triplet* ground state $O(^3P)$ and the higher-energy *singlet* state $O(^1D)$:

$$O_2 + h\nu \longrightarrow O(^3P) + O(^3P), \qquad \text{(R 13)}$$

$$O_3 + h\nu \longrightarrow O_2 + O(^1D). \qquad \text{(R 14)}$$

The stabilization of $O(^1D)$ is made by collision with abundant species (O_2 or N_2),

$$O(^1D) + M \longrightarrow O(^3P) + M. \qquad \text{(R 15)}$$

In the following, O stands for the triplet state $O(^3P)$.

4.1.2.2 Photolytic Kinetic Rate

Let us write $h\nu(\lambda_1 \leq \lambda \leq \lambda_2)$ the absorbed radiation in the wavelength interval $[\lambda_1, \lambda_2]$.

Let $k_{AB,\lambda}$ be the number of photons of wavelength λ which is absorbed by molecules AB per unit of time and per unit of volume. It depends on the absorption properties of the molecule AB, the incident radiation (wavelength and intensity) and the molecular concentration of AB,

$$k_{AB,\lambda} = \sigma_{AB}(\lambda)\, I(\lambda)[AB], \qquad (4.8)$$

where $\sigma_{AB}(\lambda)$ is the absorption cross section of AB for the radiation of wavelength λ (in cm^2 $molecule^{-1}$), and $I(\lambda)$ is the flux of photon number per unit area of surface and per unit of time, at a given wavelength λ (also referred to as *actinic flux*, expressed in photons $cm^{-2}\,s^{-1}$). Here, the concentration [AB] is expressed in number of molecules per unit of volume.

The *quantum yield* of the process j (with $j = 1, \ldots, 6$) is defined as the fraction of the excited state which takes part in the process (in molecule $photon^{-1}$). It is usually written as $\Phi_j(\lambda)$. The sum of the quantum yields is equal to 1.

The reaction rate associated to the process $j = 1$ is therefore (expressed in unit of concentration per unit of time) $k_{AB,\lambda}\Phi_1(\lambda) = \sigma_{AB}(\lambda)\, I(\lambda)\Phi_1(\lambda)[AB]$. This defines the photolytic rate at wavelength λ, expressed in s^{-1},

$$J_\lambda = \sigma_{AB}(\lambda)\, I(\lambda)\Phi_1(\lambda). \qquad (4.9)$$

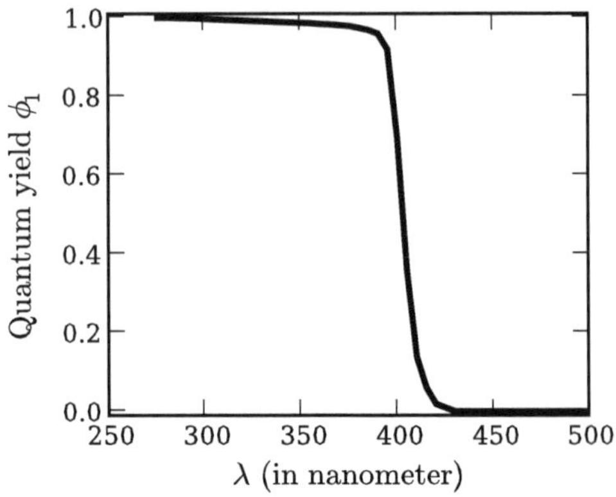

Fig. 4.2 Quantum yield for the dissociation of NO_2 as a function of wavelength

For the whole electromagnetic spectrum (considering all the wavelengths), the radiation distribution, $i(\lambda)$, is used so that $dI(\lambda) = i(\lambda)d\lambda$. The photolytic rate (associated to the process $j = 1$), usually written as J, is

$$J = \int_{\lambda_1}^{\lambda_2} \sigma_{AB}(\lambda)\, i(\lambda,t)\, \Phi_1(\lambda)\, d\lambda. \tag{4.10}$$

The calculation of the photolytic rate is based on the values of $\sigma_{AB}(\lambda)$ (given by the spectroscopic properties of AB, Sect. 2.2.1) and of the quantum yield $\Phi_1(\lambda)$ (see Fig. 4.2 for NO_2); note that σ_{AB} and Φ_1 also depend on temperature.

The actinic flux i is a function of the sun zenithal angle[3] (and therefore of the hour, day and longitude), of the atmospheric state (aerosol concentration, clouds) and of altitude. Calculating J is therefore a challenging issue.

Remark 4.1.2 (Quantum Yield) Quantum yields are very low for radiations of wavelength greater than 730 nm. As a result, terrestrial radiation is not taken into account in the calculation of photolytic rates (Chap. 2), which only depend on solar radiation. A direct consequence is that $J = 0$ during night-time.

Exercise 4.2 (Photolysis of O_2 and NO_2) The dissociation energies of O_2 in $O + O$, and of NO_2 in $NO + O$ are respectively 500 and 304 kJ mol^{-1}. Calculate the upper bound of the wavelength required for the dissociation of O_2 and NO_2.

Solution:

Let E_d be the dissociation energy. The energy associated to a photon of wavelength λ is hc/λ (Chap. 2). Thus, the dissociation of one mole requires an amount of energy $\mathcal{A}_v hc/\lambda \geq E_d$. This gives $\lambda \leq 240$ nm for O_2 and $\lambda \leq 395$ nm for NO_2.

Remark 4.1.3 (Parameterization of Photolytic Rates) Photolytic rates are often given as look-up tables of the zenithal angle, θ, as $J(\theta) = A\exp(-B/\cos\theta)$ where

[3] Angle of the solar radiation with respect to the vertical direction to the Earth's surface.

Fig. 4.3 Evolution of the photolytic rate $J(\theta)$ of $NO_2 + h\nu \rightarrow NO + O$ as a function of zenithal angle

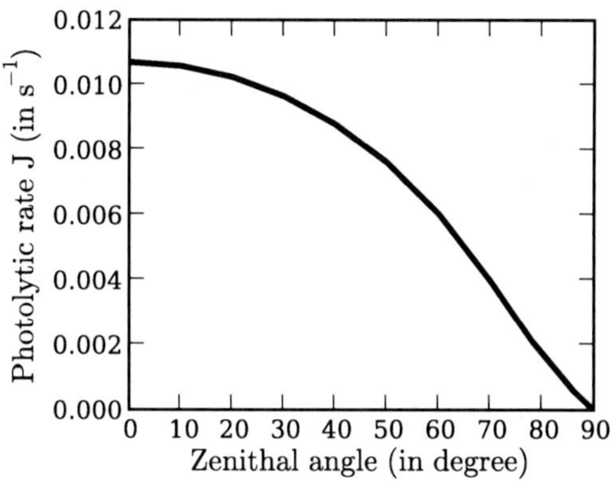

A and B are positive parameters depending on the photolytic reaction and θ is given as a function of time. The rate J is, as expected, a decreasing function of the zenithal angle (Fig. 4.3): $\theta = 0$ corresponds to the maximal value of the solar radiation ("noon"), $\theta = \pi/2$ to sunset or sunrise.

4.1.2.3 Consequences of the Radiation Dependence

The fact that chemical reactions depend on solar radiation has many consequences:

- **Night-time Chemistry Versus Day-time Chemistry**
  Because only solar radiation is efficient for photolysis, it is necessary to distinguish "day-time chemistry" from "night-time chemistry", for which photolytic reactions cannot occur.
  In the troposphere, ozone chemistry is deeply related to the hydroxyl radical OH, which is produced by the photolytic dissociation of ozone. As a result, the leading chemical reactions strongly differ between day-time and night-time.
  In the stratosphere, photolytic reactions take part in the production of chloride radicals, which play a key role for the catalysis of ozone destruction. They cannot occur during the polar night (Sect. 4.2).
- **Tropospheric Chemistry Versus Stratospheric Chemistry**
  The available part of the electromagnetic spectrum depends on altitude. This implies that the leading reactions differ depending on altitude: tropospheric chemistry may be distinguished from stratospheric chemistry.
  Oxygen and ozone filter the solar ultraviolet radiation in the stratosphere ($\lambda \leq 290$ nm). Moreover, the quantum yield for longwave radiation is negligible ($\lambda \geq 730$ nm). As a result, efficient radiation for photolysis in the troposphere occurs in the range $290\,\text{nm} \leq \lambda \leq 730\,\text{nm}$.
  This has a strong impact on sources of oxygen atoms, which are required for the formation of ozone,

$$O_2 + O + M \longrightarrow O_3 + M.$$

A possible source of oxygen atoms is the photodissociation of oxygen molecules,

$$O_2 + h\nu(\lambda \leq 242\text{ nm}) \longrightarrow O + O. \quad \text{(R 16)}$$

This reaction is possible only in the stratosphere due to the required wavelengths. In the troposphere, a possible source of oxygen atoms is the photodissociation of NO_2,

$$NO_2 + h\nu(300\text{ nm} \leq \lambda \leq 400\text{ nm}) \longrightarrow NO + O, \quad \text{(R 17)}$$

which explains the deep connection between nitrogen oxides and ozone in the troposphere (Sect. 4.3).

4.1.3 Atmosphere as an Oxidizing Reactor

4.1.3.1 Connections Between OH and O_3

The atmosphere is an oxidizing environment: an emitted hydrocarbon may be oxidized several times to form the final products CO_2 and H_2O. The three major oxidants are OH, O_3 and NO_3 (ranked by decreasing oxidation strength). Oxidation reactions play a key role to scavenge the atmosphere (the hydrocarbon concentrations would be much too high, otherwise; see Problem 4.3 for carbon monoxide).

At first approximation, the oxidizing capacity of the atmosphere is given by the concentration of OH (about 10^6 molecules cm^{-3} in the troposphere, Problem 4.1). The hydroxyl radical OH is sometimes named the "chemical scavenger" of the atmosphere. A major part of the oxidation processes take place in the tropical regions (see Table 4.2 for the oxidation of methane).

Ozone O_3 plays a decisive role for the production of OH through its photolysis,

$$O_3 + h\nu(\lambda \leq 310\text{ nm}) \xrightarrow{J_1} O_2 + O(^1D). \quad \text{(R 18)}$$

The excited state of oxygen atoms, $O(^1D)$, can either be stabilized in the form $O(^3P)$, with the reaction

$$O(^1D) + M \xrightarrow{k_2} O(^3P) + M, \quad \text{(R 19)}$$

Table 4.2 Estimation of the oxidation of CH_4 by OH as a function of altitude and latitude. Up to two thirds of oxidation takes place in the tropical region in the lower troposphere (500 hPa corresponds to an altitude of about 5 kilometers). Source: [83]

Pressure level	90°S–30°S	30°S–30°N	30°N–90°N
[85, 300] hPa	–	3%	–
[300, 500] hPa	1%	12%	2%
[500, 1000] hPa (ground)	6%	63%	13%

or produce OH with the reaction

$$O(^1D) + H_2O \xrightarrow{k_3} 2\,OH. \tag{R 20}$$

The key point is that only the excited ground state $O(^1D)$ reacts with water vapor (a stable compound). As a result, the production of OH is directly related to ozone photolysis.

The global reaction resulting from these three reactions is

$$O_3 \xrightarrow{h\nu, H_2O} 2\,OH. \tag{R 21}$$

The production rate of OH is $P_{OH} = 2k_3\,[O(^1D)][H_2O]$. Using a quasi steady-state assumption (Remark 4.1.4) for $O(^1D)$, we obtain

$$J_1[O_3] \simeq (k_2[M] + k_3[H_2O])[O(^1D)], \tag{4.11}$$

and finally, with $k_2 \gg k_3$,

$$P_{OH} \simeq \frac{2J_1 k_3[H_2O]}{k_2[M]}[O_3]. \tag{4.12}$$

The hydroxyl radical OH is a short-lived species because of its oxidation strength. It is therefore difficult to estimate its concentration by measured data. A commonly-used approach is based on the so-called *inverse modeling* of oxidized species, such as methane (CH_4) or methylchloroform (CH_3CCl_3, Problem 4.1).

4.1.3.2 Oxidation Chain

The most frequent reactions imply radicals with free electrons in the outer valence shell (which implies that they are highly reactive), such as OH or HO_2.

An oxidation chain can be partitioned in the following sequence of successive steps:

- *Initiation*
  A stable species (it is not a radical; written as X_{nonrad}) is dissociated by a photolytic reaction, which leads to the production of at least one radical (X_{rad}):

$$X_{nonrad} + h\nu \longrightarrow X_{rad}. \tag{R 22}$$

  The photolysis of ozone (R 18) is an example of initiation.
- *Propagation*
  The resulting radicals react with stable species (the notation X_{nonrad} is kept, even if it not necessarily the same species as before). New radicals (also written as X_{rad}) are produced by oxidation reactions:

$$X_{nonrad} + X_{rad} \longrightarrow X_{nonrad} + X_{rad}. \tag{R 23}$$

- *Branching Reaction*
  The stable species can be dissociated during a photolytic reaction and may generate other radicals with reactions similar to (R 22).
- *Termination*
  Radicals can react amongst them to produce a stable species, which ends the oxidation chain:

$$X_{rad} + X_{rad} \longrightarrow X_{nonrad}. \tag{R 24}$$

The stable species which are oxidized during the propagation step (R 23) should be viewed as a "fuel" for oxidation since they take part in the production of radicals. It is typically the case of VOCs, CO and CH_4.

4.1.4 Chemical Lifetime

Another key concept is the *chemical lifetime* of a given species, which is a good indicator of its chemical stability. This notion has already been introduced in Chap. 1 (Sect. 1.3).

Using the general notations presented for chemical kinetics, it is easy to check that the time derivative for species X_i can be put in the so-called *production-loss* form, namely

$$\frac{d[X_i]}{dt} = P_i - L_i\,[X_i], \tag{4.13}$$

where P_i and L_i are the nonnegative terms of production and loss, respectively. In the general case, they depend on other concentrations. This form can be obtained by summing over the reactions in which X_i is a reactant and those in which it is a product, respectively. For the first set of reactions, it is easy to prove that the reaction rate is proportional to $[X_i]$, using the mass action law.

If P_i and L_i are supposed to have constant values, the concentration has an exponential behavior,

$$[X_i](t) = \left([X_i](0) - \frac{P_i}{L_i}\right)\exp(-L_i t) + \frac{P_i}{L_i}, \tag{4.14}$$

with a characteristic timescale $1/L_i$. By extension, the characteristic timescale or *chemical lifetime* of species X_i is defined as

$$\tau_i = \frac{1}{L_i}, \tag{4.15}$$

which depends on other concentrations in the general case. Note that a rigorous definition is based on the eigenvalues of the Jacobian matrix of the source term (Sect. 6.2.2.1).

The chemical lifetime is fixed by oxidation reactions, mainly by $X_i + OH \overset{k_{OH,i}}{\rightarrow}$ ..., where $k_{OH,i}(T)$ is the kinetic rate for the oxidation of X_i by OH. The resulting loss term is $-\underbrace{k_{OH,i}[OH]}_{L_i}[X_i]$, which defines the following timescale,

$$\tau_i \simeq \frac{1}{k_{OH,i}[OH]}. \tag{4.16}$$

As expected, the chemical lifetime increases when OH concentration decreases and when $k_{OH,i}$ decreases, that is when the species is less reactive.

This lifetime defines the impact scale through atmospheric transport (Chap. 1). It depends on:

- the species reactivity: highly oxidized species are usually more reactive and have a lower lifetime;
- temperature: an increase in temperature usually results in an increase in reactivity;
- the oxidizing capacity of the environment, given by the concentration of hydroxyl radical: this motivates the focus put on the long-term evolution of the OH concentration (Problem 4.1).

Exercise 4.3 illustrates an application for the case of methane.

Exercise 4.3 (Lifetime of Methane) This exercise is a follower of Exercise 1.9. Using the kinetic data for methane oxidation, calculate the chemical lifetime of methane. Compare to the result obtained in Exercise 1.9.
Data:
– take a mean concentration of OH of about 1.2×10^6 molecule cm^{-3};
– $k_{OH,CH_4} = 2.65 \times 10^{-12} \exp(-1800/T)$ cm^3 molecule^{-1} s^{-1}.
Solution:
The mean temperature of the troposphere is about 255 K. Thus, the kinetic rate is 2.3×10^{-15} cm^3 molecule^{-1} s^{-1}, which gives a chemical lifetime of about 11.5 years.

Using an approach based on a global budget, we have obtained 8 years. The difference is related to the missing reactions not taken into account in this exercise.

Remark 4.1.4 (Quasi Steady-State Assumption) For the most reactive species (short-lived species) $L_i \gg 1$ and usually $P_i \gg 1$. The *quasi steady-state assumption* (QSSA) consists in replacing the equation evolution (4.13) by the algebraic constraint

$$[X_i]_{QSSA} \simeq \frac{P_i}{L_i}. \tag{4.17}$$

In the general case, this defines a system of nonlinear quadratic equations with respect to concentrations because the loss and production terms depend on concentrations.

The mathematical background can be illustrated by scaling P_i and L_i, supposed to be large terms. With $P_i = \tilde{P}_i/\epsilon$ and $L_i = \tilde{L}_i/\epsilon$ ($\epsilon \ll 1$), the time evolution reads

$$\epsilon \frac{d[X_i]}{dt} = \tilde{P}_i - \tilde{L}_i\,[X_i], \tag{4.18}$$

and, at the limit $\epsilon \to 0$, we obtain the approximation (4.17). We refer to Exercise 6.5 for the physical interpretation of ϵ (ratio of timescales of the fastest processes to the slowest ones).

4.1.4.1 Chemical Lifetime of a Few VOCs in the Lower Troposphere

The chemical lifetimes of a few VOCs are shown in Table 4.3. They are computed with respect to the oxidizing species OH, O_3 and NO_3, using an equation similar to (4.16).

Reactions with OH are usually the fastest ones and, as a consequence, control the chemical lifetime. It is easy to check that the increase in the "chemical complexity" results in an increase in the reactivity, and thus a decrease in the lifetime.

4.1.4.2 Local Impact Versus Regional Impact of Tropospheric Pollution

When the lifetime of an emitted species is low (highly reactive species), its spatial impact is local. In the opposite case, the species can be involved in long-range transport. A typical example is given by ozone (Table 4.4 and Sect. 4.3).

4.1.4.3 Tropospheric Impact Versus Stratospheric Impact

The timescale of the vertical transport from the troposphere to the stratosphere typically varies from 7 to 10 years (Chap. 1). A given species, which is emitted at the

Table 4.3 Chemical lifetime at a temperature of 298 K for a few VOCs (*d* for day, hr for hour, and min for minute). Oxidant concentrations have the following illustrative values: $[OH] = 10^6$ molecule cm^{-3}, $[O_3] = 10^{12}$ molecule cm^{-3} (50 ppb) and $[NO_3] = 5.4 \times 10^8$ molecule cm^{-3} (20 ppt). Source: [18]

Species	oxidation by OH	oxidation by O_3	oxidation by NO_3
methane	1837 d	–	–
ethane	48 d	–	2690 d
butane	4.8 d	–	391 d
ethene	1.4 d	6.7 d	107 d
propene	10.6 hr	1.1 d	2.3 d
isoprene	2.8 hr	20.2 hr	45 min
β-pinene	3.5 hr	17.2 hr	12 min
limonene	1.6 hr	1.3 hr	3 min

Table 4.4 Estimation of ozone lifetime in the lower troposphere (at the altitude of 5 kilometers). In summer, over the tropical regions, the lifetime is lower due to the available solar radiation. Source: [130]

Season	20°N	40°N
Summer	5 days	10 days
Winter	15 days	100 days

ground, will reach the stratosphere if its lifetime is long enough. This is the case of CFCs (see Table 4.5), which are involved in destruction reactions for stratospheric ozone.

4.1.4.4 Delayed Effect

Another interesting concept is the so-called *delayed effect*. An emitted species can have a long-term environmental impact (during tens of years) if its lifetime is long enough: this is the case of some greenhouse gases and of CFCs. Note that the substitutes of CFCs, the so-called HCFCs, have much lower lifetimes because adding an hydrogen atom (H) favors oxidation and thus increases the reactivity (Table 4.5).

4.1.4.5 Sink and Reservoir Species

Sink Species A species whose lifetime is short acts as a *chemical sink* for its precursors (the species which contribute to its formation).

The typical case is a *soluble* species because its atmospheric residence time is governed by wet scavenging (which is typically less than two weeks, Chap. 5).

A first example is nitric acid (HNO_3), a species involved in cloud acidity. Its tropospheric lifetime is about a few days due to its solubility (once it is dissolved, it is scavenged by precipitations). As it is produced from nitrogen oxides ($NO_x = NO + NO_2$), it can be viewed as a sink of NO_x.

A second example is hydrogen peroxide (H_2O_2), which is also a soluble species. This species is a sink of OH and HO_2 through the formation reaction

$$HO_2 + HO_2 + M \longrightarrow H_2O_2 + M, \qquad \text{(R 25)}$$

HO_2 is produced during the oxidation of VOCs by OH.

Reservoir Species Short-lived species may also have a non-local impact. Actually, they can be transformed into more stable species which play the role of *reservoir species*. These latter species may indeed transform back into the initial short-lived species after long-range transport under specific meteorological or chemical conditions.

- **PAN, reservoir of NO_x**

  A typical example is given by peroxyacid nitrates (PAN). The simplest PAN is peroxyacetyl nitrate, $CH_3C(O)OONO_2$, which is produced from the oxidation products of VOCs and NO_2, according to

$$CH_3CHO + OH \longrightarrow CH_3CO + H_2O, \quad \text{(R 26)}$$

$$CH_3CO + O_2 + M \longrightarrow CH_3C(O)OO + M, \quad \text{(R 27)}$$

$$CH_3C(O)OO + NO_2 + M \longrightarrow CH_3C(O)OONO_2 + M. \quad \text{(R 28)}$$

  These species are not soluble and weakly reactive. Their photolysis is not significant. On the other hand, their reactivity strongly depends on temperature and therefore on altitude. They constitute a possible reservoir for NO_2 with the thermal decomposition

$$PAN + M \longrightarrow CH_3C(O)OO + NO_2 + M. \quad \text{(R 29)}$$

  The lifetime of PAN, τ, is a decreasing function of temperature. For a pressure of 1 atm,
  - $\tau \simeq 30$ minutes at 298 K,
  - $\tau \simeq 30$ hours at 273 K,
  - $\tau \simeq 70$ days at 258 K.

  Remember that the mean moist adiabatic lapse rate is about $-6.5\ \mathrm{K\,km^{-1}}$. Thus, a temperature of 273 K corresponds to an altitude of about 2 or 3 kilometers (the dependence of pressure and of the third body concentration on altitude is omitted).

  Once formed in the ABL, the PAN is stabilized during its vertical motion to upper colder layers of the atmosphere. It may be transported to regions with low emissions of NO_x, where it may produce NO_2 by thermal decomposition after subsidence motions (inside downdrafts down to the lower layers, Chap. 3).
- **Hydrogen chloride, reservoir of chloride**

  In the stratosphere, hydrogen chloride (HCl) is a reservoir species for chloride (Cl), which takes part in the destruction reaction of ozone (Sect. 4.2).

  For example, chloride may be transformed by reacting with methane,

$$Cl + CH_4 \longrightarrow HCl + CH_3. \quad \text{(R 30)}$$

  HCl is much more stable than Cl. It can produce chloride under strong oxidizing conditions as

$$OH + HCl \longrightarrow H_2O + Cl, \quad \text{(R 31)}$$

  which makes chloride available.

4.1.5 Validity of Chemical Mechanisms

The leading chemical reactions are deeply connected to the atmospheric conditions (at the meteorological, radiative and chemical levels). The general behavior can be strongly modified.

A first example is provided by the role of NO_x in the evolution of ozone: the nitrogen oxides usually catalyze ozone production in the troposphere (Sect. 4.3) and ozone destruction in the stratosphere (Sect. 4.2).

A second example is the existence of two *chemical regimes* for tropospheric ozone (Sect. 4.3). When the environment is poor in NO_x, the emission of VOCs does not impact ozone formation. When the environment is rich in NO_x, it favors ozone production (the more common situation for urban photochemical smog).

After this general presentation of atmospheric chemistry, the focus is put first on stratospheric chemistry and second on tropospheric chemistry.

For convenience and for the sake of clarity, the most important chemical reactions are highlighted. It should be kept in mind that such an approach does not correspond to real-life chemical reactions since hundreds or even thousands of species and reactions are involved in chemical mechanisms (Fig. 4.4). For instance, the description of VOCs is particularly coarse: the so-called *model species* (*surrogate species*) represent a set of true species with similar chemical functions or reactivities with respect to OH (Exercise 4.4 for the concept of *lumping*).

Exercise 4.4 (Species Lumping) *Species lumping* is a commonly-used approach to reduce the number of chemical species and reactions in chemical mechanisms. The idea is to lump volatile organic compounds which have similar reactivity with respect to an oxidizing species (e.g. OH).

There exist many available methods which are more or less rigorously justified. This exercise aims at giving an example of such methods. Let us consider a set of species $\{X_i\}_i$ with similar reactivity with respect to OH. Let k be the kinetic rate of the oxidation reaction by OH, supposed to be the same one for all the species. Let $\{P_i\}_i$ be the oxidation products. Moreover, the production rate of X_i is $F_i(t)$ and

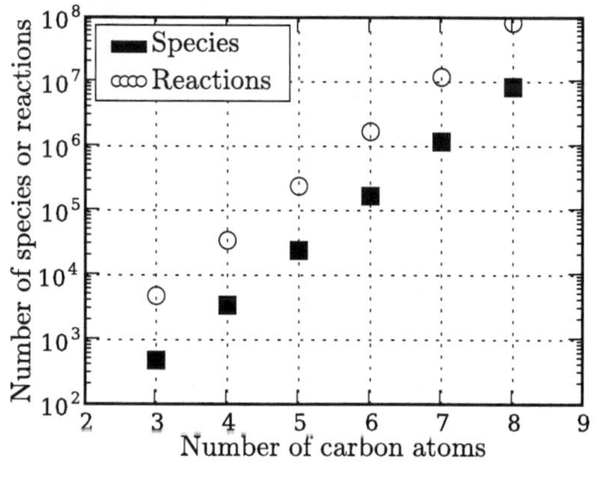

Fig. 4.4 Evolution of the number of VOCs and chemical reactions in a detailed explicit mechanism, as a function of the number of carbon atoms. Source: [11]

is supposed to be given by other chemical reactions. The detailed mechanism may therefore be written as

$$\cdots \xrightarrow{F_i(t)} \mathrm{X_i}$$

$$\mathrm{X_i} + \mathrm{OH} \longrightarrow \mathrm{P_i}.$$

1. Write the evolution equations for the "pure" species and for the "lumped" species $\mathrm{X} = \sum_i \mathrm{X_i}$.
2. Write the lumped mechanism in the form

$$\cdots \xrightarrow{\sum_i F_i(t)} \mathrm{X}$$

$$\mathrm{X} + \mathrm{OH} \longrightarrow \sum_i \alpha_i(t) P_i,$$

where the stoichiometric coefficient $\alpha_i(t)$ is a function of time. Give a closure law for α_i ($\mathrm{X_i}$ are not solved in the lumped mechanism). Use the quasi steady-state assumption for $\mathrm{X_i}$.

Solution:

1. The time evolution of the concentration of $\mathrm{X_i}$, $[\mathrm{X_i}]$, is governed by

$$\frac{\mathrm{d}[\mathrm{X_i}]}{\mathrm{d}t} = F_i(t) - k[\mathrm{X_i}][\mathrm{OH}].$$

Defining $\mathrm{X} = \sum_i \mathrm{X_i}$, we obtain

$$\frac{\mathrm{d}[\mathrm{X}]}{\mathrm{d}t} = \sum_i F_i(t) - k[\mathrm{X}][\mathrm{OH}], \qquad \frac{\mathrm{d}[\mathrm{P_i}]}{\mathrm{d}t} = k[\mathrm{X_i}][\mathrm{OH}] = \frac{[\mathrm{X_i}]}{[\mathrm{X}]} \times k[\mathrm{X}][\mathrm{OH}].$$

2. The equivalent reaction is defined with $\alpha_i = [\mathrm{X_i}]/[\mathrm{X}]$. As a result, the coefficient $\alpha_i(t)$ cannot be solved explicitly since $\mathrm{X_i}$ is not solved (only the *lumped* species are solved). Using the quasi steady-state assumption, $F_i(t) \simeq k[\mathrm{X_i}][\mathrm{OH}]$ and finally $\alpha_i(t) \simeq F_i(t)/\sum_j F_j(t)$.

To Know More ([154]):

L. WHITEHOUSE, A. TOMLIN, AND M. PILLING, *Systematic lumping of complex tropospheric chemical mechanisms: a time-scale based approach*, Atmos. Chem. Phys. Discuss., **4** (2004), pp. 3785–3834

4.2 Stratospheric Chemistry of Ozone

4.2.1 Destruction and Production of Stratospheric Ozone

As shown in Chap. 2, stratospheric ozone plays a decisive role as a shield against the solar ultraviolet radiation.

Its chemical dynamics is governed by the balance between production reactions and destruction reactions. In a first approximation, the transfer fluxes with tropospheric ozone are omitted.

The history of the understanding of stratospheric ozone chemistry may be summarized as follows:

- Sidney Chapman proposed in the 1930s a simple chemical cycle to explain the high ozone concentrations in the stratosphere;
- the surestimation of the resulting estimated concentrations with respect to real-life concentrations (given by observational data) implies that there are sinks of stratospheric ozone (namely destruction reactions catalyzed by HO_x, OH and HO_2). Reactions were proposed in the 1950s;
- in the 1960s, the increase in aircraft traffic leads to the investigation of catalysis cycles with NO_x (P. Crutzen);
- in the 1970s, other catalysis cycles by CFCs were discovered (Molina and Rowland, who will share the Nobel Prize with Crutzen in 1995);
- in the early 1980s, observational data suggest a strong decrease in stratospheric ozone concentration over the South Pole. Additional chemical reactions, namely heterogeneous reactions at the surface of the so-called *polar stratospheric clouds* under cold conditions, were then suggested.

4.2.1.1 Chapman Cycle

The only production reaction for ozone is

$$O + O_2 + M \longrightarrow O_3 + M. \qquad \text{(R 32)}$$

Oxygen atoms O are issued from the photolysis of oxygen molecules O_2,

$$O_2 + h\nu(\lambda \leq 242\ \text{nm}) \longrightarrow O + O. \qquad \text{(R 33)}$$

Oxygen atoms can also react with ozone,

$$O_3 + O \longrightarrow O_2 + O_2. \qquad \text{(R 34)}$$

These three reactions constitute the so-called *Chapman cycle* for stratospheric ozone.

4.2.1.2 Other Destruction Reactions

The Chapman mechanism is not able to reproduce the chemical composition of the stratosphere and leads to an overestimation compared to the observational data. For stratospheric ozone, two other mechanisms complete the description of the leading destruction reactions.

Destruction Catalysis by HO_x A first pathway (proposed in the 1950s) is catalyzed by hydrogen oxides (HO_x).

- The initiation step consists in the production of OH and HO_2.
- The propagation reactions are

$$O_3 + OH \longrightarrow O_2 + HO_2, \qquad (R\,35)$$

$$O_3 + HO_2 \longrightarrow 2\,O_2 + OH, \qquad (R\,36)$$

  which is summarized by the global reaction

$$2\,O_3 \xrightarrow{HO_x} 3\,O_2. \qquad (R\,37)$$

  The notation suggests that the radicals HO_x take part in the elementary reactions but do not appear in the global budget (HO_x catalysis).
- The termination step is a recombination reaction amongst radicals, e.g.,

$$OH + HO_2 \longrightarrow H_2O + O_2. \qquad (R\,38)$$

Destruction Catalysis by NO_x The second pathway is catalyzed by NO_x and has been proposed in the 1960s by P. Crutzen.

- The propagation reactions are

$$O_3 + NO \longrightarrow NO_2 + O_2, \qquad (R\,39)$$

$$NO_2 + O(^1D) \longrightarrow NO + O_2, \qquad (R\,40)$$

  with the global reaction

$$O_3 + O(^1D) \xrightarrow{NO_x} 2\,O_2. \qquad (R\,41)$$

- During day-time, the termination step is

$$NO_2 + OH + M \longrightarrow HNO_3 + M. \qquad (R\,42)$$

  During night-time, because the radical OH is not produced by photolysis, the termination is

$$NO_2 + O_3 \longrightarrow NO_3 + O_2, \qquad (R\,43)$$

$$NO_2 + NO_3 + M \longrightarrow N_2O_5 + M, \qquad (R\,44)$$

$$N_2O_5 \xrightarrow{H_2O,\ \text{aerosol}} 2\,HNO_3. \qquad (R\,45)$$

This latter reaction is a heterogeneous reaction at the aerosol surface (Chap. 5, Sect. 5.1.4.6). The species HNO_3 and N_2O_5 have high lifetimes and play the role of reservoir species.

This pathway represents the major part of the stratospheric ozone destruction. Oxygen atoms result from the photolysis of oxygen molecules and ozone. The natural source of stratospheric NO_x is the oxidation of nitrogen peroxide (N_2O) by the ground state of oxygen atom,

$$N_2O + O(^1D) \longrightarrow NO + NO. \qquad (R\,46)$$

Nitrogen peroxide takes part in the biochemical cycle of nitric acid (HNO_3). For example, it may be generated by bacteria activity in soils.

The anthropogenic source of stratospheric NO is aircraft emissions. This motivates the focus on the impact of aircraft on stratospheric ozone destruction.

4.2.1.3 Ozonolysis and Temperature Inversion

Another source of stratospheric ozone destruction is the ozonolysis reaction

$$O_3 + h\nu(240\text{ nm} \leq \lambda \leq 320\text{ nm}) \longrightarrow O_2 + O(^1D). \qquad (R\,47)$$

This photolytic reaction explains the role of the stratosphere as a shield against ultraviolet radiation. At the chemical level, this reaction is not significant for ozone because it is equilibrated by ozone production,

$$O + O_2 + M \longrightarrow O_3 + M, \qquad (R\,48)$$

following the stabilization of $O(^1D)$ to $O(^3P)$.

This reaction is an exothermic reaction ($\Delta Q = 100.3\,\text{kJ}\,\text{mol}^{-1}$). At the chemical level, the budget is null while radiative energy is transferred into thermal energy, which explains the temperature inversion in the stratosphere (Exercise 4.5).

Exercise 4.5 (Temperature Inversion) Using the data given in Exercise 2.3, calculate the photolytic rate at the altitude at which the photolysis of ozone is maximal. Calculate the resulting change in temperature for a 12-hour day. Compare with Fig. 1.1.
Data:
– $I(\infty) = 3.8 \times 10^{14}$ photons cm^{-2} s^{-1} and $\rho(50\,\text{km}) \simeq 2 \times 10^{-10}$ kg cm^{-3};
– $c_p = 1005\,\text{J}\,\text{kg}^{-1}\,\text{K}^{-1}$.

Solution:
With $-dI/ds$ the photolytic rate is $J \simeq 2.8 \times 10^8$ molecule cm^{-3} s^{-1}. The increase in temperature during a period Δt (12 hours) is given by

$$\Delta T = \frac{J \Delta t \Delta Q}{\mathcal{A}_v \rho c_p} \simeq 10\,\text{K}.$$

4.2.2 Ozone Destruction Catalyzed by Bromide and Chloride Compounds

In the 1970s, it was recognized that other chemical destruction reactions imply CFCs, which are particularly stable compounds.

4.2.2.1 Destruction Catalysis

Chloride (ClO_x) and bromide (BrO_x) oxides, which are mainly related to anthropogenic sources, play a role similar to NO_x and HO_x.

- The initiation step is related to the photolysis of CFCs, which results in the production of Cl:

$$\mathrm{CFC} + h\nu \longrightarrow \mathrm{Cl}. \tag{R 49}$$

- For example, the propagation reactions are for chloride,

$$\mathrm{O_3} + \mathrm{Cl} \longrightarrow \mathrm{O_2} + \mathrm{ClO}, \tag{R 50}$$

$$\mathrm{O(^1D)} + \mathrm{ClO} \longrightarrow \mathrm{O_2} + \mathrm{Cl}, \tag{R 51}$$

  which results in the global reaction

$$\mathrm{O_3} + \mathrm{O(^1D)} \xrightarrow{\mathrm{ClO}_x} 2\,\mathrm{O_2}. \tag{R 52}$$

- The termination step consists of conversion of radicals to more stable species, e.g.,

$$\mathrm{Cl} + \mathrm{CH_4} \longrightarrow \mathrm{HCl} + \mathrm{CH_3}, \tag{R 53}$$

$$\mathrm{ClO} + \mathrm{NO_2} + \mathrm{M} \longrightarrow \mathrm{ClNO_3} + \mathrm{M}. \tag{R 54}$$

Observational data in the antarctic region suggest that there is a correlation between the increased destruction of ozone and high concentrations of ClO. Because the concentration of oxygen atoms is not high enough, the previous mechanism cannot explain alone these measurements. Other chemical mechanisms have therefore been proposed.

- A first mechanism implies a branching step for chloride and bromide compounds (photolysis of a species leading to the production of supplementary radicals). For example, as far as chloride is concerned,

$$\mathrm{ClO} + \mathrm{ClO} + \mathrm{M} \longrightarrow \mathrm{Cl_2O_2} + \mathrm{M}, \tag{R 55}$$

$$\mathrm{Cl_2O_2} + h\nu \longrightarrow \mathrm{Cl} + \mathrm{OClO}, \tag{R 56}$$

$$OClO + M \longrightarrow Cl + O_2 + M, \tag{R 57}$$

$$Cl + O_3 \longrightarrow O_2 + ClO, \tag{R 58}$$

with the global reaction

$$2\,O_3 \xrightarrow{ClO_x} 3\,O_2. \tag{R 59}$$

- A second mechanism is based on the coupled action of bromide and chloride oxides,

$$ClO + BrO \longrightarrow Cl + Br + O_2, \tag{R 60}$$

$$Cl + O_3 \longrightarrow O_2 + ClO, \tag{R 61}$$

$$Br + O_3 \longrightarrow O_2 + BrO, \tag{R 62}$$

which results in the global reaction

$$2\,O_3 \xrightarrow{ClO_x, BrO_x} 3\,O_2. \tag{R 63}$$

Remark 4.2.1 (Destruction Rate) Using a quasi steady-state assumption for ClO and BrO radicals, it is easy to prove that the destruction rate of ozone is quadratic with respect to the concentrations of ClO and BrO.

4.2.2.2 Chloride and Bromide Sources

Most of the biogenic sources of chloride and bromide are filtered in the troposphere and do not provide significant ingoing fluxes in the stratosphere. These sources comprise:

- dissolved salts from the oceans (Cl_2 and HCl) with a lifetime of about a week;
- hydrogen chloride (HCl) related to volcanic eruptions. Hydrogen chloride can reach the stratosphere only in the case of cataclysmal eruptions with emission altitudes greater than 10 kilometers, e.g. volcanoes Krakatoa (1883), Agun (1963), Fuego (1974), El Chichon (1988) and Pinatubo Mount (1991).

The main anthropogenic sources are chlorofluorocarbons (CFCs), whose chemical symbol is $C_aH_bCl_cF_d$ (with a, b, c and d integers). These compounds have been extensively used in many industries because they are stable and not toxic.

Their tropospheric lifetime (Table 4.5) varies from a few years to tens of years. As a consequence, they can reach the stratosphere, where their photolysis produces chloride and bromide.

Alternative compounds, such as hydrochlorofluorocarbons (HCFCs) or hydrofluorocarbons (HFCs), have a shorter lifetime because adding a hydrogene atom (H) favors the oxidation by OH (Table 4.5). For example, the substitution of one hydrogen atom in CFC-12 (of lifetime 100 years) defines HCFC-22 (of lifetime 12 years); the substitution of two hydrogen atoms defines HFC-32 (of lifetime 5 years).

Table 4.5 Lifetime (in years) of a few CFCs, HCFCs and HFCs. Source: [106]

Species	Symbol	Lifetime
Chlorofluorocarbons (CFCs)		
CFC-11	CCl_3F	45
CFC-12	CCl_2F_2	100
CFC-13	$CClF_3$	640
CFC-113	CCl_2FCClF_2	85
CFC-114	CF_3CClF_2	300
Hydrochlorofluorocarbons (HCFCs)		
HCFC-21	$CHCl_2F$	2
HCHC-22	$CHClF_2$	11.9
HCFC-123	CF_3CHCl_2	1.4
Hydrofluorocarbons (HFCs)		
HFC-23	CHF_3	260
HFC-32	CH_2F_2	5
HFC-41	CH_3F	2.6

4.2.3 Antarctic Ozone Hole

The *ozone layer* is an idealized concept: the whole stratospheric ozone is assumed to be in a layer. The unit for the layer thickness is the *Dobson Unit*, which corresponds to 0.01 mm (Exercise 1.6).

In 1985, observational data of a scientific team (Farman et al., [38]) show a strong decrease in the ozone layer thickness over the Halley Bay station in the antarctic region, in October (Fig. 4.5 and Fig. 4.6).

This phenomenon results from many chemical and physical processes. It is deeply related to the very low temperatures in the antarctic stratosphere, and to the polar stratospheric clouds, which play the role of chemical reactors to produce chloride and bromide.

1. **Cold temperatures in the stratosphere over the antarctic region and formation of the polar vortex**
   From June to September, during the polar night, there is no incident solar radiation. The topography of the antarctic region (an isolated continent surrounded by oceans) leads to the formation of a circular polar vortex (winds isolate a cold polar air mass from any ingoing warm air mass).
   Due to its circular nature, the polar vortex explains the strong isolation of the antarctic region, which favors very low temperatures (from 15 to 20 degrees less than those over the North Pole).
2. **Formation of the polar stratospheric clouds**
   Low temperatures allow the formation of the so-called *polar stratospheric clouds* (PSC) at altitudes of about 20 kilometers. This formation is related to the con-

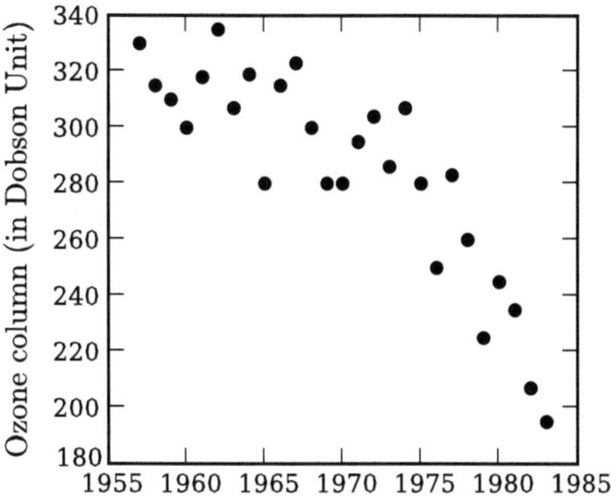

Fig. 4.5 Evolution of the ozone column at Halley Bay station (in Dobson Unit), for October for the period 1957–1983. Source: [38] (article of Farman et al.)

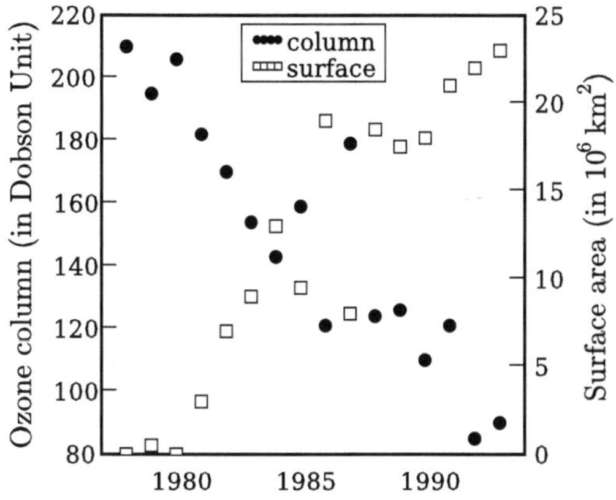

Fig. 4.6 Evolution from 1979 to 1994 of the du minimal value of the ozone column (in Dobson Unit) and of the horizontal extension of the ozone hole (in 10^6 km^2). Antarctic region (80°–90°). Months of September, October and November. Source: data from NASA, Goddard Space Flight Center

densation of water vapor and of other compounds (e.g. hydrogen chloride HCl, nitric acid HNO_3, sulfuric acid H_2SO_4), in the form of solid crystals.
While the crystallization of sulfuric acid is rather common (it can take place for temperatures below −55 K), that of hydrogen chloride requires temperatures lower than −75 K, which are likely to be reached only in the stratospheric region.

3. **Heterogeneous reactions at the PSC surface**
Heterogeneous reactions take place at the PSC surface. They mainly result in the transformation of "stable" compounds (N_2O_5, $ClNO_3$) into less stable compounds (HOCl, Cl_2), typically with

$$ClNO_3 + HCl \xrightarrow{PSC} Cl_2 + HNO_3. \tag{R 64}$$

4. **Photolytic production of chloride and bromide**
When photolysis can take place (namely after the polar night), chloride and bromide are produced,

$$Cl_2 + h\nu \longrightarrow Cl + Cl, \tag{R 65}$$

$$HOCl + h\nu \longrightarrow H + ClO, \tag{R 66}$$

which favors the catalysis of ozone destruction (Fig. 4.7).

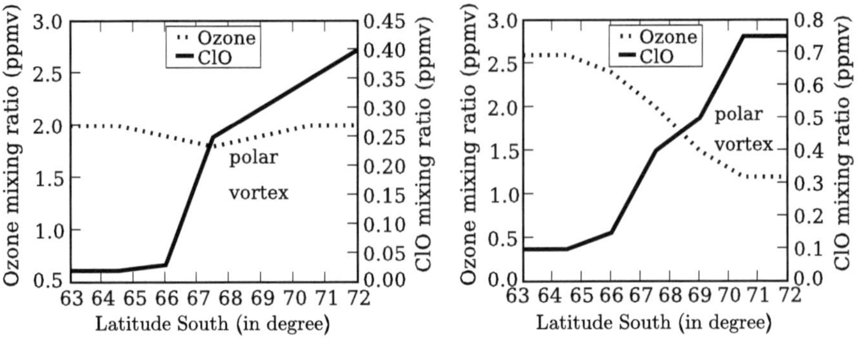

Fig. 4.7 Sketch of the latitude distribution of mixing ratios for ozone and ClO in the Antarctic region. *Left panel*: in August; *right panel*: in September

Remark 4.2.2 (Natural Variability) Understanding all the processes (meteorological, chemical and physical processes, heterogeneous processes, etc.) is a challenging issue. Note that models are required to estimate changes of a few per cent due to anthropogenic activities while the natural variability is much stronger (especially due to temperature changes).

Remark 4.2.3 (International Treaties) The discovery of the ozone hole led to a series of international treaties in the 1980s to reduce antropogenic emissions, such as those of CFCs (Montreal Protocol in 1987). We refer to Problem 1.5.

Remark 4.2.4 (Feedbacks and Climate Change) A decrease in the stratospheric temperature favors ozone destruction. A few studies suggest that a decrease in temperature of 1 K implies an increase in ozone destruction of about 5%.

Satellite data show a cooling of the stratosphere of about −0.5 K per decade since the 1960s (except specific heatings due to volcanic eruptions such as El Chichon or Mont Pinatubo). Two reasons may be invoked:

- the first effect is relatively direct: the destruction of stratospheric ozone leads to a decrease in the stratospheric temperatures since less solar radiation is absorbed. In the lower stratosphere, ozone destruction results in a decrease of infrared radiation absorption because ozone is a greenhouse gas;
- the second effect is much more complicated and is related to greenhouse gases. The increase in the concentrations of greenhouse gases blocks infrared radiation in the layers just above the Earth's surface: in other words, a smaller fraction of infrared radiation reaches the stratosphere. Meanwhile, the energy budget related to the radiative behavior of CO_2 has also a cooling effect.

This example illustrates the coupling of atmospheric issues. Understanding the interactions between stratospheric chemistry and the greenhouse effect is still a prevailing issue.

4.3 Tropospheric Chemistry of Ozone

The focus is now put on tropospheric chemistry for ozone. A key point is the existence of two chemical regimes: when the environment is NO_x-poor, the emission of VOCs does not impact ozone concentration, whereas when the environment is NO_x-rich, the emission of VOCs favors ozone formation (this corresponds to the classical case of ozone peaks during a photochemical smog event).

4.3.1 Basic Facts for Combustion

Emissions of CO and VOCs, Mixture Richness For the sake of clarity, let us consider a hydrocarbon whose chemical symbol is C_xH_y (not yet oxidized). Combustion in the air (3.8 moles of N_2 per mole of O_2), under *stoichiometric conditions*, is described by the *global* chemical reaction

$$C_xH_y + \left(x + \frac{y}{4}\right)(O_2 + 3.8N_2) \longrightarrow x\ CO_2 + \frac{y}{2}\ H_2O + \left(x + \frac{y}{4}\right)\ N_2. \qquad (R\,67)$$

To oxidize one mole of hydrocarbon, $x + y/4$ air moles are thus required (2 for methane CH_4, 3.5 for ethane C_2H_6, 11 for heptane C_7H_{16}, etc.).

Because combustion does not actually take place under stoichiometric conditions, (R 67) needs to be modified. The mixture *richness* is defined as the ratio of the number of hydrocarbon moles to the number of oxygen moles, normalized with respect to stoichiometric conditions, namely

$$\phi = \left(x + \frac{y}{4}\right)\frac{C_{C_xH_y}}{C_{O_2}}. \qquad (4.19)$$

Other products can be produced, depending on the mixture richness.

- In a rich case ($\phi > 1$), CO and H_2 are produced and the reaction becomes

$$C_xH_y + n(O_2 + 3.8N_2) \longrightarrow aCO_2 + (1-a)CO + bH_2O + \left(\frac{y}{2} - b\right)H_2 \qquad (R\,68)$$

  with $n = (1 + y/4)/\phi$. There are also $3.78n$ molecules of N_2 in the products. The coefficients a and b can be calculated as a function of n (that is to say of hydrocarbon and of richness; Exercise 4.6).
- In a poor case ($\phi < 1$), oxygen is also produced,

$$C_xH_y + n(O_2 + 3.8N_2) \longrightarrow CO_2 + \frac{y}{2}H_2O + \left(n - 1 - \frac{y}{4}\right)O_2 + 3.78nN_2. \qquad (R\,69)$$

Exercise 4.6 (CO Emission as a Function of Richness) In the rich case, CO is produced from the reaction

$$CO_2 + H_2 \rightleftharpoons CO + H_2O. \tag{R 70}$$

In practice, the temperature is nearly constant at about 1700 K and the reversible reaction is at equilibrium (with an equilibrium constant $K \simeq 3.8$).

Calculate the proportion of produced CO (coefficient a in (R 68)).

Solution:

Oxygen conservation implies

$$2n = 2a + (1-a) + b = a + b + 1.$$

The equilibrium reads

$$K = \frac{[CO][H_2O]}{[CO_2][H_2]} = \frac{(1-a)b}{a(\frac{y}{2} - b)}.$$

This leads to a two-equation system with two unknowns (a and b). Solving for a, we obtain a quadratic equation

$$(K-1)a^2 + \left[K\left(\frac{y}{2} - 2n + 1\right) + 2n\right]a - (2n-1) = 0.$$

The discriminant is positive and the product of roots is negative (due to the values of K and n). There exists therefore a unique positive root for a, which can be computed as a function of n (and, as a result, of richness ϕ).

The mixing ratio of CO_2 is then $a/(1 + y/2 + 3.78n)$; the mixing ratio of CO is $(1 - a)/(1 + y/2 + 3.78n)$.

The evolution of mixing ratios as a function of richness (including the poor case) is shown in Fig. 4.8.

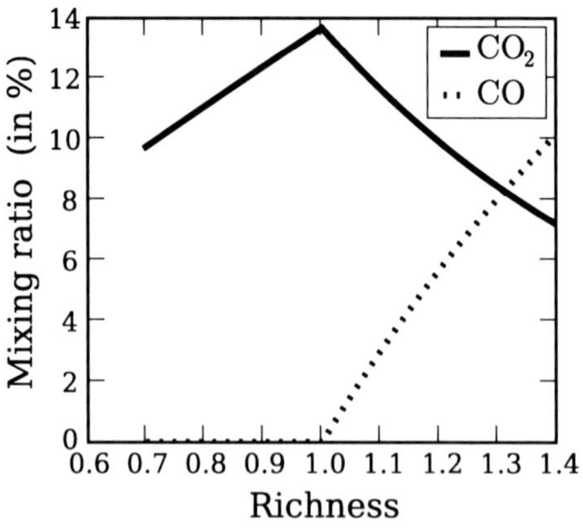

Fig. 4.8 Theoretical evolution of CO and CO_2 emissions for combustion of $CH_{1.75}$ (similar to a "classical" fuel for a gasoline vehicle). Source: [49]

NO_x Emission Two chemical mechanisms are usually distinguished for the formation of NO. Zeldovitch (1946) has proposed the following mechanism for NO formation from N_2 of air:

$$N_2 + O \longrightarrow NO + N, \quad (R\,71)$$

$$N + O_2 \longrightarrow NO + O, \quad (R\,72)$$

$$N + OH \longrightarrow NO + H. \quad (R\,73)$$

Taking into account the first two reactions, the global reaction is,

$$N_2 + O_2 \longrightarrow 2\,NO. \quad (R\,74)$$

These reactions can only occur at very high temperatures (the activation energy of the first reaction is very high), typically above 1700 K. As a consequence, the produced nitrogen monoxide is usually referred to as *thermal* NO.

A second mechanism, sometimes referred to as the Fenimore mechanism, implies radicals which include hydrogen atoms (e.g. CH). The produced nitrogen monoxide is said to be *prompt* NO. The initiation step

$$CH + N_2 \longrightarrow HCN + N \quad (R\,75)$$

has a rather low activation energy. It can thus take place at lower temperatures (about 1000 K). HCN leads thereafter to the formation of CN with the reactions

$$HCN + H \longrightarrow CN + H_2, \quad (R\,76)$$

$$HCN + OH \longrightarrow CN + H_2O. \quad (R\,77)$$

Radicals HCN and CN lead to NO formation (reactions not detailed here).

A third mechanism implies nitrogen peroxide (N_2O) and takes place at high pressures.

Finally, the typical ratio of NO concentration to that of NO_2 is about 10% in emissions.

Evolution of Emissions as a Function of Richness Figure 4.9 shows a typical evolution of emissions of NO_x, CO and hydrocarbons as a function of richness. NO_x emissions have a maximum for a slightly poor mixture ($\phi \in [0.9, 0.95]$). Actually, the formation rate of NO depends exponentially on the combustion temperature and on the oxygen concentration and follows a bell curve. Below the maximum, the combustion temperature is the limiting factor; above, the oxygen concentration is the limiting factor.

Emissions cannot be reduced simultaneously for all chemical components (VOCs, CO and NO_x). This remark motivates that targeting emission reductions have to be chosen (see Sect. 4.3.5) when considering traffic-induced emissions.

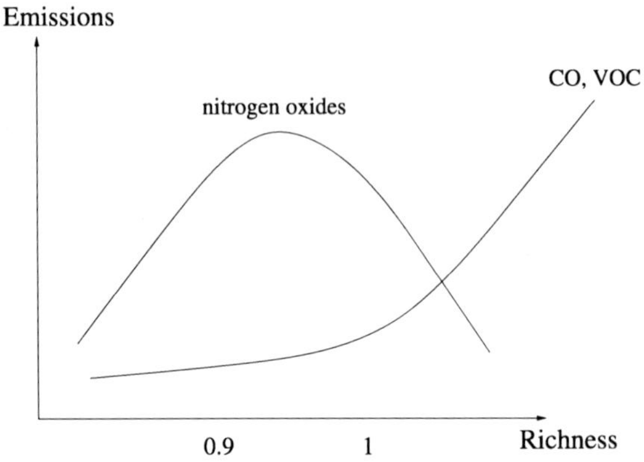

Fig. 4.9 Schematic evolution of NO, CO and hydrocarbon emissions as a function of richness. Source: [50]

4.3.2 Photostationary State of Tropospheric Ozone

The only reaction for ozone production implies oxygen atoms O, namely

$$O_2 + O + M \longrightarrow O_3 + M. \qquad \text{(R 78)}$$

A possible source for oxygen atoms is provided by the photolysis of oxygen molecules,

$$O_2 + h\nu(\lambda \leq 242 \text{ nm}) \longrightarrow O + O. \qquad \text{(R 79)}$$

Such a reaction can take place only in the stratosphere because the radiation is short-wave. In the troposphere, the source is the photolysis of NO_2,

$$NO_2 + h\nu(300 \text{ nm} \leq \lambda \leq 400 \text{ nm}) \longrightarrow NO + O, \qquad \text{(R 80)}$$

which explains the close interactions between ozone and nitrogen oxides in the troposphere.

Ozone destruction through the oxidation by NO,

$$NO + O_3 \longrightarrow NO_2 + O_2, \qquad \text{(R 81)}$$

equilibrates the chemical cycle made of reactions (R 78) and (R 80). This reaction is sometimes referred to as the *titration reaction* of ozone by nitrogen monoxide.

These reactions replace the Chapman cycle in the troposphere. Using a quasi steady-state assumption for NO, the *photostationary equilibrium* is defined from reactions (R 80) and (R 81),

$$[O_3] \simeq \frac{J[NO_2]}{k[NO]},$$

with J the photolytic rate of (R 80) and k the kinetic rate of (R 81).

The possible increase in ozone is induced by an increasing oxidation of NO into NO_2 by other oxidants than ozone, which alters the cycle in favor of reaction (R 78).

The oxidants may be peroxy radicals HO_2 and RO_2, which result from chemical reactions affecting emitted species in the troposphere (VOCs, methane CH_4 and carbon monoxide CO).

4.3.3 *Oxidation Chains of VOCs*

In this section, the oxidation of a generic hydrocarbon (RH) is detailed. Thereafter, the focus is put on carbon monoxide (CO) and methane (CH_4). These oxidation chains are initiated by HO_x and *catalyzed* by NO_x, and result in ozone production.

The difference with the stratosphere (where NO_x catalyzes ozone destruction) is the low concentration of ground state oxygen atom $O(^1D)$ in the troposphere (due to the lower ozone concentrations). Reaction (R 40) can then be omitted in a first approximation.

4.3.3.1 Hydrocarbon Oxidation

Let RH be a generic hydrocarbon. Oxidation by OH leads to the formation of compounds with many chemical functions, such as *peroxy* radicals (RO_2), *oxy* radicals (RO) and *carbonyl/aldehyde* radicals (R′CHO where R′ stands for a chain with a lower number of carbon atoms than the initial hydrocarbon).

The oxidation of this chain is catalyzed by NO_x and contributes to ozone production. The propagation reactions are

$$RH + OH \longrightarrow R + H_2O, \qquad \text{(R 82)}$$

$$R + O_2 + M \longrightarrow RO_2 + M, \qquad \text{(R 83)}$$

$$RO_2 + NO \longrightarrow RO + NO_2, \qquad \text{(R 84)}$$

$$RO + O_2 \longrightarrow R'CHO + HO_2, \qquad \text{(R 85)}$$

$$HO_2 + NO \longrightarrow OH + NO_2. \qquad \text{(R 86)}$$

There are two molecules of nitrogen dioxide produced from one molecule of RH with reactions (R 84) and (R 86). As a result, there is ozone production with

$$NO_2 + h\nu \longrightarrow NO + O, \qquad \text{(R 87)}$$

$$O_2 + O + M \longrightarrow O_3 + M. \qquad \text{(R 88)}$$

The global budget is therefore

$$RH \xrightarrow{NO_x} R'CHO + 2\,O_3. \qquad \text{(R 89)}$$

The termination reactions are given by production of H_2O_2 and HNO_3 (to be viewed as sink species, Sect. 4.1.4.5),

$$HO_2 + HO_2 \longrightarrow H_2O_2 + O_2, \quad \text{(R 90)}$$

$$NO_2 + OH + M \longrightarrow HNO_3 + M. \quad \text{(R 91)}$$

Actually, this chain can be "amplified" by branching reactions related to the photolysis of R′CHO. In the case of methane, this generates three supplementary ozone molecules, as detailed in the following. Moreover, other oxidation reactions originate from R′CHO.

4.3.3.2 Methane Oxidation

Similarly to the general case, with R taken as CH_3 (methyl radical), the oxidation mechanism is:

$$CH_4 + OH \longrightarrow CH_3 + H_2O, \quad \text{(R 92)}$$

$$CH_3 + O_2 + M \longrightarrow CH_3O_2 + M, \quad \text{(R 93)}$$

$$CH_3O_2 + NO \longrightarrow CH_3O + NO_2, \quad \text{(R 94)}$$

$$CH_3O + O_2 \longrightarrow HCHO + HO_2. \quad \text{(R 95)}$$

At this stage, the branching reaction is the photolysis of formaldehyde (HCHO) with the chemical mechanism

$$HCHO + h\nu \xrightarrow{O_2} CHO + HO_2, \quad \text{(R 96)}$$

$$CHO + O_2 \longrightarrow CO + HO_2, \quad \text{(R 97)}$$

$$CO + OH \longrightarrow CO_2 + H, \quad \text{(R 98)}$$

$$H + O_2 + M \longrightarrow HO_2 + M. \quad \text{(R 99)}$$

Four molecules of HO_2 are produced from reactions (R 95), (R 96), (R 97) and (R 99). They can thereafter take part in the production reaction of NO_2,

$$HO_2 + NO \longrightarrow NO_2 + OH. \quad \text{(R 100)}$$

The produced nitrogen dioxide (one molecule with (R 94) and four molecules with (R 100), thus five molecules from one molecule of methane), contributes to produce ozone:

$$NO_2 + h\nu \longrightarrow NO + O, \quad \text{(R 101)}$$

$$O_2 + O + M \longrightarrow O_3 + M. \quad \text{(R 102)}$$

The global budget resulting from this cycle is

$$CH_4 \xrightarrow{NO_x} CO_2 + 5\,O_3 + 2\,OH. \quad \text{(R 103)}$$

4.3.3.3 Carbon Monoxide Oxidation

The oxidation of carbon monoxide is similar but ozone production is weaker because there is no hydrogen atom in CO. The propagation reactions are

$$CO + OH \longrightarrow CO_2 + H, \quad \text{(R 104)}$$

$$H + O_2 + M \longrightarrow HO_2 + M. \quad \text{(R 105)}$$

HO_2 oxides NO into NO_2, which is available for ozone production with the reactions

$$HO_2 + NO \longrightarrow OH + NO_2, \quad \text{(R 106)}$$

$$NO_2 + h\nu \longrightarrow NO + O, \quad \text{(R 107)}$$

$$O_2 + O + M \longrightarrow O_3 + M, \quad \text{(R 108)}$$

whose global budget is

$$CO \xrightarrow{NO_x} CO_2 + O_3. \quad \text{(R 109)}$$

4.3.4 *NO_x-Limited Versus VOC-Limited Chemical Regimes*

The ability to produce oxidants (OH, HO_2) is therefore related to NO_x concentration. In the previous oxidation chains, NO_x concentration was implicitly assumed to be high enough (the regime is sometimes said to be *high*-NO_x), so that the key reaction is

$$HO_2 + NO \longrightarrow OH + NO_2, \quad \text{(R 110)}$$

which leads to an enhanced production of ozone.

In a *low*-NO_x regime, the termination reaction

$$HO_2 + HO_2 \longrightarrow H_2O_2 + O_2 \quad \text{(R 111)}$$

dominates (R 110), which inhibits ozone production because the environment has a lower oxidizing capacity.

The conclusion of the competition between reactions (R 110) and (R 111) is actually driven by the ratio [VOC]/[NO_x] (Problem 4.2).

Figure 4.10 shows a sketch of the isopleths of ozone concentrations as a function of NO_x and VOC concentrations. There are two regimes:

- in the *high*-NO_x regime, also referred to as *VOC-limited* regime, a decrease in NO_x concentration results in higher ozone production on the contrary to a decrease in VOC concentration;
- in the *low*-NO_x regime, also referred to as NO_x-*limited* regime, a decrease in NO_x results in lower ozone concentration while changes in VOC concentrations do not affect ozone concentrations.

Remark 4.3.1 (Source of OH Radical During Day-time) As seen in the previous sections, the hydroxyl radical OH plays a leading role in the initiation steps of oxidation chains. Assessing the oxidizing capacity of the atmospheric environment is then related to an accurate estimation of sources and sinks for the hydroxyl radical. Note that the sinks are related to the oxidation reactions of VOCs. The sources may strongly vary:

- the major source is ozone photolysis (Sect. 4.1.3);
- in the context of air quality, another important source is the photolysis of aldehydes and HONO,

$$HONO + h\nu(\lambda \leq 390 \text{ nm}) \longrightarrow NO + OH; \qquad \text{(R 112)}$$

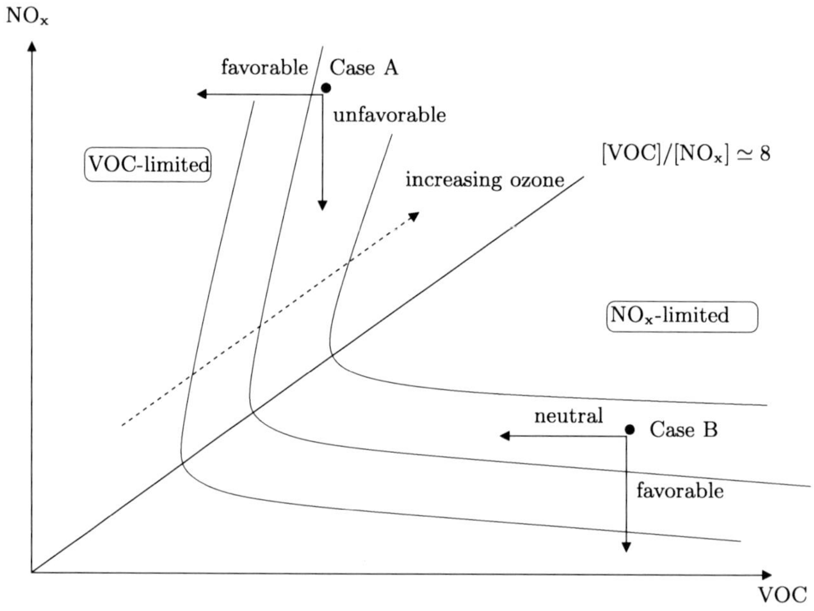

Fig. 4.10 Chemical regimes of tropospheric ozone. Sketch of *isopleths* for ozone concentrations as a function of NO_x and VOC concentrations. This figure is usually based on the so-called EKMA model (*empirical kinetic modeling approach*). The ratio $[VOC]/[NO_x] \simeq 8$ is typical of North-American conditions

HONO is produced by heterogeneous reactions at the surface of aerosols and water drops,

$$NO_2 + NO_2 + H_2O \longrightarrow HONO + HNO_3. \qquad \text{(R 113)}$$

During night-time, as there is no photolysis of NO_2, HONO can accumulate. At sunrise, it constitutes therefore a reservoir species to produce OH, which is rather produced by ozone photolysis during day-time;

- even if photolysis does not occur during night-time, "nocturnal" chemistry can have a significant impact, especially because of the activity of the nitrate radical NO_3. During day-time, nitrate is quickly consumed by photolysis,

$$NO_3 + h\nu(400\ \text{nm} \leq \lambda \leq 640\ \text{nm}) \longrightarrow NO_2 + O. \qquad \text{(R 114)}$$

During night-time, reactions

$$RO_2 + NO_3 \longrightarrow RO + NO_2 + O_2, \qquad \text{(R 115)}$$

$$HO_2 + NO_3 \longrightarrow OH + NO_2 + O_2 \qquad \text{(R 116)}$$

result in an environment which is rich in OH radicals and NO_2.

4.3.5 Emission Reduction Strategies for Ozone Precursors

Because there are different chemical regimes for tropospheric ozone, the impact of emission reduction strategies, applied to NO_x and VOCs, on ozone concentrations is particularly difficult to assess *a priori*.

In the 1960s and 1970s, the North-American strategies devoted to the reduction of photochemical smog were based on reductions of local VOC emissions. An illustrative example of the regulatory "spirit" is shown in Fig. 4.11: the required reduction of VOC emission is calculated as a function of the ambient ozone peak, depending on the regulatory target (here 160 $\mu g\ m^{-3}$ for the hourly average of ozone concentration).

In the 1980s, the significant role of NO_x was recognized, as illustrated by the NRC report (National Research Council, [144]) on air quality:

> To substantially reduce O_3 concentrations [...] the control of NO_x emissions will probably be necessary in addition to, or instead of, the control of VOCs.

As seen in the previous sections, the regimes are particularly complicated because they may vary with space and time.

Many studies have shown that urban areas are usually VOC-limited. In this case, a decrease in NO_x emissions should result in an increase in ozone concentration (Fig. 4.10). This fact is sometimes referred to as the NO_x *disbenefit*.

A first illustration is provided by the evolution of air quality over Canada in the 1990s. While the traffic-induced NO_x emissions and the resulting concentrations in

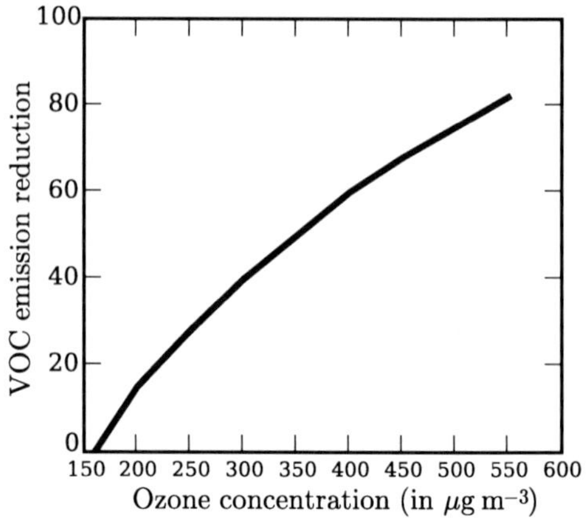

Fig. 4.11 Required reduction (in %) of VOC emissions as a function of the ozone peak to reach the NAAQS target (*National Ambient Air Quality Standard*) of 160 μg m^{-3} (US EPA recommendation, 1971). Source: [3]

urban areas were reduced, the VOC emissions and the resulting concentrations in urban areas were not modified. The observational data have shown an increase in ozone concentrations in urban areas, which may be explained by the VOC-limited regime.

A second illustration is given by the occurrence of ozone peaks at the end of the week in a few large cities because of the decrease in the traffic-induced NO_x emissions (studies in Los Angeles, Toronto, Brussels, etc.).

The impact of biogenic emissions of VOCs (especially of isoprene, C_5H_8) may also significantly affect ozone concentration. Having an accurate estimation of biogenic emissions is therefore required to forecast the impact of anthropogenic emission reductions. For example, the reevaluation to a higher value of isoprene emissions in the USA during the 1990s resulted in a modified assessment of emission reduction of VOCs (weaker impact) because the real-life atmospheric environment was likely to be in a NO_x-limited regime (Fig. 4.10).

To conclude, the following key facts can be summarized:

- a decrease in ozone concentrations in an urban area should be based *a priori* on a decrease in VOC emissions;
- a decrease in ozone concentrations at the regional scale, over areas with high biogenic emissions of VOC, should be based *a priori* on a decrease in NO_x emissions.

Figure 4.12 shows an indirect estimation of the chemical regimes over Europe for summer 2001. The respective impacts on ground ozone concentration of a 35% reduction of NO_x and VOC emissions are compared. Note that the regime is likely to be VOC-limited over polluted urban areas and NO_x-limited in the southern part of Europe (with high biogenic emissions of VOCs).

Note that the role of meteorological conditions has also to be taken into account (Chap. 3) to give a global assessment of air quality: actually, meteorological conditions can impact the long-range transport of pollutants, the dilution state (mixing height in the atmospheric boundary layer), ventilation or stagnation of air masses,

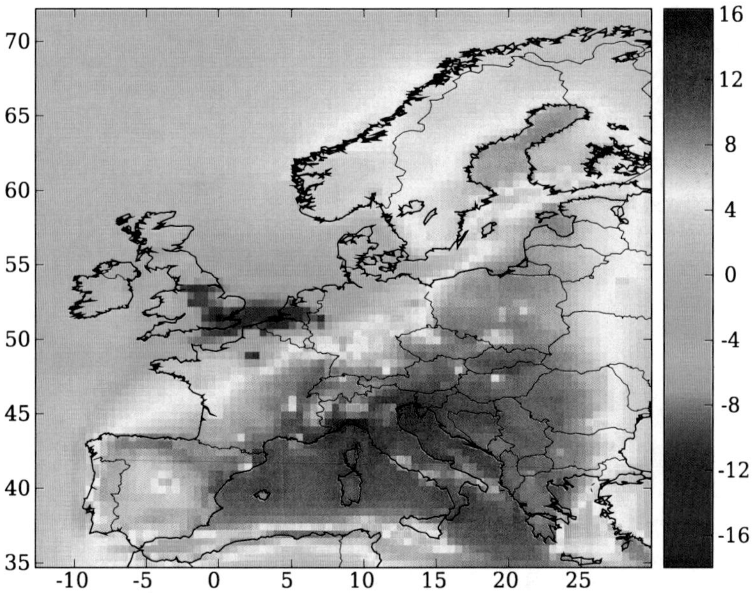

Fig. 4.12 Mean difference (in $\mu g\,m^{-3}$), for summer 2001, of the resulting impact on ozone concentration at ground, of a 35% decrease in NO_x and VOC emissions. A positive value (a negative value, respectively) suggests an NO_x-limited regime (a VOC-limited regime, respectively). Simulation with the POLYPHEMUS system. Credit: Yelva Roustan, CEREA

etc. We refer to the next section devoted to the investigation of air quality over Paris (Île-de-France region).

To Know More ([3]):
An assessment of Tropospheric Ozone pollution: a North American perspective. NARSTO Assessment, 2006. Available at http://www.narsto.org

More generally, the topics related to the assessment of emission reduction strategies are decisive for decision-making purposes. A typical example is the assessment of the impact of modified emissions, for instance for traffic-induced emissions. Changes in emissions can be induced by new fuels or new engines. We refer to Exercise 4.7 for the case of "biofuels".

Exercise 4.7 (Biofuel Impact on Air Quality) Traffic-induced emissions can be strongly modified by the introduction of new fuels and new engines. The estimation of the impact of photochemistry is a challenging issue. An illustrative case is related to *biofuels* (e.g., ethanol, E85).

Using the data given in Table 4.6, estimate the *a priori* impact of the introduction of E85 on photochemistry.

Solution:
In an urban area, the increase of emitted VOCs and the decrease of emitted NO_x should result *a priori* in an increase in ozone concentration. Numerical simulations over Los Angeles ([64]) indicate a possible increase in the daily average of ozone concentration of about 3 ppb, and

Table 4.6 Evolution of emissions for a vehicle: comparison between a gasoline vehicle (with a "classical" fuel) and a vehicle using biofuels (here E85, ethanol). *PM* stands for *particulate matter*. Source: [64]

Species	E85 *versus* gasoline	Sign
VOCs (total)	[+34, +95]%	+
Methane	[+43, +340]%	+
Formaldehyde (HCHO)	[+7, +228]%	+
Acetaldehyde (CH_3CHO)	[+1250, +4340]%	+
Carbon monoxide (CO)	[−38, +320]%	?
NO_x	[−59, +17]%	−
PM (mass)	+31%	+
PM (number)	+100%	+

in the peaks of 4 ppb. These preliminary studies are still limited (to date) to the case of the USA and further studies are still required.

Notice that there is also a fear about the radiative impact of biofuels. Using biofuels should result in an increase of agriculture-induced emissions, especially of nitrogen peroxide (N_2O). The impact could be that "N_2O release from agro-biofuel production negates global warming reduction by replacing fossil fuels" (from the title of [28], see below for the reference).

To Know More ([28, 64, 65]):

M. JACOBSON, *Effects of Ethanol (E85) versus Gasoline Vehicles on Cancer and Mortality in the United States*, Environ. Sci. Tech. (2007)

M. JACOBSON, J. SEINFELD, G. CARMICHAEL, AND D. STREETS, *The effect on photochemical smog of converting the US fleet of gasoline vehicles to modern diesel vehicles*, Geophys. Res. Lett., **31** (2004)

P. CRUTZEN, A. MOSLER, K. SMITH, AND W. WINIWARTER, N_2O *release from agro-biofuel production negates global warming reduction by replacing fossil fuels*, Atmos. Chem. Phys. Discuss., **7** (2007), pp. 11191–11205

4.3.6 Example of Photochemical Pollution at the Regional Scale: Case of Île-de-France Region

The Île-de-France region is a good example of the genesis of photochemical pollution events with the joint effects of chemical and meteorological conditions. The resulting ozone concentrations are driven by long-range transport and ozone titration by NO_x emissions (sketch of the processes in Fig. 4.13).

- **Continental Transport**
  Because the lifetime of ozone is high enough (Table 4.4), a significant fraction of pollution events is related to continental transport of polluted air masses (typically coming from the north-eastern direction). Figure 4.15 illustrates the "correlation" between the ozone concentrations at 3 p.m. (illustrative of the daily peak) for a rural station in the northeast of Paris and an urban station (Chatelet-Les Halles,

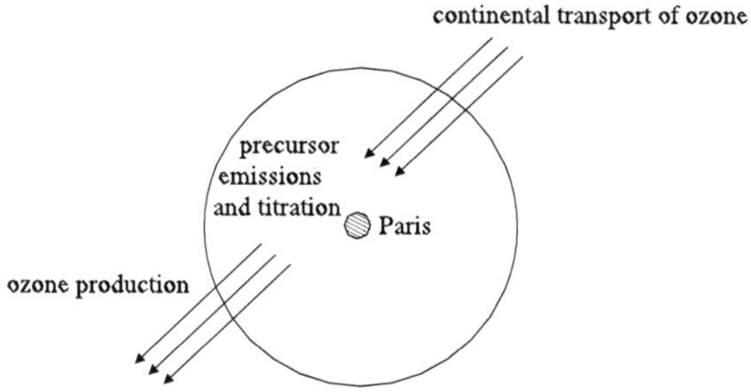

Fig. 4.13 Sketch of the development of photochemical pollution events over Île-de-France region: combined effects of continental transport, ozone production and titration by NO_x emissions

center of Paris), for the year 2005. The significant titration of ozone is noticeable over Paris.

- **Precursor Emission and Ozone Production in Rural Regions**
  The emissions of ozone precursors in the Paris region contribute to ozone formation in the air masses downwind Paris. The maximal values of ozone concentrations are actually observed in rural areas, especially in the southwest of Paris (Fig. 4.15).
- **Titration by NO_x Emissions Close to Strong Emission Areas**
  Previous figures illustrate the ozone titration in urban areas. The evolution of ozone concentrations in the Paris region for the period 1992–2005 (Fig. 4.16) also shows the impact of titration. The background concentrations, as measured by rural stations, show an increase of about 15% while the concentrations measured by urban stations have increased by about 84%. A likely reason is the technical improvement of vehicles with a strong decrease of emitted NO (decrease in NO urban concentrations of 218% and Fig. 4.14).

4.3.7 Transcontinental Transport

In summertime, the lifetime of ozone in the lower troposphere can be estimated to range from one to two weeks (Table 4.4). Trans-pacific and trans-atlantic transport of ozone may therefore be significant. A few studies (e.g. [56]) suggest that the increase in ozone concentration at measurement stations in California may vary from 5 to 10 ppb (in April and May), due to Asian emissions. The transported pollutants can be ozone or PAN. Similarly, the contribution of North-American pollution to surface ozone concentration over Europe is estimated to be up to a few ppb (from 3 to 5 ppb on average for summertime, with peaks ranging from 10 to 20 ppb for specific events, [48] and [87]). The strongest increase takes place in the free troposphere (from 10 to 30 ppb).

The underlying issue related to these assessments is to check that local emission reductions would not be negated by long-range transcontinental transport of pollutants. In [87], up to 20% of the violations of the regulatory thresholds for European ozone are likely to be directly connected to North-American pollution. In [61], the expected increase in the Asian emissions for the period 1985–2010

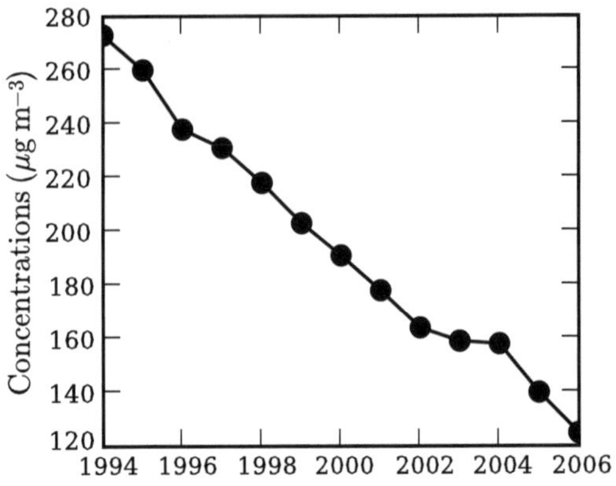

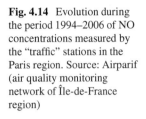

Fig. 4.14 Evolution during the period 1994–2006 of NO concentrations measured by the "traffic" stations in the Paris region. Source: Airparif (air quality monitoring network of Île-de-France region)

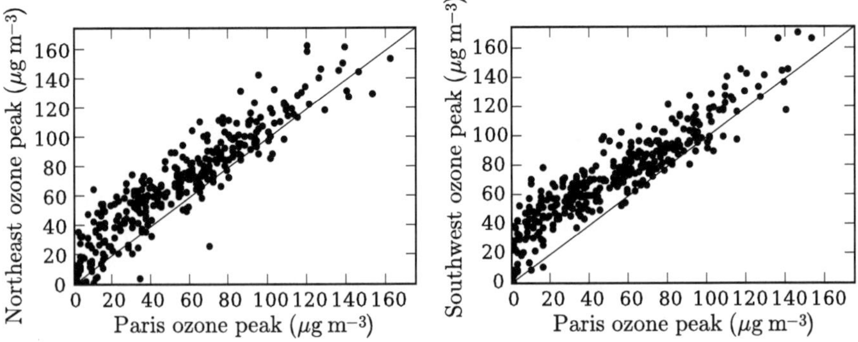

Fig. 4.15 Correlation between the peaks in ozone concentration (supposed to occur at 15:00 p.m), during the year 2005, for a station in Paris center (Chatelet-Les Halles) and: *left*, a rural station in the northeast of Paris (Montgé-en-Gole), *right*, a rural station in the southwest of Paris (Rambouillet). Source: Airparif

should negate the North-American emission reduction of NO_x and VOCs (of about 25%) for North-American ozone.

4.4 Brief Introduction to Indoor Air Quality

The focus of air quality studies is usually put on *outdoor* air quality while we spend most of our time inside buildings ! *Indoor air quality* is therefore a key topic.

Indoor air quality has its own characteristics. It depends, in a complicated way, both on indoor "atmospheric" physics and chemistry, and on air transfer with outdoor.

- Photolysis has a role which is less significant than outdoor because part of the ultraviolet and visible radiation is filtered. For example, the photolytic rate of NO_2 should be divided by a scaling factor ranging from 3 to 5, according to a few studies, as compared to the outdoor rate.

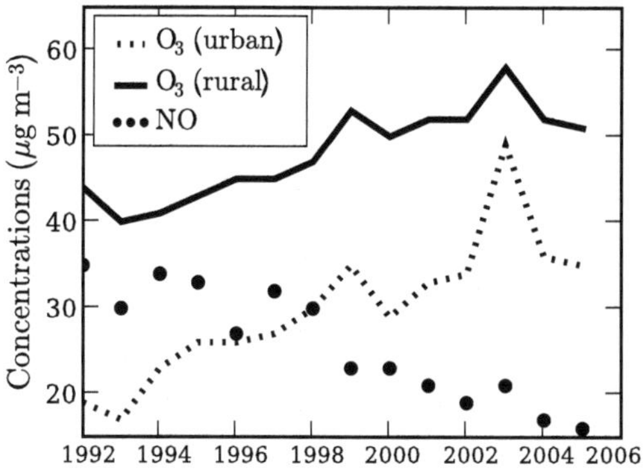

Fig. 4.16 Evolution during the period 1992–2005 of measured concentrations of ozone and nitrogen monoxide in the Paris region. The concentrations are averaged over the monitoring network, which is representative of rural and urban conditions (constant network during the period of interest). Note that the anomaly of the year 2003 is related to the heat wave of summer 2003. Source: Airparif

- The indoor environment is yet characterized by its strong oxidizing ability with high levels of the hydroxyl radical OH (only one order of magnitude lower than outdoor). The minor role of photolysis implies that there are other chemical pathways to produce oxidants.
- Heterogeneous chemical reactions (e.g. at the surface of painted walls) play a leading role, which explains the concentration levels for VOCs (e.g. aldehydes) and HONO (available for the production of OH, see Remark 4.3.1).
- A few studies devoted to health impact (e.g. allergies, asthma, etc.) suggest the role of oxidation products of VOCs by OH, O_3 and NO_3. The resulting products can lead thereafter to secondary organic aerosols (see Chap. 5).
- Deposition processes are not well described in indoor environments. It is usually assumed that the dry deposition velocities should be divided by one or two orders of magnitude, as compared to outdoor values.
- Last, the understanding of the interactions between indoor air quality and outdoor air quality is still a challenging issue. These interactions stongly depend on the ventilation conditions and on the outdoor meteorological conditions in the lower atmospheric boundary layer (Fig. 4.17).

Exercise 4.8 introduces a simplified box model for indoor air quality.

Exercise 4.8 (Indoor Air Quality Modeling) Let us consider a room of volume V, of surface A, with a given ventilation rate $\lambda_{in/out}$ (supposed to be inversely proportional to a characteristic time). Let X_i be a species with a mass emission rate Q_i (in $kg\,s^{-1}$), a deposition velocity v_i^{dep} and a chemical source term χ_i. Formulate a model for the time evolution of the indoor mass concentration C_i^{in}. Write C_i^{out} the outdoor mass concentration.
Data: for illustration, $A/V \simeq 3\,m^{-1}$, $v_{O_3}^{dep} \sim 0.04\,cm\,s^{-1}$.

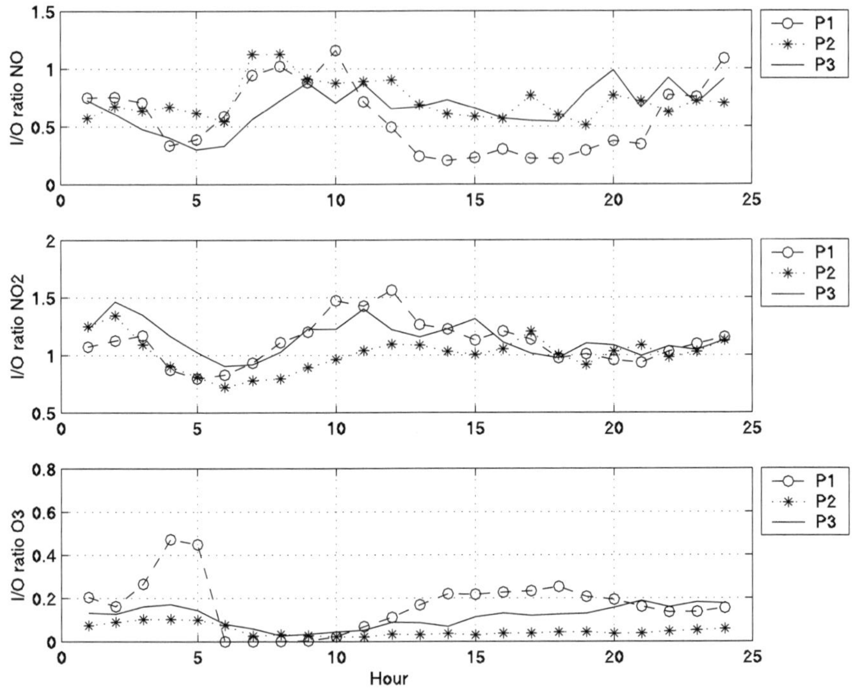

Fig. 4.17 Evolution of the ratio of *indoor* concentration (I for *indoor*) to *outdoor* concentration (O for *outdoor*) for NO, NO_2 and O_3 (three-day observational data in a flat of Paris suburb, June 2000: P1, P2 and P3). Credit: Stephanie Lacour (CEREA) from measured data of Centre scientifique et technique du batiment (CSTB)

Solution:
The time evolution is governed by

$$\frac{dC_i^{in}}{dt} = -v_i^{dep}\frac{A}{V}C_i^{in} - \lambda_{in/out}(C_i^{in} - C_i^{out}) + \frac{Q_i}{V} + \chi_i$$

To Know More ([21, 153]):

N. CARSLAW, *A new detailed chemical model for indoor air pollution*, Atmos. Env., **41** (2007), pp. 1164–1179

C. WESCHLER, *Ozone-initiated reaction products indoors may be more harmful than ozone itself*, Atmos. Env., **38** (2004), pp. 5715–5716

Problems Related to Chap. 4

Problem 4.1 (Inverse Modeling of OH) The evolution of OH concentration is often estimated from observations of oxidized species, such as methane. Another

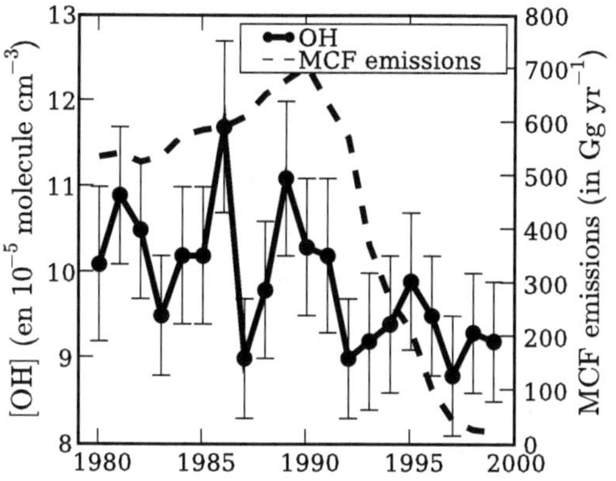

Fig. 4.18 Evolution during the period [1980–2000] of OH concentration and MCF emission, obtained by inverse modeling of MCF observations. The error bars indicate the uncertainties, as estimated by the inversion procedure. Source: [17]

commonly-used species for this purpose is methyl-chloroform (CH_3CCl_3, written as MCF).

1. MCF is a solvent used in industrial applications. Its lifetime varies from 5 to 6 years. Its emissions have been strongly reduced (from more than 700 $\mathrm{Gg\,yr^{-1}}$ in 1991 to about 20 $\mathrm{Gg\,yr^{-1}}$ in 2000). Why?
2. Give the main chemical reactions which govern the evolution of MCF. Formulate a model to describe the time evolution of MCF concentration.
3. Suppose there is a set of stations (actually more than ten) which measure MCF concentrations. Writing the model as $C_{\mathrm{MCF}} = f(E_{\mathrm{MCF}}, C_{\mathrm{OH}})$, give an estimation of C_{OH}.

Solution:

1. Because its lifetime is high enough, MCF can reach the stratosphere. Upon dissociation, it may produce chloride and, thus, take part in ozone destruction. It has therefore been included in the species regulated by the Montreal Protocol.
2. MCF undergoes photolysis and oxidation by OH:

$$\mathrm{MCF} + \mathrm{OH} \longrightarrow \mathrm{CH_2CCl_3} + \mathrm{OH}$$

$$\mathrm{MCF} + h\nu \longrightarrow \mathrm{CH_3CCl_2} + \mathrm{Cl}.$$

The evolution of MCF concentration is given by a reactive dispersion model (Chap. 6),

$$\frac{\partial C_{\mathrm{MCF}}}{\partial t} + \mathrm{div}(V C_{\mathrm{MCF}}) = \mathrm{div}(K \nabla C_{\mathrm{MCF}}) + E_{\mathrm{MCF}} - J C_{\mathrm{MCF}} - k_{\mathrm{OH,MCF}} C_{\mathrm{OH}} C_{\mathrm{MCF}},$$

with V the wind field, K the eddy diffusion coefficient, E_{MCF} the emissions, J the photolytic rate and $k_{\mathrm{OH,MCF}}$ the oxidation rate by OH.
3. Inverse modeling can be used (Chap. 6) by minimizing the so-called *cost function* (**J** is not a photolytic rate here)

$$\mathbf{J}(u) = \frac{1}{2}(obs - Hf(u))^T R^{-1}(obs - Hf(u)) + \frac{1}{2}(u - u_b)^T B^{-1}(u - u_b)$$

where $u = (E_{MCF}, C_{OH})$. The observation operator H maps the simulated concentrations onto the observations, written as *obs*; R and B are, respectively, the observation and background error covariance matrices; u_b is the background (prior estimation of u).

The results are shown in Fig. 4.18. A key point is the evaluation of uncertainties. Strong decrease of MCF emissions could be noticed after the Montreal Protocol. OH concentrations have a marked interannual variability: they tend to increase in the 1980s and then to decrease in the 1990s.

Because the MCF concentrations are likely to be much lower in the future, using these observational data will be much more challenging. Similar studies are therefore carried out with other indicators of atmospheric oxidation, such as methane.

To Know More ([17, 84]):

P. BOUSQUET, D. HAUGLUSTAINE, P. PEYLIN, C. CAROUGE, AND P. CIAIS, *Two decades of OH variability as inferred by an inversion of atmospheric transport and chemistry of methyl chloroform*, Atmos. Chem. Phys., **5** (2005), pp. 2635–2656

J. LELIEVELD ET AL., *Watching over tropospheric hydroxyl* (OH), Atmos. Env., **40** (2006), pp. 5741–5743

Problem 4.2 (Chemical Regimes of Tropospheric Ozone) This problem, taken from [58], aims at illustrating the chemical regimes of ozone in the troposphere. Let us consider the oxidation chain of RH (branching reactions are omitted for the sake of clarity):

$$RH + OH \xrightarrow{O_2,\, k_1} RO_2 + H_2O, \qquad (R\,117)$$

$$RO_2 + NO \xrightarrow{k_2} RO + NO_2, \qquad (R\,118)$$

$$RO + O_2 \xrightarrow{k_3} R'CHO + HO_2, \qquad (R\,119)$$

$$HO_2 + NO \xrightarrow{k_4} NO_2 + OH, \qquad (R\,120)$$

$$HO_2 + HO_2 \xrightarrow{k_5} H_2O_2 + O_2, \qquad (R\,121)$$

$$NO_2 + OH + M \xrightarrow{k_6} HNO_3 + M. \qquad (R\,122)$$

1. Reactions (R 117)–(R 118)–(R 119)–(R 120) are assumed to be fast reactions. Using a quasi steady-state assumption for species RO_2, RO and HO_2, calculate the reaction rates. It is assumed that ω_5 can be neglected with respect to ω_4.
2. Express ozone production P_{O_3}, which is supposed to be directly related to that of NO_2, as a function of $[HO_2]$ and [NO].

3. The concentration of HO_2 can be calculated from the production of radicals HO_x, P_{HO_x}, supposed to be given (defined by the oxidizing power of the environment). Calculate P_{HO_x} by using a quasi steady state assumption for HO_x. Show that $[HO_2]$ can be calculated as the root of a quadratic equation.
   Hint: calculate [OH] as a function of $[HO_2]$, [NO] and [RH].
4. It is not desired to solve the quadratic equation. Distinguish the *low* and *high* NO_x regimes and calculate ozone production.

Solution:

1. The reaction rates are equal: $\omega_1 \simeq \omega_2 \simeq \omega_3 \simeq \omega_4$.
2. Ozone production is equal to that of NO_2 and is therefore given by

$$P_{O_3} \simeq \omega_2 + \omega_4 \simeq 2\omega_4 = 2k_4[HO_2][NO].$$

3. The budget for HO_x due to the propagation reactions (R 117)–(R 118)–(R 119)–(R 120) is null:

$$-\omega_1 + \omega_3 - \omega_4 + \omega_4 = -\omega_1 + \omega_3 \simeq 0.$$

   There is a loss in the HO_x radicals due to reactions (R 121)–(R 122): $-\omega_5 - \omega_6$. The equilibrium between loss and production (using QSSA) reads

$$P_{HO_x} \simeq \omega_5 + \omega_6 = 2k_5[HO_2]^2 + k_6[NO_2][OH].$$

   Using QSSA for the propagation reactions and $\omega_1 \simeq \omega_4$ yields

$$[OH] \simeq \frac{k_4[HO_2][NO]}{k_1[RH]},$$

   and finally

$$P_{HO_x} \simeq 2k_5[HO_2]^2 + \frac{k_4k_6[NO_2][NO]}{k_1[RH]}[HO_2].$$

   This defines a quadratic equation in $[HO_2]$. It is easy to check with $P_{HO_x} > 0$ that there exists a unique positive root.
4. Under *low*-NO_x conditions, $[HO_2] \propto \sqrt{P_{HO_x}}$, which leads for ozone production to

$$P_{O_3} \propto \sqrt{P_{HO_x}}[NO].$$

   This regime is also referred to as NO_x-limited regime because ozone production is limited by NO_x concentrations (case B in Fig. 4.10).
   Under *high*-NO_x conditions, $[HO_2] \propto P_{HO_x}\frac{[RH]}{[NO][NO_2]}$, and thus for ozone production

$$P_{O_3} \propto P_{HO_x}\frac{[RH]}{[NO_2]}.$$

   This regime is referred to as VOC-limited regime, because ozone production is limited by VOC concentration (case A in Fig. 4.10).

To Know More ([77]):

L. KLEINMAN, *Low and high* NO_x *tropospheric photochemistry*, J. Geophys. Res., **99** (1994), pp. 16831–16838

Problem 4.3 (Oxidizing Power of the Atmosphere and CO) Carbon monoxide (CO) plays a leading role for the oxidizing power of the atmosphere. It is usually estimated that more than one half of the tropospheric sinks of OH is given by the oxidation reaction of CO,

$$CO + OH \xrightarrow{O_2} CO_2 + HO_2.$$

1. Using the data, calculate the lifetime of CO.
2. The oxidation reaction of OH is assumed to be the only sink of OH. Calculate, in a first approximation, the impact of a division by a factor 2 of the "technological" sources of CO.
3. Actually, detailed models which take into account the whole oxidation processes show that the decrease in CO concentration is only of about 15%. Why?

Indicative data:

- primary sources of CO: 1185 $Tg\,yr^{-1}$ (including 500 related to "technological" sources);
- secondary sources associated to VOC oxidation: 1270 $Tg\,yr^{-1}$;
- atmospheric burden of CO: 350 Tg.

Solution:

1. Writing the global budget of CO, $\tau = m_{CO}/E \simeq 50$ days, where E is the annual mass emissions and m_{CO} is the atmospheric burden of CO.
2. Let us neglect the photochemical feedbacks, which means that τ and the secondary sources of CO are supposed to be constant. Thus,

$$\delta m_{CO}/m_{CO} = -\delta E/E \simeq -20\%.$$

3. The decrease in the concentration of CO results in an increase of OH concentrations (from 3 to 4%). There is therefore an increasing oxidation of VOCs, which leads to an increase in the secondary chemical sources of CO. Thus, the decrease in CO concentration is lower.

To Know More ([73, 111]):

M. Kanakidou and P. Crutzen, *The photochemical source of carbon monoxide: importance, uncertainties and feedbacks*, Chemosphere: Global Change Sci., **1** (1997), pp. 91–109

N. Poisson, M. Kanakidou, and P. Crutzen, *Impact of non-methane hydrocarbons on tropospheric chemistry and the oxidizing power of the global troposphere: 3-dimensional modelling results*, J. Atmos. Chem., **36** (2000), pp. 157–230

Chapter 5
Aerosols, Clouds and Rains

Chapter 4 has introduced the main aspects of gas-phase atmospheric chemical kinetics. Many processes actually imply condensed matter: for instance, heterogeneous processes govern the chemical kinetics of stratospheric ozone.

Condensed matter consists of *suspended particles*, in liquid or solid forms. One usually distinguishes the condensed forms of water (raindrops, cloud drops, graupel), sometimes referred to as *hydrosols*, from the remaining part of the condensed matter (liquid and solid particles), referred to as *aerosols*.

A few multiphase processes play a leading role for air quality:

- Aerosols can have a direct adverse effect on human health. This is the case for the finest particles, not filtered by the respiratory system (Table 5.1).
- Aerosols interact with the gaseous phase through condensation and evaporation. This is a key process for the fate of organic atmospheric matter. Volatile organic compounds are oxidized in the gaseous phase (Chap. 4), which leads to the formation of semi-volatile organic compounds that can condense on aerosols. Once bound to aerosols, they are thereafter scavenged by precipitations. The organic matter would otherwise accumulate in the atmosphere, which would result in toxic concentrations.
- Aerosols govern the formation of clouds by providing *cloud condensation nuclei* for water vapor.
- Aerosols modify the radiative properties of the atmosphere either in a *direct* way or in an *indirect* way (Chap. 2).
- Clouds can be viewed as chemical reactors: mass transfer occurs between the gas phase and the cloud droplets and the aqueous-phase chemical reactions can provide major pathways for the formation of some species (such as sulfate).
- Rain is both a vector of pollution (*acid rains*) and a key tool for the *scavenging* of the atmosphere (*wet removal* of gases and aerosols).

The objective of this chapter is not to describe the microphysics of clouds and aerosols. This is indeed the main part of atmospheric physics! Only a few key facts strongly related to air quality are presented, concerning aerosol dynamics (evolution of the chemical composition and of the size distribution of aerosols), interactions with clouds and the role of rains (acid rains and scavenging).

B. Sportisse, *Fundamentals in Air Pollution*,

Table 5.1 Filtration of atmospheric particles by the human respiratory system as a function of the aerosol diameter d_p. A few studies have shown the "lung pollution" in a polluted area (a lung pollution of $PM_{10} \simeq 70\,\mu g\,m^{-3}$ in Mexico City to be compared to $PM_{10} \simeq 15\,\mu g\,m^{-3}$ in Vancouver; [19]). Source: [94]

nose and pharynx	filtration of particles with $d_p \geq 5\,\mu m$
trachea and bronchial tube	filtration of particles with $d_p \in [1, 5]\mu m$
lung alveoli	residence of fine particles with $d_p \leq 1\,\mu m$

This chapter should be viewed as an introduction to multiphase processes in the atmosphere. It is organized as follows. The aerosols are briefly presented in Sect. 5.1. The major chemical and physical processes occurring inside clouds are detailed in Sect. 5.2 with a focus on cloud condensation nuclei. The role of rains is investigated in Sect. 5.3 with two examples: formation of acid rains and wet scavenging by clouds and rains.

The radiative properties of the condensed matter (aerosols and clouds) have already been studied in Chap. 2.

5.1 Aerosols and Particles

5.1.1 General Facts

The liquid and solid particles that are suspended in the air are called *aerosols*. They are usually split from the condensed forms of the atmospheric water (cloud drops, raindrops, graupel), as detailed in Sect. 5.2.3.2.

Aerosols are classified according to different factors such as emissions, chemical composition, size distribution and mixing state.

5.1.1.1 Emissions

As for the gas-phase species one distinguishes the *primary* aerosols (directly emitted) from the *secondary* aerosols (formed in the atmosphere). Emissions can be either related to anthropogenic sources (for instance, soot issued from combustion) or to biogenic processes (sea salts, mineral aerosols—dust—, vegetal particles). The formation of secondary aerosols mainly results from the condensation of semi-volatile gas-phase species onto preexisting particles.

The estimation of aerosol emissions into the atmosphere is given in Table 5.2. The biogenic sources (dust and sea salts) are dominant as far as the aerosol mass is concerned. One can notice the well-balanced repartition of emissions related to biomass burning between the two hemispheres; on the other hand, the aerosols related to fossil fuel combustion are mostly emitted in the Northern hemisphere. These emissions are highly uncertain, especially for sea salts or mineral aerosols.

Table 5.2 Estimation of aerosol emissions (in Tg year^{-1}) for the year 2000: total and hemispheric repartition. The range of uncertainties is indicated inside brackets. Source: [106]

Type	total		North	South
Sea salt	3344	[1000, 6000]	43%	57%
diameter < 1 μm	54	[18, 100]		
diameter in [1, 16] μm	3290	[1000, 6000]		
Mineral aerosol (dust)	2150	[1000, 3000]	84%	16%
diameter < 1 μm	110			
diameter < [1, 2] μm	290			
diameter in [2, 20] μm	1750			
Organic aerosol				
biomass burning	54	[45, 80]	50%	50%
fossil fuel combustion	28	[10, 30]	98%	2%
biogenic	56	[0, 90]	98%	2%
Elemental carbon	12.3			
biomass burning	5.7	[5, 9]	50%	50%
fossil fuel combustion	6.6	[6, 8]	98%	2%
Industrial emissions	100	[40, 130]		

Biogenic processes are difficult to represent. For example, sea-salt aerosols are generated over oceans by wind stress at the surface, according to two mechanisms: bubble bursting during whitecap formations and spume drops torn directly from the wave crest. The first mechanism, the so-called *indirect* mechanism, generates particles whose dry radius ranges from 0.5 μm to 5 μm. The second mechanism, the so-called *direct* mechanism, generates particles with a dry radius above 5 μm. The emitted particles are mainly composed of sodium and chloride, but also of sulfate, potassium, magnesium and calcium. The averaged emissions of sea salts over Europe for the year 2001 are shown in Fig. 5.1, as calculated by a commonly-used parameterization.

5.1.1.2 Chemical Composition

Aerosols that contain carbon are referred to as *organic aerosols* as opposed to *inorganic aerosols*. In practice, atmospheric aerosols are a complicated mixture of organic and inorganic species, soot (elemental carbon) and, eventually, liquid water (Fig. 5.2).

The averaged composition of the atmospheric aerosol column is shown in Table 5.3. The typical composition of urban aerosols is presented in Table 5.4. Note that the composition of a significant part of the mass is not identified by current measurement methods. Measurement artefacts, linked to surface processes on the filters or to enhancement of evaporation, can also alter the aerosol composition.

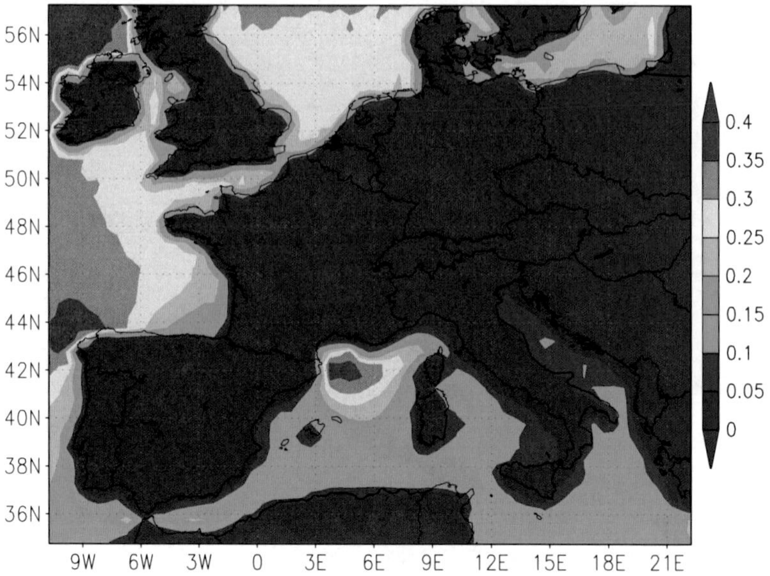

Fig. 5.1 Averaged sea salt emissions over Europe for the year 2001 (in $\mu g\,m^{-2}\,s^{-1}$). Parameterizations of the POLYPHEMUS system ([93]). Credit: Karine Sartelet, CEREA

Table 5.3 Chemical composition of the atmospheric aerosol column. Source: [8] (1995)

Species	Column (in $mg\,m^{-2}$)
sea salt	7.0
mineral (crustal)	36.1
sulfate	6.8
nitrate	1.3
biomass burning (soot)	3.9
organic fraction	4.5

The composition of marine aerosols is given in Table 5.4, showing, as expected, a major fraction of sodium and chloride.

5.1.1.3 Size

The size of atmospheric aerosols ranges from a few nanometers to tens of micrometers. We will not describe the coarsest aerosols ("vegetal" aerosols) and the focus is put on aerosols whose diameter is less than about 10 micrometers. Gravitational settling plays a leading role for the coarsest aerosols, which are usually quickly removed from the atmosphere (Exercise 5.1 and Fig. 5.3).

The notation PM_x stands for aerosols whose *aerodynamic* diameters are less than or equal to x µm. The aerodynamic diameter is defined as the diameter of a sphere

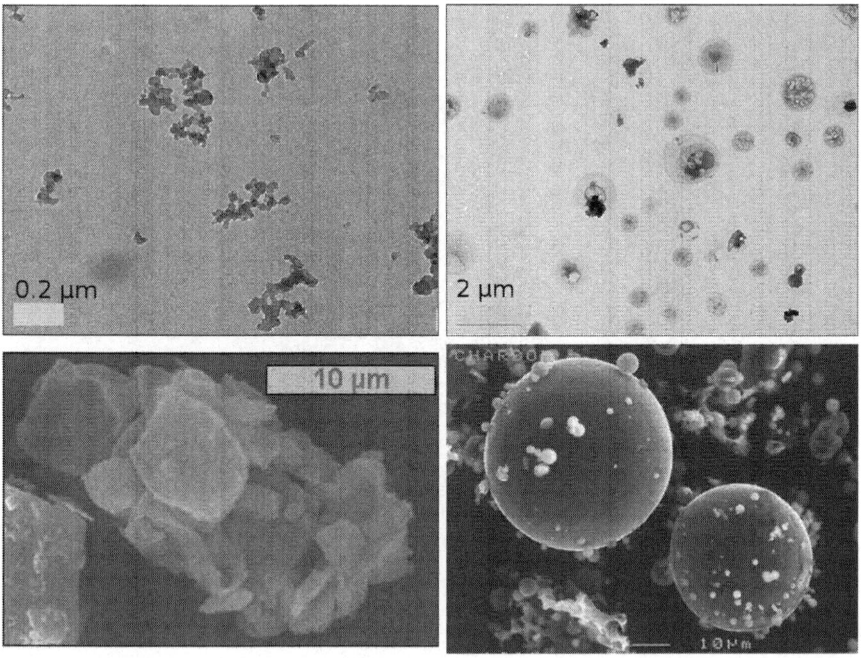

Fig. 5.2 Images of aerosols obtained by an electron microscope. *Top left*: urban aerosol (fine mode; measured in Paris, Primequal-Puffin project, 2007). *Top right*: urban aerosol (coagulation mode; same source). *Bottom left*: mineral aerosol. *Bottom right*: coal ash (measured in the Paris suburb). Credit: Annie Gaudichet, LISA (joint laboratory CNRS/Universities Paris XII and Paris VII)

Table 5.4 Left: averaged composition of the urban aerosols over Europe (PM_{10}). "Others" stand for the fraction that cannot be alloted to given species. Sources left: [113], right: averaged composition of the marine aerosols [130]

Urban aerosol		Marine aerosol	
Species	Composition (%)	Species	Composition (%)
organic compounds	20	chloride	55.0
nitrate	15	sodium	30.6
sulfate	13	sulfate	7.7
mineral	9	magnesium	3.7
ammonium	7	others	3.0
soot	5		
sea salt	4		
others	27		

with a density (mass per volume) of $1\,\mathrm{g\,cm^{-3}}$ and with a settling velocity equal to that of the particle. The aerodynamic diameter is very close to the diameter for small aerosols.

For regulatory purposes, it is usually desired to estimate $PM_{2.5}$ and PM_{10}.

Exercise 5.1 (Settling Velocity of a Particle) We investigate the settling of a spherical particle of diameter d_p and of density ρ_p. The forces applied to the particle are the gravity force and the *drag* force exerted by air. The drag force is upward oriented and is given, in a first approximation, by the Stokes law: $F \simeq 3\pi \mu_{air} d_p v_s$ with $v_s > 0$ the settling velocity of the particle (s for *sedimentation*) and μ_{air} the dynamic air viscosity.

Using the formulation of Newton's law, estimate v_s as a function of the particle diameter by calculating the stationary value. What is the value obtained for $d_p = 10\,\mu\text{m}$ and for $d_p = 100\,\mu\text{m}$ in the case of a water droplet?

Data: $\mu_{air} \simeq 1.8 \times 10^{-5}\,\text{kg}\,\text{m}^{-1}\,\text{s}^{-1}$.

Solution:

The first principle of dynamics (Newton's law) leads to

$$m_p \frac{\mathrm{d}v_s}{\mathrm{d}t} = m_p g - 3\pi \mu_{air} d_p v_s,$$

with $m_p = \rho_p \times \pi d_p^3/6$ the particle mass. A positive velocity is downward oriented. With the initial condition $v_s(0) = 0$ one easily gets

$$v_s(t) = \tau g \left(1 - \exp\left(-\frac{t}{\tau}\right)\right),$$

where $\tau = m_p/(3\pi \mu_{air} d_p) = d_p^2 \rho_p/(18\mu_{air})$. At equilibrium, this defines the so-called Stokes velocity,

$$v_s(\infty) = \tau g = \frac{d_p^2 \rho_p g}{18\mu_{air}}.$$

For a water droplet ($\rho_p = 1000\,\text{kg}\,\text{m}^{-3}$) with $d_p = 10\,\mu\text{m}$, we get $\tau \simeq 10^{-4}$ s and $\tau g \simeq 3 \times 10^{-3}\,\text{m}\,\text{s}^{-1}$. For a drop of diameter $d_p = 100\,\mu\text{m}$, $\tau \simeq 10^{-2}$ s and the Stokes velocity is $\tau g \simeq 0.3\,\text{m}\,\text{s}^{-1}$. This corresponds to a velocity of about 1 $\text{km}\,\text{h}^{-1}$. The coarsest particles are therefore quickly removed from the atmosphere.

However, the Stokes law only provides an approximation of the drag force, to be used for diameters ranging from 1 to 20 μm. For the smallest aerosols, splipping effects have to be taken into account and the drag force is much larger than the one computed with the Stokes law (three times larger for a diameter of 0.1 μm, 200 times larger for a diameter of 1 nm). One refers to Fig. 5.3 for the evolution of v_s as a function of particle diameter, as calculated by the appropriate laws.

To know more ([102]):

E. NASLUND AND L. THANING, *On the settling velocity in a nonstationary atmosphere*, Aerosol Sci. and Technol., **14** (1991), pp. 247–256

The partitioning of aerosols according to their size is described by a *distribution function*. Let d_p be the aerosol diameter. For example, the number distribution can be represented as a function $n(d_p)$, so that the number of aerosols per air volume with a diameter in $]D_{min}, D_{max}[$ is $\int_{D_{min}}^{D_{max}} n(d_p)\mathrm{d}(d_p)$. The unit of the function

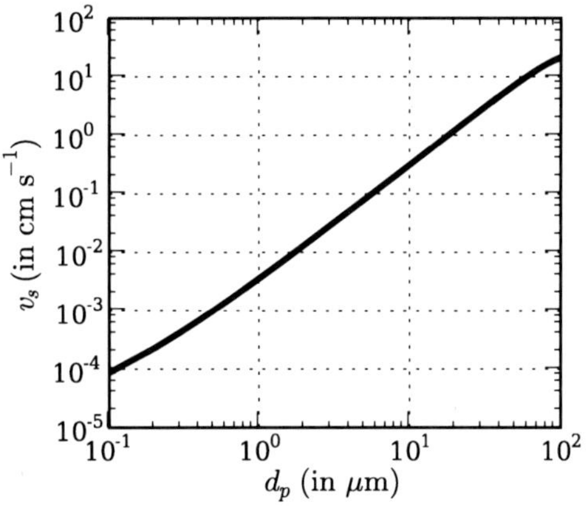

Fig. 5.3 Settling velocity of a water drop as a function of diameter. Source: [135]

n is then a number of aerosols per air volume and per aerosol diameter (usually in $\text{cm}^{-3}\,\text{nm}^{-1}$). Other size variables can be chosen for the distribution functions, e.g. the aerosol volume or the aerosol mass.

The measured distributions often reveal a superposition of a few *modes*, usually fitted to log-normal functions of the aerosol diameter (Fig. 5.4),

$$n(d_p) = \sum_{l=1}^{N_m} \frac{N_l}{\sqrt{2\pi}\,\ln(\sigma_l)} \exp\left[-\frac{1}{2}\left(\frac{\ln^2(d_p/D_l)}{\ln^2(\sigma_l)}\right)\right], \tag{5.1}$$

where, for the mode l, N_l is the number concentration, D_l is the median diameter and σ_l is the variance of the mode. The number of modes is typically 3 or 4. They are usually defined as:

- the *nucleation* mode ($D_1 \in [1, 10]$ nm),
- the *accumulation* mode ($D_2 \in [10, 100]$ nm),
- the *Aitken* mode ($D_3 \in [100, 1000]$ nm),
- and the *coarse* mode ($D_4 \in [1, 10]$ μm).

Each mode is governed by a given physical process that differs according to the mode's size (see Sect. 5.1.3 and Fig. 5.6).

5.1.1.4 Mixing State

The mixing state is a key feature of an aerosol distribution:

- *single internal mixing* refers to aerosols in which the chemical components are completely well mixed: there is only one aerosol "family" in this case.
- *multiple internal mixing* refers to the case where one considers a few families of well mixed aerosols. Each family is defined by its composition and size distribution.
- *external mixing* refers to aerosols in which the species are not mixed.

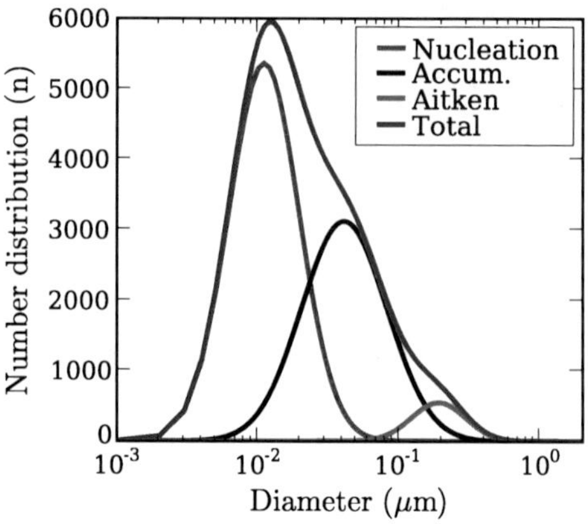

Fig. 5.4 Aerosol number distribution. Measured data in Leipzig (Germany). Source: [113]

Except near emissions, the assumption of external mixing is not realistic because there are processes, such as coagulation, that mix aerosols. Field measurements confirm that atmospheric aerosols are internally mixed. The assumption of single internal mixing is also not realistic: the aerosol chemical composition results from ageing processes acting on it (from aerosol "history"). For instance, the coagulation of a soot and of a sulfate aerosol is likely to lead to the formation of an aerosol with a carbon *core* surrounded by an inorganic liquid film (*shell*). The most realistic assumption is then probably multiple internal mixing (Fig. 2.17, Chap. 2).

The mixing state is a decisive property for estimating the radiative impact of aerosols (Sect. 5.1.1.4).

5.1.2 Residence Time and Vertical Distribution

The atmospheric residence time of aerosols is governed by the loss processes, especially dry deposition and rain scavenging. The magnitude of these processes strongly depends on the size distribution (Sect. 5.3.2 and Fig. 3.24).

The *empirical* formula of Jaenicke ([68]) gives the residence time τ as a function of the aerosol diameter d_p with

$$\frac{1}{\tau(d_p)} = \frac{1}{\tilde{\tau}}\left(\frac{d_p}{d_{max}}\right)^2 + \frac{1}{\tilde{\tau}}\left(\frac{d_{max}}{d_p}\right)^2 + \frac{1}{\tau_{wet}}, \tag{5.2}$$

where $\tilde{\tau} = 1.28 \times 10^8$ s and $d_{max} = 0.6\,\mu$m. The residence time associated to wet scavenging, τ_{wet}, is a function of altitude.

It is easy to prove that τ is first an increasing function and then a decreasing function of d_p. Its maximal value is obtained at d_{max} and is about τ_{wet} (Fig. 5.5). The parameter τ_{wet} varies with the altitude: τ_{wet} is about 1 day in the lower troposphere, 10 days in the free troposphere and 100 days in the lower stratosphere.

The residence time can be estimated on the basis of observational data. For example, the atmospheric aerosol column can be used for marine aerosols (Problem 5.1),

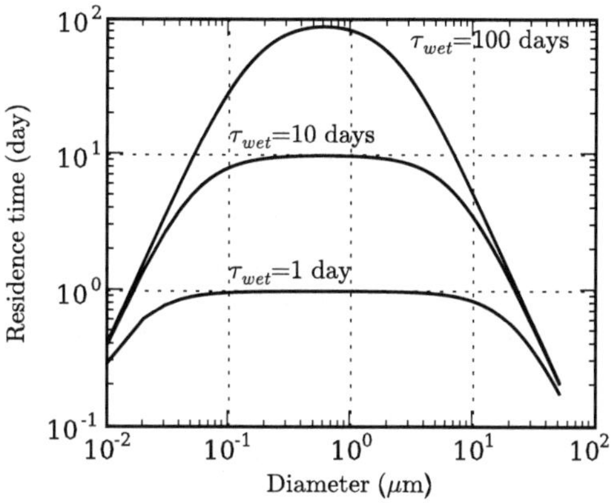

Fig. 5.5 Evolution of the residence time of aerosols as a function of the aerosol diameter for a few altitudes (fixing the value of τ_{wet}). Source: [68]

which leads to a residence time of about one day. Another approach is to investigate atmospheric *tracers* bound to aerosols. One refers to Problem 1.4 with the example of the Radon radioactive chain (lead, Bismuth and Polonium). The residence time of the continental tropospheric aerosol is then estimated to be about ten days.

Because of the low residence times, the vertical aerosol distribution strongly decreases with altitude. However, observational data indicate that there is a layer of sulfate aerosols in the lower stratosphere (at the altitude of 20 kilometers): this layer is usually referred to as the Junge layer.[1] Because the residence time of tropospheric aerosols is about ten days, the formation of this layer can only be explained by an *in situ* production of sulfate aerosols in the stratosphere. The reservoir species of SO_2 (Sect. 4.1.4.5) is carbonyl sulfide (OCS), whose residence time is large enough for reaching the stratosphere (Exercise 5.2). Its photolysis leads to the formation of SO_2 according to the mechanism

$$OCS + h\nu \longrightarrow CO + S, \quad \text{(R 123)}$$

$$O + OCS \longrightarrow CO + SO, \quad \text{(R 124)}$$

$$S + O_2 \longrightarrow SO + O, \quad \text{(R 125)}$$

$$SO + O_2 \longrightarrow SO_2 + O, \quad \text{(R 126)}$$

$$SO + NO_2 \longrightarrow SO_2 + NO. \quad \text{(R 127)}$$

The oxidation of SO_2 leads thereafter to the formation of sulfuric acid (H_2SO_4) that will condense to form sulfate aerosols.

The lack of scavenging processes in the stratosphere contributes to the stability of the Junge layer. Direct injections of SO_2 into the stratosphere, during volcanic events (Pinatubo, El Chichon, etc.), can strengthen the Junge layer.

[1]Named from the atmospheric scientist that measured it in the late 1950s and early 1960s.

Exercise 5.2 (Atmospheric Residence Time of Sulfur)

1. A maximal bound for the tropospheric residence time of sulfur dioxide (SO_2) can be estimated by investigating dry deposition. An averaged value for the dry deposition of SO_2 is $v_{dep}^{SO_2} \simeq 0.8\,\mathrm{cm\,s^{-1}}$. What is the residence time of SO_2 in the lower troposphere? Take a characteristic height $H \simeq 5\,\mathrm{km}$.
2. The mixing ratio of carbonyl sulfide (OCS) is about 500 pptv and the yearly emissions are estimated to be $E \simeq 0.7\,\mathrm{Tg\,year^{-1}}$. Estimate its atmospheric residence time. Conclude for the Junge layer.

Data: $M_{\mathrm{OCS}} = 60\,\mathrm{g\,mol^{-1}}$.

Solution:

1. The corresponding residence time is $H/v_{dep}^{SO_2}$ (Exercise 1.8), which leads to a value of about one week. This estimation is a maximal bound because one does not take into account other loss processes, such as oxidation (see Exercise 5.10 for the case of acid rains).
2. The methodology is similar to that used in Exercise 1.9. The total mass of OCS is $0.5 \times 10^{-12} \times (M_{\mathrm{OCS}}/M_{air})m_{atm} \simeq 5.4\,\mathrm{Tg}$. The residence time is therefore 5.4/0.7, namely 7 or 8 years. OCS is therefore a reservoir of sulfur and can reach the stratosphere, providing the main source for stratospheric sulfate.

5.1.3 Aerosol Dynamics

5.1.3.1 Phenomenology of Processes

The evolution of size distribution and chemical composition of aerosols is governed by many processes (Fig. 5.6).

Nucleation The formation of the smallest aerosols is related to the aggregation of gaseous molecules. The resulting *clusters*, when thermodynamically stable, have a characteristic size in [0.1, 1] nm. This process is referred to as *nucleation*.

The binary nucleation of water and sulfuric acid (H_2O–H_2SO_4), and the ternary nucleation of water, sulfuric acid and ammonia (H_2O–H_2SO_4–NH_3) are relatively well understood and often taken into account in models.

Coagulation The coagulation of aerosols is governed by aerosol motions. For atmospheric aerosols one usually only describes Brownian coagulation, which is related to thermal motions. Other motions can be taken into account: they are referred to as *phoretic effects*, namely those related to gradients of temperature and to electromagnetic fields.

Aerosol coagulation can be neglected for aerosols whose diameters are above a few micrometers.

Condensation and Evaporation Gas-phase compounds with low saturation vapor pressure can condense onto existing aerosols. Condensed matter can also evaporate from the particulate phase to the gaseous phase. This mass transfer is governed

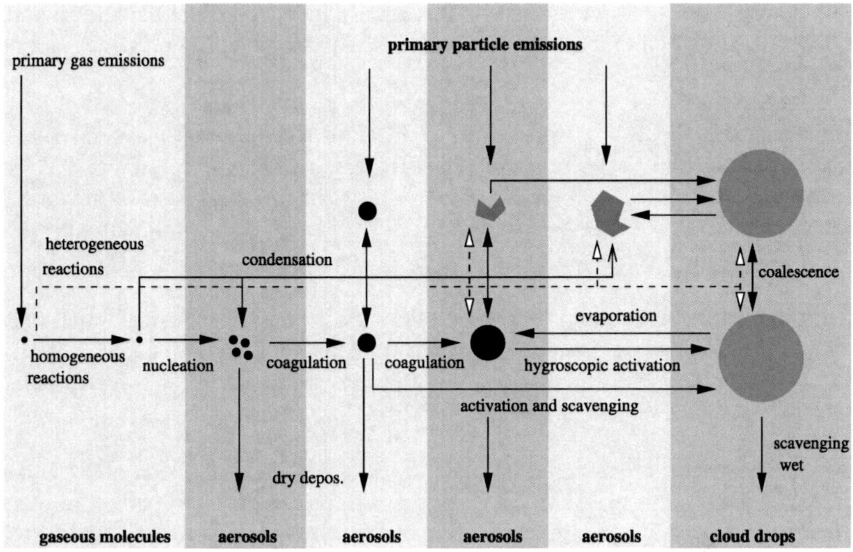

Fig. 5.6 Sketch of aerosol dynamics. Source: [117]

by the concentration gradient between the gaseous phase far from the aerosol and at the aerosol surface. The gas-phase concentration at the aerosol surface is supposed to be at *local thermodynamic equilibrium* with the aerosol mixture.

Condensation is a fast process for submicronic aerosols and a slow process for coarser aerosols. Aerosols growing by condensation tend to accumulate between diameters of a few nanometers to one micrometer. They define the so-called *accumulation mode*.

The composition of continental inorganic aerosols is deeply related to the interaction between sulfate, nitrate and ammonium. The gaseous precursors are respectively:

- sulfur dioxide (SO_2) which is oxidized in sulfuric acid (H_2SO_4) whose saturation vapor pressure is about zero (H_2SO_4 is therefore only in the particulate phase);
- nitric acid (HNO_3) which results from the oxidation of NO_x;
- ammonia (NH_3), mainly related to agriculture-induced emissions.

Inside aerosols nitrate and sulfate compete with respect to ammonium for the formation of ammonium nitrate ($NH_4NO_3 = NH_4^+ + NO_3^-$) or of ammonium sulfate ($(NH_4)_2SO_4 = 2NH_4^+ + SO_4^{2-}$). A key point is that the competition is in favor of the second salt.

One usually defines two regimes according to the value of the so-called *ammonium richness* (other indicators can also be used):

- if the environment is ammonium-rich, ammonium sulfate is formed and then ammonium nitrate with the remaining part of ammonium. This leads to weakly acidic aerosols (even to neutral aerosols).

- if the environment is ammonium-poor, there is not enough ammonium to neutralize the sulfate and the resulting aerosols are acid. Ammonium nitrate cannot be formed.

Mass Transfer to Cloud Drops Part of the aerosol distribution plays the role of condensation nuclei for water vapor (CCN for *Cloud Condensation Nuclei*). These aerosols are referred to as *activated aerosols*. This is a major process for the formation of cloud drops because the condensation of water vapor cannot occur without a condensation surface (Sect. 5.2).

Heterogeneous Reactions Reactions at the surface of condensed matter (particles and cloud or fog drops) can have a significant impact on gas-phase photochemistry and on aerosols. These processes, still highly uncertain, are usually described by first-order chemical mechanisms.

Loss Processes Aerosols are scavenged by rains and deposited at ground. *Gravitational settling* (or *sedimentation*) is also a major process for the coarsest aerosols (Exercise 5.1).

Aerosol dynamics is detailed in the sequel with the following processes: nucleation, condensation/evaporation and coagulation. Aerosol activation is described in Sect. 5.2 and loss processes in Sect. 5.3.

5.1.3.2 Notations

Aerosols are supposed to be spherical (see Exercise 5.3 for the case of soot).

Internal Mixing and External Mixing Aerosols are supposed to be *internally mixed* with a single family of aerosols. This means that there exists a unique chemical composition for a fixed size. Real-life aerosol populations are more complicated: there exist several chemical compositions for a given size, due to the different ageing processes leading to the formation of aerosols (primary aerosols, aerosols resulting from coagulation, aerosols resulting from condensation of semi-volatile gases, etc.).

Aerosol Distributions Let $\{X_i\}_{i=1,n_s}$ be a set of n_s species that constitute an internal mixture of aerosols. Aerosols are described by:

1. a size distribution, let us say $n(\star, t)$ for the number distribution, a function of a size variable $\star$ (volume v, diameter d_p or dry mass m) and of time. The unit of n is a number of aerosols per unit of air volume and per unit of size variable.
2. a chemical composition as a function of size, $\{q_i(\star, t)\}_{i=1,n_s}$, where $q_i(\star, t)$ is the mass distribution of species X_i $(1 \le i \le n_s)$ inside aerosols of size $\star$. The unit of q_i is a unit of mass per unit of air volume and per unit of size variable.

The mass distributions satisfy

$$\sum_{i=1}^{n_s} q_i = mn. \tag{5.3}$$

We also define the mass $m_i(m,t)$ of species X_i in aerosols of mass m with

$$m_i(m,t) = \frac{q_i(m,t)}{n(m,t)}. \tag{5.4}$$

Mass conservation for aerosols of mass m is given by

$$\sum_{i=1}^{n_s} m_i(m,t) = m. \tag{5.5}$$

Chemical Composition The following compounds are taken into account:

- liquid water;
- inert species (or species supposed to be inert): crustal species (MD for *Mineral Dust*), soot or elemental carbon (EC for *Elemental Carbon*; this can be also referred to as *Black Carbon*) and heavy metals (such as lead or cadmium);
- for liquid aerosols, inorganic compounds in ionic form in an aqueous-phase solution: sodium (Na^+), sulfate (SO_4^{2-}), ammonium (NH_4^+), nitrate (NO_3^-), chloride (Cl^-);
- for solid aerosols, salts such as sodium chloride NaCl, ammonium chloride NH_4Cl or ammonium nitrate NH_4NO_3;
- organic compounds: primary organic matter (POA for *Primary Organic Aerosol*) and secondary organic matter (SOA for *Secondary Organic Aerosol*).

Exercise 5.3 (Fractal Dimension of Soot) Assuming that soot particles are spherical is not realistic, especially for soots emitted by diesel engines.

1. Measurements indicate that the surfacic mass of soot is up to 100 g m^{-2}. Calculate the surfacic mass of an equivalent spherical particle. Suppose that the aerosol density is $\rho = 2\,\mathrm{g\,cm^{-3}}$ and the radius is 100 nm. Comment.
2. Observational data for soot (for instance with an electron microscope) show that soot is made of spherule clusters (Fig. 5.7). The fractal dimension $D_f \in [2,3]$ is defined from the surface (s) and the volume (v) as $s \sim v^{D_f/3}$. Calculate the ratio of the surface of a linear cluster ($D_f = 3$) to the surface in the spherical case ($D_f = 2$) for an aggregate of 1000 spherules with a fixed unitary volume ($v = 1$).

Solution:

1. The surfacic mass of a spherical particle of radius r and of density ρ is

$$4\pi r^2/(4/3\pi r^3 \times \rho) \simeq 3/(\rho r).$$

We obtain 15 g m^{-2}. This value is much smaller than the measured data. The discrepancy can be explained either by a smaller radius (one gets 150 g m^{-2} for particles with a radius of 10 nm) or by a fractal dimension, resulting in a larger surface for a fixed volume.

2. Let s_{D_f} be the surface associated to a fractal dimension D_f and V the total volume generated by the 1000 spherules ($V = 1000 \times v = 1000$). We obtain

$$\frac{s_3}{s_2} = \frac{V}{V^{2/3}} = V^{1/3} = 10.$$

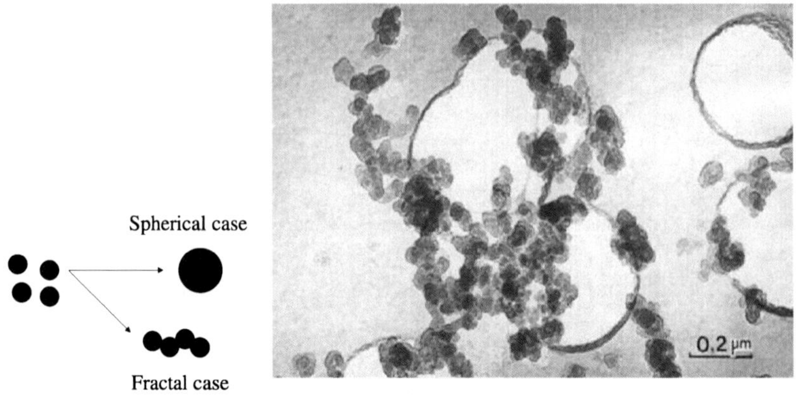

Fig. 5.7 *Left panel*: clustering of soot particles in the spherical case (fractal dimension $D_f = 2$) and in the fractal case (with $D_f = 3$), Exercise 5.3. *Right panel*: observation of micro soots with an electron microscope. Credit: Annie Gaudichet, LISA (joint laboratory CNRS/Universities Paris XII and Paris VII)

The available surface (especially for condensation) is much larger in the fractal case: the surface linearly increases with volume and similarly with the number of spherules.

To know more ([133, 109]):
C. SORENSEN, *Light scattering by fractal aggregates: a review*, Aerosol Sci. and Technol., **35** (2001), pp. 648–687 (for physics and measurement techniques)
S. PARK, S. ROGAK, W. BUSHE, J. WEN, AND M. THOMSON, *An aerosol model to predict size and structure of soot particles*, Combust. Theory Model., **9** (2005), pp. 499–513 (for modeling)

5.1.3.3 General Dynamic Equation for Aerosols

The time evolution of the number distribution and of the mass distributions of chemical compounds in the particulate phase is governed by the *General Dynamic Equation* (GDE). The size variable is the dry mass m.

The following notations are used (the parameterizations are detailed below for each process).

- The nucleation threshold, m_0, is the dry mass of the smallest aerosol that is thermodynamically stable, in practice fixed by nucleation.
- $J_0(t)$ is the nucleation rate, expressed in terms of number of (nucleated) aerosols per unit of time.
- $I_i(m,t)$ is the condensation/evaporation (c/e) rate for semi-volatile species X_i in an aerosol of dry mass m. It is expressed in unit of mass per unit of time.
- The total condensation rate, I_0, is defined by

$$I_0(m,t) = \sum_{i=1}^{n_s} I_i(m,t). \tag{5.6}$$

- The coagulation kernel $K(m_1, m_2)$ describes the coagulation between two aerosols of dry mass m_1 and m_2, and is expressed in unit of volume per unit of time.

Number Distribution The equation for the number distribution is given by

$$\frac{\partial n}{\partial t}(m,t) = \underbrace{\theta(m \geq 2m_0)\frac{1}{2}\int_{m_0}^{m-m_0} K(\tilde{m}, m-\tilde{m})n(\tilde{m},t)n(m-\tilde{m},t)\,\mathrm{d}\tilde{m}}_{\text{coagulation gain}}$$
$$\underbrace{- n(m,t)\int_{m_0}^{\infty} K(m,\tilde{m})n(\tilde{m},t)\,\mathrm{d}\tilde{m}}_{\text{coagulation loss}} - \underbrace{\frac{\partial(I_0 n)}{\partial m}}_{\text{c/e}} + \underbrace{\delta_{(m,m_0)}J_0(t)}_{\text{nucleation}}. \quad (5.7)$$

The function $\theta(\mathbf{A})$ is dimensionless. Its value is 1 if the logical expression $\mathbf{A}$ is met, 0 otherwise. Moreover, δ stands for the Dirac function and is expressed in inverse unit of mass.

Chemical Composition Similarly, we obtain for species labelled by $i = 1, \ldots, n_s$,

$$\frac{\partial q_i}{\partial t}(m,t) = \underbrace{\theta(m \geq 2m_0)\int_{m_0}^{m-m_0} K(\tilde{m}, m-\tilde{m})q_i(\tilde{m},t)n(m-\tilde{m},t)\,\mathrm{d}\tilde{m}}_{\text{coagulation gain}}$$
$$\underbrace{- q_i(m,t)\int_{m_0}^{\infty} K(m,\tilde{m})n(\tilde{m},t)\,\mathrm{d}\tilde{m}}_{\text{coagulation loss}}$$
$$- \underbrace{\frac{\partial(I_0 q_i)}{\partial m}}_{\text{advection by c/e}} + \underbrace{(I_i n)(m,t)}_{\text{mass transfer by c/e}} + \underbrace{\delta_{(m,m_0)}m_i(m_0,t)J_0(t)}_{\text{nucleation}}. \quad (5.8)$$

Note that, for condensation/evaporation, the advection term (corresponding to a shift along the distribution) should be distinguished from the mass transfer term which actually corresponds to a flux between the gaseous and particulate phases.

5.1.4 Parameterizations

Readers who are not interested in modeling may omit this (rather technical) section.

5.1.4.1 Coagulation

The expression of the coagulation kernel is simpler when the aerosol diameter d_p is chosen as the size variable. The computation of the wet mass (the dry mass plus the mass of the aerosol liquid water) is given by

$$m_p = \pi \frac{d_p^2}{6} \rho_p, \quad (5.9)$$

where ρ_p is the aerosol density (supposed to be fixed or depending on the chemical composition).

Let K_{12} be the coagulation kernel between aerosols of diameters d_{p_1} and d_{p_2}. In the case of Brownian coagulation, different regimes are distinguished depending on the aerosol size. The *mean free path* of a gas is defined as the mean distance between two collisions of gaseous molecules. The air mean free path, written as λ_{air}, is given by

$$\lambda_{air} = \frac{2\mu_{air}}{P} \sqrt{\frac{\pi RT}{8M_{air}}}, \quad (5.10)$$

where μ_{air} is the dynamic viscosity of air. For $T = 298$ K and $P = 1$ atm, one gets $\lambda_{air} \simeq 65$ nm.

The coagulation regimes are defined in the following way:

- **Continuum Regime**
  If d_{p_1} and d_{p_2} are much larger than λ_{air}, the coagulation kernel is

$$K_{12} = 2\pi(D_1 + D_2)(d_{p_1} + d_{p_2}). \quad (5.11)$$

  D_1 and D_2 are the diffusion coefficients in the air for aerosols of diameters d_{p_1} and d_{p_2}, respectively, given by

$$D_i = \frac{k_B T}{3\pi \mu_{air} d_{p_i}}, \quad (5.12)$$

  where k_B is the Boltzmann constant.
- **Free Molecular Regime**
  If d_{p_1} and d_{p_2} are much lower than λ_{air}, the coagulation kernel is

$$K_{12} = \frac{\pi}{4}(d_{p_1} + d_{p_2})^2 \sqrt{(\bar{v}_1^2 + \bar{v}_2^2)}, \quad (5.13)$$

  where m_{p_i} and $\bar{v}_i$ stand for the mass and the thermal velocity of the aerosol i, respectively, with

$$\bar{v}_i = \sqrt{\frac{8k_B T}{\pi m_{p_i}}}. \quad (5.14)$$

- **Transition Regime**
  If d_{p_1} and d_{p_2} have a value similar to the value of λ_{air}, the coagulation kernel of the continuum regime has to be corrected with a multiplying coefficient ([41]). One refers to [130] for the analytical formula (not appropriate for this presentation).

The coagulation kernel has a maximal value when the colliding aerosols have very different sizes. Because the small aerosols actually have a large thermal velocity but the surface available for collision is low, the best trade-off is obtained for collisions between smallest and coarsest aerosols (see Fig. 5.8).

5.1.4.2 Condensation and Evaporation

The mass flux I_i is parameterized by

$$I_i = \underbrace{2\pi D_i^g d_p f(K_{n_i}, \alpha_i)}_{\text{condensation kernel}} \big(c_i^g - c_i^s(d_p, t)\big) \tag{5.15}$$

where D_i^g and c_i^g stand for the air diffusivity and the gas-phase concentration of species X_i, respectively. The concentration c_i^s corresponds to the concentration at the aerosol surface and is supposed to be at *local* thermodynamic equilibrium with the aerosol chemical composition, namely

$$c_i^s(d_p, t) = \eta(d_p)\, c_i^{eq}(m_1, \ldots, m_{n_s}, RH, T). \tag{5.16}$$

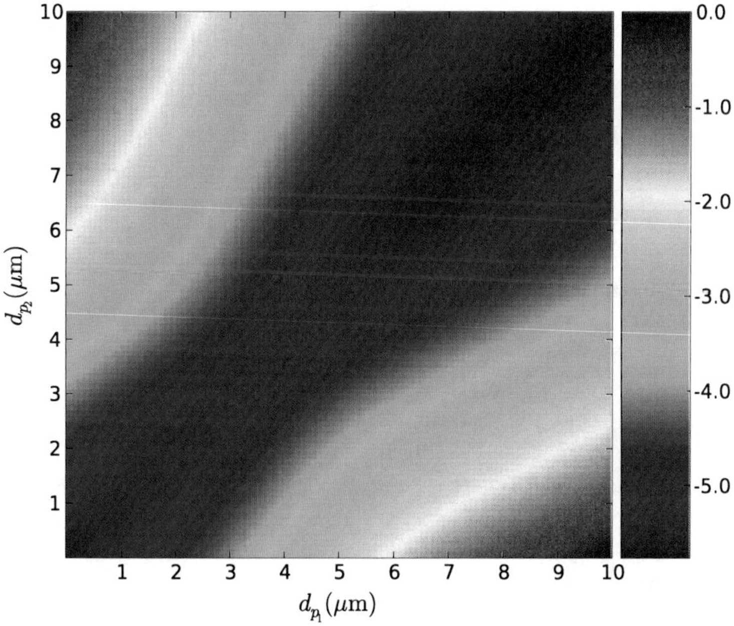

Fig. 5.8 Distribution of the coagulation kernel $K(d_{p_1}, d_{p_2})$ as a function of the diameters of the colliding particles. The map corresponds to $\log_{10}(K(d_{p_1}, d_{p_2})/K_{max})$ where K_{max} is the maximal value obtained for a pair of diameters (expressed in μm) $(d_{p_1}, d_{p_2}) \in [0.01, 10] \times [0.01, 10]$. Note the symmetry: $K(d_{p_1}, d_{p_2}) = K(d_{p_2}, d_{p_1})$

RH is the relative humidity (defined in Sect. 5.2) and T is the temperature. The coefficient $\eta(d_p)$ describes the curvature effect of aerosols (the so-called Kelvin effect, Sect. 5.2). It is given by

$$\eta(d_p) = \exp\left(\frac{4\sigma M_p}{k_B T d_p \rho_p}\right), \tag{5.17}$$

where σ is the surface tension. M_p is the molecular weight of the aerosol, calculated from its composition (see Sect. 5.2 and (5.42)).

The function f is the Fuchs-Sutugin function. It describes noncontinuum effects ([29]). It depends on the Knudsen number for species X_i, K_{n_i}, and on the accommodation coefficient α_i,

$$f(K_{n_i}, \alpha_i) = \frac{1 + K_{n_i}}{1 + 2K_{n_i}(1 + K_{n_i})/\alpha_i}, \qquad K_{n_i} = \frac{2\lambda_i}{d_p}, \tag{5.18}$$

with λ_i the air mean free path for species X_i.

For large values of K_{n_i}, f is nearly equal to $\alpha_i / 2K_{n_i}$, which leads to the following expression for mass flux in the free molecular regime,

$$I_i = \alpha_i \frac{\bar{v}_i^g}{4} \pi d_p^2 \left(c_i - c_i^s(d_p)\right). \tag{5.19}$$

The thermal velocity of a gas-phase species X_i in the air, $\bar{v}_i^g$, satisfies

$$D_i^g = \frac{\lambda_i \bar{v}_i^g}{2}. \tag{5.20}$$

Typical values of these parameters are summarized in Table 5.5.

5.1.4.3 Local Thermodynamic Equilibrium

Solving the mass transfer between the gaseous and particulate phases requires to compute the concentration at thermodynamic equilibrium, namely $c_i^{eq}(m_1, \dots, m_{n_s}, RH, T)$.

This local thermodynamical equilibrium is defined by chemical equilibria of the aerosol compounds. Most state-of-the-art models do not take into account mixing

Table 5.5 Thermal velocity and gas-phase diffusivity for a few species at $T = 300$ K and $P = 1$ atm

Species	$\bar{v}_i^g$ (m s^{-1})	D_i^g (m^2 s^{-1})
sulfate	254.58	1.07×10^{-5}
ammonium	611.24	2.17×10^{-5}
nitrate	317.51	1.47×10^{-5}
chloride	417.15	1.72×10^{-5}

between organic and inorganic species. Thermodynamic equilibria are usually defined independently for the organic fraction and the inorganic fraction.

This assumption is a weakness of current models because the aerosol behavior with respect to water is not well described: the issue is indeed to distinguish the hydrophilic (the species that favor condensation of water vapor) from the hygroscopic species. When the organic and the inorganic species are not thermodynamically coupled, one often assumes that only the inorganic species are hydrophilic. Recent studies indicate the role of hydrophilic secondary organic aerosols.

The aerosol state, liquid or solid, depends on the relative humidity and on the properties of chemical compounds with respect to water condensation. The *deliquescence relative humidity* of a given compound (*DRH*) is defined as the relative humidity above which water vapor begins to condense on the aerosol, which becomes therefore liquid. Compounds of low *DRH* enhance the condensation of water.

When the humidity decreases, the aerosol does not become solid when *RH* reaches *DRH*. The aerosol is still a liquid aerosol until the so-called *crystallization relative humidity*, *CRH*. This hysteresis phenomenon is illustrated in Fig. 5.9. One also refers to Table 5.6 for a few values of *DRH* and *CRH*. A simple computation of equilibrium conditions is given in Exercise 5.4 for the case of a solid salt of ammonium nitrate.

The deliquescence properties of a mixture of salts can be obtained from those of the individual salts. For example, the *DRH* of a mixed salt (usually referred to as *MDRH*, *M* for *mutual*) is always lower than the lowest *DRH* of the individual salts.

Exercise 5.4 (Ammonium Nitrate at Thermodynamic Equilibrium) This exercise aims at illustrating an equilibrium condition in a simple case, with a salt of ammonium nitrate (NH_4NO_3).

Consider an environment with the following mixing ratios (in ppb) for ammonia, $C^o_{NH_3}$, and nitric acid, $C^o_{HNO_3}$, at $T = 298$ K and $RH = 40\%$.

1. Formulate thc condition required for having a solid aerosol.
2. The solid aerosol of ammonium nitrate is derived from the reaction

$$NH_3 + HNO_3 \rightleftharpoons NH_4NO_3,$$

whose equilibrium constant is $K_p = C_{NH_3} C_{HNO_3}$. Show that the salt can only be formed when an excess of ammonium and of nitrate is available.

Table 5.6 Estimation of *DRH* and *CRH* for a few salts at 298 K. Source: [67]

Salt	*DRH* (in %)	*CRH* (in %)
NH_4Cl	77	47
NH_4NO_3	61.8	[25, 32]
NH_4HSO_4	40	[0.05, 22]
$(NH_4)_2SO_4$	79.9	–

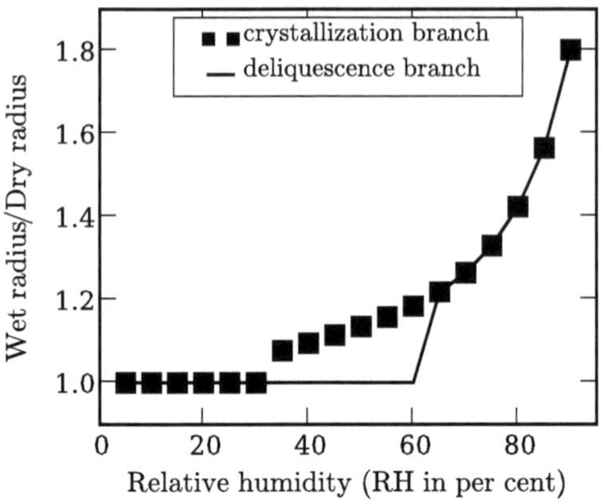

Fig. 5.9 Evolution of the ratio of the wet radius to the dry radius for an aerosol made of ammonium nitrate (NH_4NO_3), as a function of relative humidity. When the relative humidity increases, the aerosol absorbs water for $RH \geq DRH$; when the relative humidity decreases, the aerosol state is solid only when $RH \leq CRH$. Computed with ISORROPIA ([100]) at $T = 300\,\mathrm{K}$ with $[NH_3] = 0.34\,\mu\mathrm{g\,m^{-3}}$ and $[HNO_3] = 1.26\,\mu\mathrm{g\,m^{-3}}$. Credit: Édouard Debry, CEREA

Solution:

1. As $RH < DRH$ ($40 < 61.8$), the aerosol of ammonium nitrate is solid.
2. It is assumed that a molar fraction x of nitrate has taken part in the formation of the salt. Due to the stoichiometric relations, a molar fraction x of ammonium has also been consumed. Thus one gets at thermodynamic equilibrium,

$$K_p = (C^o_{\mathrm{NH_3}} - x)(C^o_{\mathrm{HNO_3}} - x).$$

It is easy to show, for instance by plotting in a figure, that this equation has two roots:

$$x_1 \leq \min(C^o_{\mathrm{NH_3}}, C^o_{\mathrm{HNO_3}}), \qquad x_2 \geq \max(C^o_{\mathrm{NH_3}}, C^o_{\mathrm{HNO_3}}).$$

The root x_2 does not have a physical meaning. The root x_1 can be accepted only if it has a positive value. This implies that the product of the roots has to be positive. The condition is therefore

$$C^o_{\mathrm{NH_3}} C^o_{\mathrm{HNO_3}} > K_p.$$

This can be interpreted as an excess of ammonium and sulfate.

These properties have a significant impact on the estimation of the radiative impact of aerosols (Fig. 2.18, Chap. 2).

5.1.4.4 Nucleation

The binary nucleation between sulfuric acid and water is described by parameterizations of the nucleation rate J, given as a function of temperature, relative humidity and sulfuric acid concentration $C_{\mathrm{H_2SO_4}}$ (usually expressed in molecule $\mathrm{cm^{-3}}$). The parameterizations for the ternary nucleation also include the ammonia concentration (NH_3).

The nucleation diameter is typically of a few nanometers. The magnitude of J for ternary nucleation can be many tens times larger than the values obtained in the case of binary nucleation.

An estimation of the nucleation rate J is shown in Fig. 5.10 for binary nucleation ([145]). The nucleation rate has larger values for low temperatures and high relative humidities.

5.1.4.5 Secondary Organic Aerosols

Principle A specific treatment has to be applied to an organic species. Commonly-used gas-phase chemical mechanisms are actually designed for photochemistry and they do not have a fine description of the volatile organic compounds (VOCs). The key point is indeed to represent the compounds with a low saturation vapor pressure, which can condense onto existing aerosols.

The oxidation of VOCs leads to the formation of species which can themselves be oxidated to lead to complicated chemical functions with high polarizations. This oxidation process results in a decrease of saturation vapor pressure: the produced organic species are usually referred to as *semi-volatile organic compounds* or SVOCs (Fig. 5.11). The formation of SVOCs is usually described by adding new products in the oxidation reactions of VOCs by OH, O_3 and NO_3, the so-called *double products*. The oxidized VOCs are often biogenic species (such as terpens or limonen) or aromatic anthropogenic species.

The generic oxidation reaction

$$\mathrm{COV} + \mathrm{Ox} \longrightarrow \mathrm{P}$$

with Ox an oxidant to be chosen among OH, O_3 or NO_3, is then modified to

$$\mathrm{COV} + \mathrm{Ox} \longrightarrow \mathrm{P} + \alpha_1 \mathrm{P}_1 + \alpha_2 \mathrm{P}_2$$

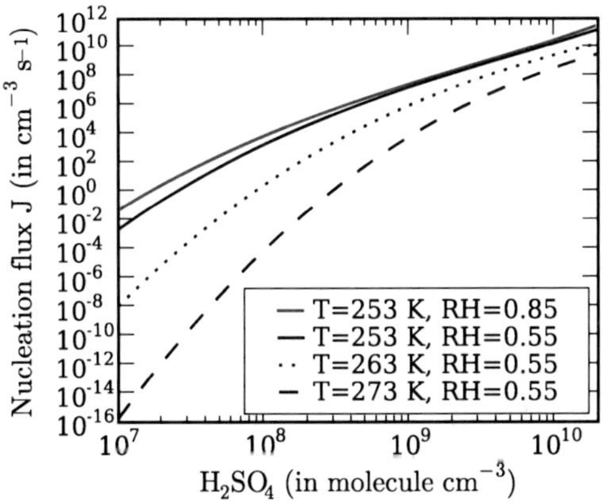

Fig. 5.10 Evolution of the nucleation rate as a function of sulfuric acid concentration at temperature $T = 253$, 263 and 273 K for a relative humidity $RH = 55\%$, and at $T = 253$ K for $RH = 85\%$. Source: [145]

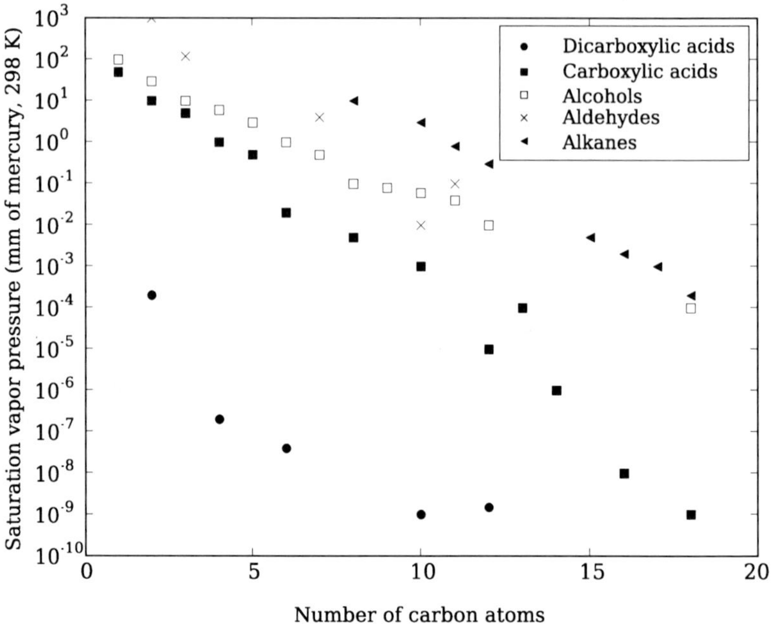

Fig. 5.11 Evolution of the saturation vapor pressure of VOCs as a function of the carbon number and of the level of "chemical complexity". By increasing order of chemical complexity: alkanes, aldehydes, alcohols, carboxylic acids and dicarboxylic acids. Sources: [47] and private communication of Christian Seigneur (AER and CEREA)

with P_1 and P_2 the double products (SVOCs). The stoichiometric coefficients α_i (the so-called *yields*) are usually derived from measurements in atmospheric *smog chambers* and are characterized by large uncertainties. Moreover, the conditions are rather different from real-life atmospheric conditions.

To summarize, organic aerosols are difficult to measure and understanding the processes that govern their atmospheric fate is still an open question (on the contrary to inorganic aerosols).

Partitioning Assume that organic species and inorganic species are not well mixed (this may not be the case in practice). The partitioning of organic species between the gaseous and particulate phases is then carried out in the following way.

Let n_{OM} be the number of organic compounds in the particulate phase (including both primary and secondary species). We assume that the organic species constitute an *ideal* solution. Thus one gets for species X_i

$$(q_i)_g = (x_i)_a \, q_i^{sat}, \tag{5.21}$$

with q_i^{sat} the saturation mass concentration of species X_i in a pure solution and $(x_i)_a$ the molar fraction of species X_i in the particulate organic phase. The activity coefficient in the organic phase is supposed to be equal to 1.

The molar fraction $(x_i)_a$ is given by

$$(x_i)_a = \frac{\frac{(q_i)_a}{M_i}}{\frac{q_{OM}}{M_{OM}}} = \frac{\frac{(q_i)_a}{M_i}}{\sum_{j=1}^{n_{OM}} \frac{(q_j)_a}{M_j} + \frac{(q_{POA})_a}{M_{POA}}}, \tag{5.22}$$

where q_{OM} stands for the total concentration of the organic matter (primary and secondary species) in the particulate phase, providing an absorbing medium. The molar mass M_i of species X_i is expressed in µg mol^{-1} (with the same units as the mass concentrations q_i); M_{OM} is the mean molar mass of organic matter (in µg mol^{-1}). *POA* stands for primary organic aerosols, which are supposed to be composed of species that do not evaporate to the gas phase.

The saturation vapor pressure, p_i^{sat}, can be computed from q_i^{sat} by

$$p_i^{sat} = \frac{RT}{M_i} q_i^{sat}. \tag{5.23}$$

It is also given as a function of temperature with the Clausius-Clapeyron law,

$$p_i^{sat}(T) = p_i^{sat}(298\,\mathrm{K}) \exp\left[-\frac{\Delta H_{vap}}{R}\left(\frac{1}{T} - \frac{1}{298}\right)\right], \tag{5.24}$$

where ΔH_{vap} is the enthalpy of vaporization.

5.1.4.6 Heterogeneous Reactions

The reactions at the surface of condensed matter (particles and cloud drops) can have a significant impact on ozone and aerosols. These processes are usually parameterized as first-order chemical reactions. For example, for the following mechanism ([59]),

$$HO_2 \xrightarrow{\text{condensed matter}} 0.5\,H_2O_2, \tag{R 128}$$

$$NO_2 \xrightarrow{\text{condensed matter}} 0.5\,HONO + 0.5\,HNO_3, \tag{R 129}$$

$$NO_3 \xrightarrow{\text{condensed matter}} HNO_3, \tag{R 130}$$

$$N_2O_5 \xrightarrow{\text{condensed matter}} 2\,HNO_3, \tag{R 131}$$

the kinetic rate of the heterogeneous reaction associated to consumption of the gas-phase species X_i over the aerosols of radius r is

$$k_i(r) = \left(\frac{r}{D_i^g} + \frac{4}{\bar{v}_i^g \gamma}\right)^{-1} S(r), \tag{5.25}$$

with D_i^g the air diffusivity and $\bar{v}_i^g$ the thermal velocity of species X_i. The coefficient γ is the reaction probability and $S(r)$ the available surface of condensed matter, integrated over all particles of radius r, per air volume.

The coefficient γ strongly depends on the composition and size distribution of aerosols. Its value is highly uncertain (Table 5.7).

5.2 Aerosols and Clouds

Aerosols play a leading role in the formation of clouds by providing condensation nuclei to water vapor. This section details a few related processes.

Another interaction between clouds and air pollution is given by aqueous-phase chemical reactions inside cloud drops. Clouds can actually be viewed as chemical tanks. The atmospheric aqueous-phase chemistry is also described in Sect. 5.3 with a focus on acid rains.

5.2.1 Primer for Clouds

5.2.1.1 A Few Data

- Clouds are difficult to define in a rigorous way. The *liquid water content* (LWC) is often used. It is defined as the mass of liquid water per air volume. The environment may be supposed to be a cloud for values of LWC beyond $0.1\,\text{g}\,\text{m}^{-3}$. The liquid water content can strongly differ from one type of clouds to another one (Table 5.8).

  The integrated columns of liquid water content typically vary from $50\,\text{g}\,\text{m}^{-2}$ (at high latitudes) to $200\,\text{g}\,\text{m}^{-2}$ (at the Equator).
- The mean global cloud coverage is about 60%. The volume generated by clouds is about 5–10% of the tropospheric volume (up to 15% in the lower troposphere). The cloud mass is up to 10% of the total mass.
- The global efficiency of precipitations is about 50% : this means that up to one half of the cloud water is scavenged during a rain event.
- Clouds are schematically classified in 10 types of clouds (cumulonimbus, cumulus, stratus, etc.), with three classes of altitude: high (from 5 to 13 km), medium

Table 5.7 Range of variation for the reaction probability γ (heterogeneous reactions). Source: [59]

	Default value	Range of variation
γ_{HO_2}	0.2	[0.1, 1]
γ_{NO_2}	10^{-4}	$[10^{-6}, 10^{-3}]$
γ_{NO_3}	10^{-3}	$[2 \times 10^{-4}, 10^{-2}]$
$\gamma_{N_2O_5}$	0.03	[0.1, 1]

Table 5.8 Typical values of microphysical parameters for a few cloud types

Cloud type	Density of drops (in cm^{-3})	Liquid water content (in $g\,m^{-3}$)
Stratus	100–500	0.1–0.3
Nimbostratus	300	0.5
Altostratus	50–500	0.05–0.2
Cumulus	100	0.5
Cumulonimbus	70	1–2
Mean	–	0.3

(from 2 to 5 km) and low (from 0 to 2 km). The microphysical properties can strongly differ from one type to another (Table 5.8).

5.2.1.2 Characteristic Timescales

Clouds are formed by condensation of water vapor on particles, the so-called *cloud condensation nuclei* (CCN). Note that the evaporation of a cloud drop leads to the "formation" of an aerosol. As the residence time of CCN is about a week and the duration of a rain event is about a few hours, one can consider that the CCN take part in several condensation/evaporation cycles (typically, from 5 to 10).

A few characteristic timescales are given in Table 5.9. One refers to Exercise 5.5 for an estimation of the atmospheric residence time of water.

Exercise 5.5 (Atmospheric Residence Time of Water) The mass of atmospheric water is about 1.3×10^{16} kg. The mean precipitation flux at ground is about 2 mm per day. Estimate the atmospheric residence time of water.

Solution:

Let m be the mass of atmospheric water and p the daily rain rate. The water mass that is scavenged during one day is $\dot{m} = \rho_{H_2O} \times 4\pi R_t{}^2 \times p$ with $\rho_{H_2O} = 1000\,\text{kg}\,\text{m}^{-3}$ the density of liquid water and $R_t = 6400\,\text{km}$ the Earth radius. This corresponds to an atmospheric residence time $m/\dot{m} \simeq 12.6$ days.

5.2.2 Saturation Vapor Pressure of Water, Relative Humidity and Dew Point

The *partial pressure* of a gas-phase species X_i is defined as $P_{X_i} = C_{X_i} P$. The partial pressures meet $\sum_i P_{X_i} = P$.

The *saturation vapor pressure* of water, $P^{sat}_{H_2O}$, is defined as the equilibrium pressure between liquid water and water vapor. The formation of a cloud occurs when

Table 5.9 A few timescales associated to clouds

Property	Characteristic timescale
Cumulus lifetime	hour
Scavenging of a soluble species	week
CCN lifetime	week
Residence time of water	two weeks

the partial pressure of water is above $P^{sat}_{H_2O}$. Similarly, the liquid water drops evaporate once the partial pressure of water is below $P^{sat}_{H_2O}$.

An increase of temperature leads to an enhancement of evaporation because $P^{sat}_{H_2O}$ is an increasing function of temperature (Fig. 5.12). Many parameterizations are available, for instance ([16]),

$$P^{sat}_{H_2O}(T) = 6.11 \times \exp\left(17.3\frac{T-273.15}{T-35.86}\right), \tag{5.26}$$

with T in K and $P^{sat}_{H_2O}$ in hPa.

The relative humidity, RH (in %), is defined as

$$RH = 100 \times \frac{P_{H_2O}}{P^{sat}_{H_2O}}. \tag{5.27}$$

The condensation of water occurs when $RH \geq 100\%$ and the environment is then said to be *supersatured*. The *supersaturation* is defined by $S = RH - 100$ (in %). Note that the notation could be confused with the notation used for entropy (but the context clarifies the variable).

For a fixed partial pressure of water vapor, the *dew point temperature*, T_d, is defined as the temperature at which condensation occurs, so that

$$P_{H_2O} = P^{sat}_{H_2O}(T_d). \tag{5.28}$$

Exercise 5.6 (Dew Point Temperature) Calculate the dew point temperature for a given pressure.
Solution:
Using (5.26) and expressing P in hPa yield

$$T_d(P) = \frac{273 - 2.1\ln(P/6.11)}{1 + 0.028\ln(P/6.11)}.$$

5.2.3 Condensation Nuclei

Aerosols play a key role in the formation of cloud drops. We briefly introduce the fundamentals of cloud physics in the following.

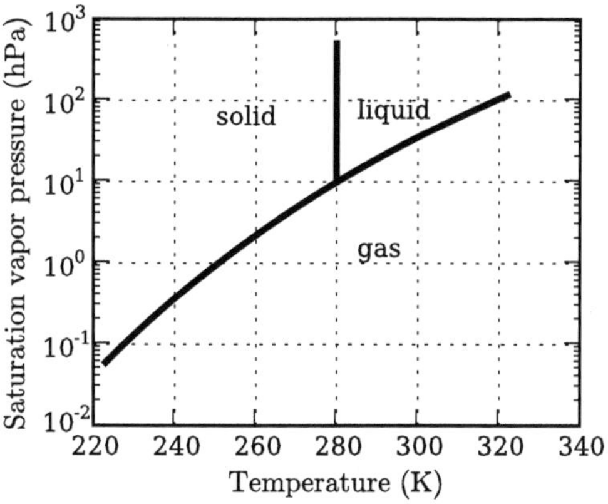

Fig. 5.12 Evolution of the saturation vapor pressure of water as a function of temperature

5.2.3.1 Formation of a Pure Water Droplet

Case of a Flat Surface The first law of thermodynamics can be written for a reversible process with molecular variables (written with lowercase letters),

$$du = T\,ds - P\,dv + g_l dn_l + g_v dn_v, \tag{5.29}$$

where $dq_{rev} = T\,ds$ is the reversible heat flux and n_l (respectively n_v) is the number of liquid molecules (respectively gaseous molecules), u is the molecular internal energy, v is the molecular volume and s is the molecular entropy. g_l (respectively g_v) is the chemical potential or the free energy of a liquid water molecule (respectively of a gaseous water molecule).

The appropriate thermodynamic variable for describing the formation of a water drop is the Gibbs free energy, defined by $G = U + PV - TS$. One easily gets for the molecular free energy

$$dg = v\,dP - s\,dT + g_l dn_l + g_v dn_v. \tag{5.30}$$

The equilibrium between the vapor phase and the liquid phase is defined by a minimum of g ($dg = 0$). For a change in the state of one molecule ($dn_l = -dn_v$), at given pressure and temperature (namely $dP = dT = 0$), one gets

$$dg = (g_l - g_v)dn_l. \tag{5.31}$$

The equilibrium condition defines the free energy at the saturation vapor pressure, g_v^{sat}, so that

$$g_l = g_v^{sat}. \tag{5.32}$$

The condensation of n molecules of water vapor, at fixed pressure and temperatures, is then associated to a change in Gibbs free energy

$$\Delta G = n(g_v^{sat} - g_v). \tag{5.33}$$

At a fixed temperature, the evolution from the gas-phase free energy, g_v, to the saturation free energy, g_v^{sat}, is given by the relation $\mathrm{d}g = v\mathrm{d}P$, obtained from (5.30) for $\mathrm{d}T = \mathrm{d}n_l = \mathrm{d}n_v = 0$. For a perfect gas, the molecular volume is defined by $Pv = k_B T$, with k_B the Boltzmann constant. Thus one gets

$$g_v^{sat} - g_v = k_B T \int_P^{P_{\mathrm{H_2O}}^{sat}} \frac{\mathrm{d}P}{P} = k_B T \ln\left(\frac{P_{\mathrm{H_2O}}^{sat}}{P}\right), \tag{5.34}$$

and

$$\Delta G = n k_B T \ln\left(\frac{P_{\mathrm{H_2O}}^{sat}}{P}\right). \tag{5.35}$$

Note that equilibrium is attained at the interface between liquid water and water vapor,

$$P = P_{\mathrm{H_2O}}^{sat}. \tag{5.36}$$

Case of a Spherical Drop As a drop is not a flat surface but a sphere, in a first approximation, the energy required for the formation of one drop is actually higher. This "overhead" is estimated by the so-called *Kelvin effect.*

Let us consider a spherical drop of radius r, resulting from the transformation of n molecules of water vapor to n molecules of liquid water,

$$n = \frac{4/3\pi r^3}{M_w/\rho_w}, \tag{5.37}$$

with M_w the molecular mass of water and ρ_w the volumic mass of liquid water (M_w/ρ_w is the molecular volume of liquid water). The formation of one drop is associated to the change in Gibbs free energy given by

$$\Delta G = n(g_l - g_v) + 4\pi r^2 \sigma, \tag{5.38}$$

with σ the surface tension at the interface between the gaseous phase and the liquid phase. Surface tension is defined as the work required for creating one surface unit of the drop and is expressed in work units per surface unit, namely in Newtons per length unit. For water, $\sigma = 72\,\mathrm{dyn\,cm^{-1}}$ (with $1\,\mathrm{dyn} = 10^{-5}$ N).

At the gas/liquid interface, one has $g_l = g_v^{sat}$ and one obtains

$$\Delta G = n(g_v^{sat} - g_v) + 4\pi r^2 \sigma. \tag{5.39}$$

At a given temperature, the change of free energy due to the formation of a drop of radius r is then, using (5.34) and (5.37),

$$\Delta G(r) = \frac{4/3\pi r^3}{M_w/\rho_w} k_B T \ln\left(\frac{P_{\mathrm{H_2O}}^{sat}}{P}\right) + 4\pi r^2 \sigma, \tag{5.40}$$

where $\Delta G(r)$ can be written as $\alpha r^2 + \beta r^3$, with $\alpha > 0$. The sign of β is the sign of $\ln(P_{\mathrm{H_2O}}^{sat}/P)$.

- When the air is not supersaturated ($P^{sat}_{H_2O} > P$), both α and β are positive and ΔG is an increasing function of r. It is therefore not possible to reach an equilibrium (associated to a minimum of free energy): the formation of drops is not favored because a drop has to evaporate (decrease of r) in order to decrease free energy.
- When the air is supersaturated ($P^{sat}_{H_2O} < P$), there exists a critical radius given by $r^\star = -2\alpha/(3\beta)$. This is the so-called Kelvin equation:

$$r^\star = \frac{2\sigma M_w}{\rho_w k_B T \ln\left(\frac{P}{P^{sat}_{H_2O}}\right)}. \tag{5.41}$$

For a radius below $r^\star$, the drops evaporate (so that the free energy decreases), while for a radius above $r^\star$, the drops grow. This is therefore an unstable equilibrium.

Rearranging the Kelvin equation leads to

$$P = P^{sat}_{H_2O} \exp\left(\frac{2\sigma M_w}{\rho_w k_B T r^\star}\right). \tag{5.42}$$

In comparison to the equilibrium condition obtained for a flat surface, in (5.36), the correcting coefficient is important for small droplets (with a radius of a few nanometers) and can be neglected for larger aerosols. A similar formula has been introduced for condensation onto aerosols with (5.17).

Figure 5.13 gives the evolution of supersaturation S required for reaching a critical size. Note that the curve exactly corresponds to the Kelvin effect. For example, the equilibrium of a pure water droplet with a diameter of 10 nm requires a supersaturation of about 12%. Such values of supersaturation are not realistic in the atmosphere and this process of "spontaneous" formation of water droplets is then not efficient. The fact that there are preexisting particles (containing species) tends to decrease the required supersaturation.

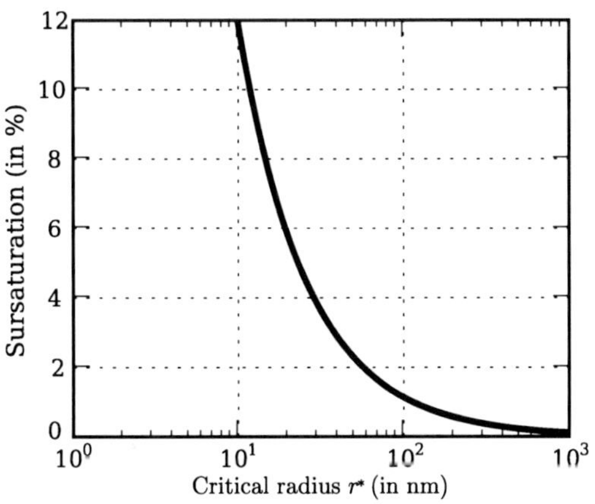

Fig. 5.13 Evolution of supersaturation as a function of the critical radius, at $T = 278$ K, for a pure water drop

5.2.3.2 Condensation of Water onto Aerosols

We now consider a water droplet that contains salts (it is not a pure water droplet). For the sake of clarity, we assume that the species are soluble in water. They constitute the so-called *solute*. The key point is that the water saturation vapor pressure is then decreased, which will favor condensation. The Raoult law is written for a mixture, supposed to be an *ideal* mixture, as

$$P^{sat,solution} = x_w \times P^{sat,pure}, \tag{5.43}$$

with $P^{sat,solution}$ the saturation vapor pressure of the mixture (the solution), $P^{sat,pure}$ the saturation vapor pressure of a pure water drop (computed in the previous sections) and x_w the molar fraction of liquid water.

Let m_s be the solute mass, with a molar mass M_s. Typically, the aerosol would be a marine aerosol (NaCl), a continental aerosol or an urban (polluted) aerosol (e.g., $(NH_4)_2SO_4$). The dissolved salt is dissociated into several ions (NaCl into Na^+ and Cl^-; $(NH_4)_2SO_4$ into $2\times NH_4^+$ and SO_4^{2-}). Let i be the resulting number of ions after dissociation. Thus, we obtain $i \times m_s/M_s$ moles of solute ions.

The molar fraction of liquid water is defined for n_w liquid water moles with

$$x_w = \frac{n_w}{n_w + i \times m_s/M_s}. \tag{5.44}$$

This can be rewritten as

$$\frac{1}{x_w} = 1 + i\frac{m_s/M_s}{n_w}. \tag{5.45}$$

Moreover, the total volume of the drop of radius r is

$$\frac{4}{3}\pi r^3 = n_w \frac{M_w}{\rho_w} + \frac{m_s}{M_s} v_s, \tag{5.46}$$

with v_s the molar volume of the solute.

We consider, as a first approximation, that the main part of the drop volume is made of liquid water. One has therefore

$$\frac{4}{3}\pi r^3 \simeq n_w \frac{M_w}{\rho_w}, \tag{5.47}$$

and

$$\frac{1}{x_w} = 1 + i\frac{3 m_s M_w}{4\pi M_s \rho_w r^3}. \tag{5.48}$$

Using the Raoult law rewritten as

$$\ln\left(\frac{P^{sat,solution}}{P^{sat}_{H_2O}}\right) = \ln\left(\frac{P^{sat,pure}}{P^{sat}_{H_2O}}\right) - \ln\left(\frac{1}{x_w}\right), \tag{5.49}$$

one gets after an asymptotic expansion based on $\ln(1+\epsilon) \simeq \epsilon$ for small values of ϵ,

$$\ln\left(\frac{P^{sat,solution}}{P^{sat}_{H_2O}}\right) = \frac{2\sigma M_w}{\rho_w k_B T r} - i\frac{3m_s M_w}{4\pi M_s \rho_w r^3}. \tag{5.50}$$

This equation is one possible expression of the Köhler equation *for an ideal solution with small quantities of solute*.

There are two competing effects: the curvature effect tends to increase the saturation vapor pressure while the solute effect tends to decrease it. Let us write (5.50) as $\alpha/r - \beta/r^3$; the critical radius (also referred to as the *activation radius*), is therefore $r^\star = \sqrt{3\beta/\alpha}$.

Figure 5.14 gives the evolution of supersaturation as a function of the droplet radius for a pure water drop and for a water drop containing solutes (NaCl and $(NH_4)_2SO_4$) in different quantities (10^{-18} or 10^{-19} kg). For increasing sizes, the solute effect can be neglected and the curvature effect is the dominant one: for large radius, all water drops behave as pure water drops. Note that an increasing mass of solute (e.g., NaCl) tends to decrease the maximal values of supersaturation. The evolution of supersaturation also depends on the solute type (with the values of i and of M_s).

Let us consider an environment with a supersaturation of 0.25% (the black horizontal line in Fig. 5.14).

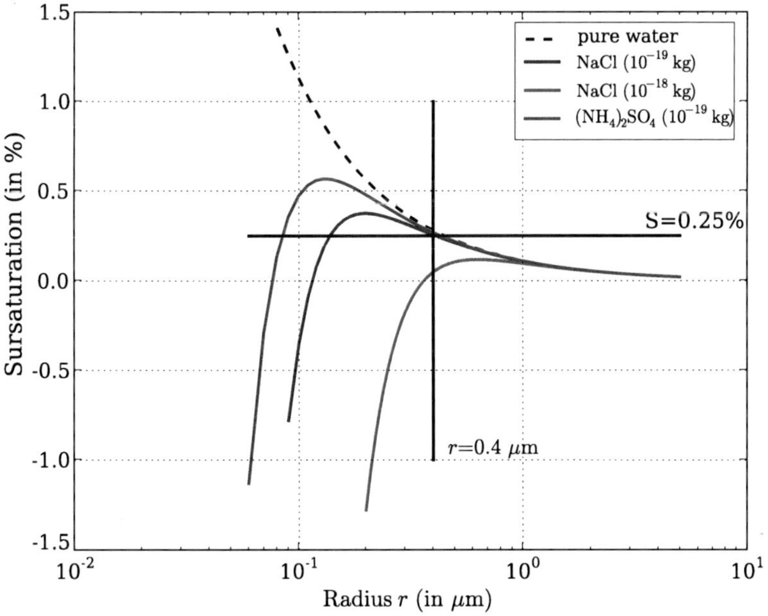

Fig. 5.14 Evolution of supersaturation as a function of the drop radius at $T = 278$ K: case of a pure water drop and of a drop containing solutes (NaCl or $(NH_4)_2SO_4$). Source: [150]

- A particle that contains enough solute mass (e.g., NaCl, 10^{-18} kg) will grow by condensation of water vapor. At the drop surface, the supersaturation will be always lower than the ambient supersaturation: the drop will grow up to become a cloud drop. This particle is then called an *activated particle*.
- In the opposite case, when the particle does not contain enough solute mass (e.g., 10^{-19} kg for NaCl), the surface supersaturation will be equal to the ambient supersaturation during evolution. Growth is therefore not possible. Once the ambient supersaturation is reached, the particle is in a stable equilibrium. The particle is then referred to as a *nonactivated particle*. Note that a particle with an initial radius greater than 0.4 μm (in the case depicted in the figure), can grow by condensation.

The parameterization of the activated fraction of the aerosol population is still an open question.

5.2.4 Mass Transfer Between the Gaseous Phase and Cloud Drops

Inside clouds, the mass transfer between the gaseous phase (the so-called *interstitial air*) and the aqueous phase defined by cloud drops affects the atmospheric chemical composition. The following processes have to be taken into account (Fig. 5.15):

- molecular diffusion of gas-phase species to cloud drops,
- mass transfer at the drop surface,
- molecular diffusion of dissolved species inside drops,
- aqueous-phase chemical reactions.

These phenomena have a significant impact for atmospheric chemistry, especially for the production of sulfate (Sect. 5.3 for the investigation of acid rains).

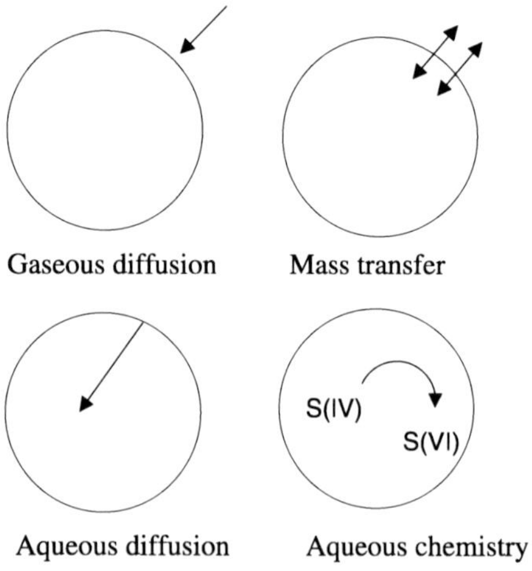

Fig. 5.15 Mass transfer in a cloud drop. The notations S(IV) and S(VI) stand for the dissolved sulfur dioxide and for the oxidized sulfur SO_4^{2-}, respectively (Sect. 5.3.1.2)

5.2.4.1 Henry's Law

The dissolution of a soluble gaseous species in a liquid drop of water

$$X \xrightleftharpoons{\text{dissolution}} (X)_{aq}, \tag{R 132}$$

is defined by Henry's law

$$H_X = \frac{[(X)_{aq}]}{p_X}, \tag{5.51}$$

with $[(X)_{aq}]$ the aqueous-phase concentration with respect to the liquid solution (in $mol\,l^{-1}$, also written as M) and p_X the partial pressure (in atm). The customary units of Henry's constant H_X are then $M\,atm^{-1}$ (Table 5.10).

In the standard thermodynamic conditions for pressure ($P = 1$ atm), one gets $p_X = 10^{-6} \times C_X$ (expressed in ppmv).

When $(X)_{aq}$ can dissociate, a so-called *effective* Henry's constant, which depends on the acidity of the water drop, is defined (Exercise 5.7).

Exercise 5.7 (Effective Henry's Constant) Consider a soluble species XH (H is the hydrogen atom) and assume that the dissolved species $(XH)_{aq}$ can be dissociated into X^- and H^+ with the reversible reaction (R 133). The equilibrium constant of this reaction is given by (5.54), namely

$$K = \frac{[(XH)_{aq}]}{[X^-][H^+]},$$

with [.] the concentration expressed in $mol\,l^{-1}$ (with respect to the solution). Calculate the *effective* Henry's constant, defined by

$$H_{XH}^{eff} = \frac{[(XH)_{aq}^{tot}]}{p_{XH}},$$

where $(XH)_{aq}^{tot} = (XH)_{aq} + X^-$ lumps all the dissolved forms of XH.

Table 5.10 Henry's constant for a few species. The value at temperature T is calculated with $H(T) = H(298)\exp(\Delta H/R(1/298 - 1/T))$. Source: [130]

Species	H ($M\,atm^{-1}$) at 298 K	Heat of dissolution ΔH ($kJ\,mol^{-1}$)
SO_2	1.23	−26.2
CO_2	3.4×10^{-2}	−20.3
NH_3	62	−34.2
H_2O_2	7.45×10^4	−60.7
O_3	1.1×10^{-2}	−21.1
HNO_3	2.1×10^5	−72.2

Solution:
The equation $[(XH)_{aq}^{tot}] = [(XH)_{aq}] + [X^-]$ can be rewritten as

$$[(XH)_{aq}^{tot}] = \left(1 + \frac{1}{K[H^+]}\right)[(XH)_{aq}] = \underbrace{\left(1 + \frac{1}{K[H^+]}\right)H_{XH}}_{H_{XH}^{eff}} \times p_{XH}.$$

The effective Henry's constant is a function of the acidity of the water drop (through H^+). This formula can be easily generalized when X^- can also be dissociated.

5.2.4.2 Mass Transfer

Mass transfer between gas-phase interstitial air and cloud drops is usually parameterized for soluble species.

Let us consider a cloud drop of radius a (supposed to be a sphere). We assume that the aqueous-phase concentration (inside the water drop), c_a, is homogeneous. Let c_g be the gas-phase concentration in the interstitial air. The time evolution of these concentrations is given by

$$\begin{cases} \frac{dc_g}{dt} = \chi_g(c_g) - Lk_{mt}\left(c_g - \frac{c_a}{HRT}\right) \\ \frac{dc_a}{dt} = \chi_a(c_a) + k_{mt}\left(c_g - \frac{c_a}{HRT}\right), \end{cases} \tag{5.52}$$

where χ_a and χ_g are the chemical source terms in the gaseous and aqueous phase, respectively, and L is the liquid water content. The coefficient k_{mt} parameterizes the mass transfer (Problem 5.2),

$$k_{mt} = \left(\frac{a^2}{3D_g} + \frac{4a}{3\bar{v}\alpha}\right)^{-1}, \tag{5.53}$$

with D_g the gas-phase diffusivity, $\bar{v}$ the thermal velocity and α the accommodation coefficient. The accommodation coefficient describes the efficiency of dissolution of a gaseous molecule after a collision with the drop (typically $\alpha \in [0.01, 1]$).

More complicated models account for the case in which the aqueous-phase diffusion and the aqueous-phase chemical reactions are limiting steps of mass transfer.

5.3 Acid Rains and Scavenging

We briefly investigate two interactions between rain and air quality: we first detail the formation of acid rains (from this viewpoint, rain contributes to environmental damage) and then scavenging of particles and soluble gases by falling raindrops (from this viewpoint, rain favors a decrease in atmospheric pollution).

5.3.1 Acid Rains

In the 1960s and 1970s, observational data in the northeastern United States of America and in northern Europe indicate that the acidity of precipitation was strongly increasing, not only near industrial regions. For a few monitoring stations, the concentration of hydrogen ion H^+ had increased 30 or 40 times over twenty years of monitoring. Simultaneously, measured concentrations of sulfate (SO_4^{2-}) and of nitrate (NO_3^-) were also increasing.

A strong reduction in SO_2 emissions, especially those related to thermal power plants, changed the situation in the 1980s. For example, the emissions were divided by a factor of 2 in western Europe (Table 0.5, Introduction), which contributes to decrease the rainwater acidity.

The continental impact of acid rains is investigated in Exercise 5.10.

5.3.1.1 Acidic Solutions

The equilibrium constant of the following dissociation reaction in water,

$$(\mathrm{XH})_{\mathrm{aq}} \rightleftharpoons \mathrm{H}^+ + \mathrm{X}^- \qquad \text{(R 133)}$$

is defined by

$$K = \frac{[(\mathrm{XH})_{aq}]}{[\mathrm{X}^-][\mathrm{H}^+]}, \qquad (5.54)$$

with [.] the concentration expressed in $\mathrm{mol\,l^{-1}}$ (with respect to the solution).

The pH of water is defined by $\mathrm{pH} = -\log_{10}[\mathrm{H}^+]$. It can be easily calculated for pure water by considering the self-ionization of water, $(\mathrm{H_2O})_{aq}$,

$$(\mathrm{H_2O})_{aq} \rightleftharpoons \mathrm{H}^+ + \mathrm{OH}^-. \qquad \text{(R 134)}$$

We assume that H_2O concentration is constant in a pure water solution. By incorporating $[H_2O]$ in the equilibrium constant, one gets $[\mathrm{H}^+][\mathrm{OH}^+] \simeq 10^{-14}$. The pH of neutral pure water ($[\mathrm{H}^+] = [\mathrm{OH}^-]$) is then equal to 7.

The dissociation of acidic species in solution contributes to formulation of hydrogen ions and therefore to a decrease in the pH of rainwater. A first example is provided by the dissociation of carbon dioxide, a leading process for the pH of "natural" rainwater (far from polluted areas). One refers to Exercise 5.8.

Exercise 5.8 (pH of "Natural" Rainwater) When dissolved, carbon dioxide, CO_2, forms carbonate ions (CO_3^{2-}) and bicarbonate ions (HCO_3^-) according to the following chemical reactions:

$$CO_2 \xrightleftharpoons{\text{dissolution}} (CO_2)_{aq}, \quad \text{(R 135)}$$

$$(CO_2)_{aq} \xrightleftharpoons{K_{a_1}} HCO_3^- + H^+, \quad \text{(R 136)}$$

$$HCO_3^- \xrightleftharpoons{K_{a_2}} CO_3^{2-} + H^+. \quad \text{(R 137)}$$

The equilibrium constants are $K_{a_1} = 4.5 \times 10^{-7}$ M and $K_{a_2} = 4.7 \times 10^{-11}$ M at 298 K. Henry's constant is $H_{CO_2} = 3.4 \times 10^{-2}$ M atm^{-1}. Calculate the pH of rainwater by considering that the only dissolved species is CO_2 ($C_{CO_2} = 365$ ppm for the "current" atmosphere). Use the electroneutrality equation.

Solution:

The partial pressure of CO_2 is $p_{CO_2} = C_{CO_2} \times P$ with P the atmospheric pressure (supposed to be 1 atm). Thus $[(CO_2)_{aq}] = H_{CO_2} p_{CO_2} \simeq 1.24 \times 10^{-5}$ M.

Let us first neglect CO_3^{2-} and OH^- in the electroneutrality equation. This gives $[HCO_3^-] = [H^+]$. The dissociation equilibrium leads to

$$[H^+] = \sqrt{K_{a_1}[(CO_2)_{aq}]} \simeq 2.4 \times 10^{-6}\ \text{M},$$

and thus pH $\simeq 5.6$. We check, *a posteriori* (Fig. 5.16), that CO_3^{2-} can be neglected for this value (this is clear for OH^-). Otherwise, we have to solve an algebraic equation of the third order in $[H^+]$ ... .

5.3.1.2 Aqueous-Phase Chemistry

The dissolution of gas-phase species can form hydrogen ions H^+, which results in an increased acidity of water drops (clouds, rains, fogs). Values of rainwater pH below

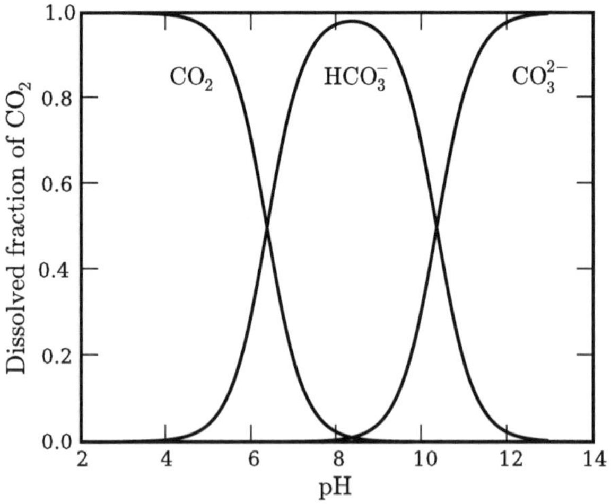

Fig. 5.16 Evolution of the dissolution of CO_2 as a function of pH (Exercise 5.8)

5 (or 5.6, Exercise 5.8) usually indicate *acid rains*. The main chemical reactions to be taken into account in the gas and aqueous phases are summarized in Table 5.11, with a focus on sulfate and nitrate.

Sulfate When dissolved in water drops, sulfur dioxide takes part in dissociation reactions similar to CO_2 (Exercise 5.8):

$$SO_2 \xrightleftharpoons{\text{dissolution}} (SO_2)_{aq}, \tag{R 138}$$

$$(SO_2)_{aq} \rightleftharpoons HSO_3^- + H^+, \tag{R 139}$$

$$HSO_3^- \rightleftharpoons SO_3^{2-} + H^+. \tag{R 140}$$

These reactions contribute to the acidity of water drops (Exercise 5.9).

Exercise 5.9 (Cloud pH) We consider a cloud over a polluted area. The mixing ratio of gas-phase sulfur dioxide is supposed to be $C_{SO_2} = 10\,\text{ppb}$. We assume that the acidity of cloud drops is governed by (R 139). Calculate the cloud pH.
Data
– The equilibrium constant is $K_{a_1} = 1.2 \times 10^{-2}\,\text{M}$ at 298 K.
– Henry's constant is $H_{SO_2} = 1.2\,\text{M}\,\text{atm}^{-1}$.
Solution:
The calculations are similar to Exercise 5.8. One easily gets $[H^+] = \sqrt{K_{a_1} H_{SO_2} p_{SO_2}}$, thus $pH = 4.9$.

Table 5.11 Main chemical reactions leading to the formation of nitrate and sulfate. Source: [128]

Gas-phase reactions	
	$SO_2 + OH \rightarrow H_2SO_4 + HO_2$
	$NO_2 + OH \rightarrow HNO_3$
	$NO_2 + O_3 \rightarrow NO_3$
	$NO_2 + NO_3 \rightarrow N_2O_5$
	$N_2O_5 + H_2O \rightleftarrows 2HNO_3$
	$COV + NO_3 \rightarrow HNO_3 + \ldots$
Heterogeneous reactions	
	$NO_3 \rightarrow HNO_3$
	$N_2O_5 + H_2O \rightleftarrows 2HNO_3$
Aqueous-phase reactions	
	$S(IV) + H_2O_2 \rightarrow S(VI)$
	$S(IV) + O_3 \rightarrow S(VI)$
	$S(IV) + O_2 \overset{\text{Fe, Mn}}{\rightarrow} S(VI)$
	$NO_3 \rightarrow NO_3^- + H^+$
	$N_2O_5 + H_2O \rightleftarrows 2(NO_3^- + H^+)$

The oxidation of dissolved SO_2 will actually increase the acidity of water drops. The ion HSO_3^- is oxidized by dissolved oxidants, such as ozone O_3 or hydrogen peroxide H_2O_2. For example, a decisive reaction inside clouds is

$$HSO_3^- + (H_2O_2)_{aq} + OH \longrightarrow SO_4^{2-} + H^+ + H_2O + O_2. \qquad (R\,141)$$

This reaction is catalyzed by acidic species and can be rewritten as

$$HSO_3^- + (H_2O_2)_{aq} + H^+ \longrightarrow SO_4^{2-} + 2H^+ + H_2O. \qquad (R\,142)$$

This process is sometimes referred to as the oxidation of sulfur in its oxidation state 4 (S(IV), namely SO_2) to its oxidation state 6 (S(VI), namely SO_4^{2-}).

Trace metals (e.g., iron or manganese) can also catalyse this oxidation.

Nitrate Nitric acid (HNO_3) is highly soluble in water (Table 5.10). Nitric acid also contributes to the acidity of water drops through the following chemical reactions:

$$HNO_3 \xrightleftharpoons{\text{dissolution}} (HNO_3)_{aq}, \qquad (R\,143)$$

$$(HNO_3)_{aq} \rightleftharpoons NO_3^- + H^+. \qquad (R\,144)$$

Let us recall that the mechanisms of formation for gas-phase acid nitric are particularly complicated (Chap. 4):

- during daytime, NO_2 is oxidized with

$$NO_2 + OH + M \longrightarrow HNO_3 + M, \qquad (R\,145)$$

- during night-time, acid nitric is mainly produced by heterogeneous reactions at the aerosol surface (Sect. 5.1.4.6) and by homogeneous and heterogeneous reactions of N_2O_5.

Acid-Neutralization by Bases The acidity can be decreased by the presence of bases and the neutralization of hydrogen ions H^+. A strong base is ammonia (NH_3) (related to agriculture-induced emissions). In a solution, ammonia is dissociated to produce ammonium ions (NH_4^+),

$$NH_3 \xrightleftharpoons{\text{dissolution}} (NH_3)_{aq}, \qquad (R\,146)$$

$$(NH_3)_{aq} \rightleftharpoons NH_4^+ + OH-, \qquad (R\,147)$$

which results in a decrease of the acidity of water drops.

Other bases are alkaline ions (Ca^{2+}, Na^+, K^+), associated to the crustal matter and Aeolian erosion of dust. A comparison between two European sites is shown in Table 5.12. The major difference is related to the Ca^{2+} concentration (with high values for the "non-acidic" site).

Salts of ammonium nitrate $(NH_4)NO_3$ and of ammonium sulfate $(NH_4)_2SO_4$ or $(NH_4)HSO_3$ can also be found in raindrops. This explains why the observed values for the ratio of ammonium to sulfate typically range from 1 to 2 (Fig. 5.17 and Fig. 5.18, data from the EMEP network for Europe, 2004).

Exercise 5.10 (Long-Range Transport of Acid Rains) We use a Lagrangian description of transport (Sect. 6.1.5.3, Chap. 6) in order to investigate the impact of dioxide sulfur emissions.

A plant emits sulfur dioxide (SO_2). The emitted plume is advected by a horizontal wind, whose velocity u is supposed to have a constant value. We assume that SO_2 has a timescale τ_{SO_2} fixed by its gas-phase oxidation to H_2SO_4 and that H_2SO_4 has a timescale $\tau_{H_2SO_4}$ fixed by rain scavenging. Calculate the distance at which H_2SO_4 concentration will have a maximal value.
Hint: transform the time variable t to the space variable x (distance from the emission point).
Data: $u = 5\,\mathrm{m\,s^{-1}}$; $\tau_{SO_2} = 2$ days and $\tau_{H_2SO_4} = 5$ days.

Table 5.12 Mean composition (for the year 2004) of rainwater for two European sites: an "acidic" site (Virolahti, Finland) and a "non-acidic" site (Campisabalos, Spain). Except for pH, the units are $\mathrm{mg\,l^{-1}}$ (e.g. $0.2\,\mathrm{mg\,l^{-1}}$ of nitrogen, N, for NH_4^+ in the acidic site). Source: [39]

Ion	"Non-acidic" site	"Acidic" site
Ca^{2+}	1.21	0.2
K^+	0.1	0.19
Mg^{2+}	0.07	0.05
NH_4^+	0.2	0.26
Na^+	0.29	0.26
Cl^-	1.00	0.45
NO_3^-	0.33	0.34
SO_4^{2-}	0.42	0.46
pH	6.19	4.65

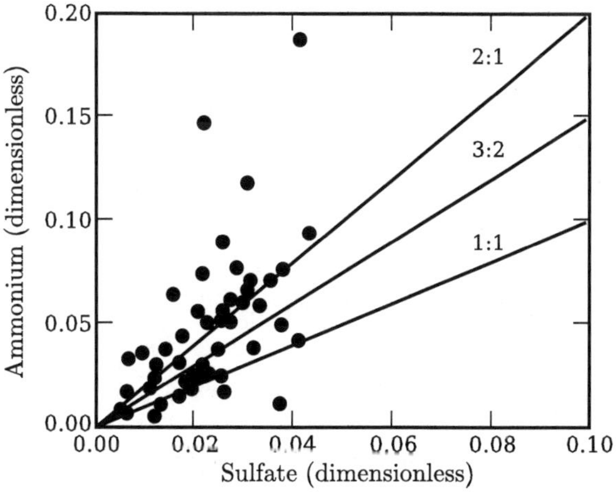

Fig. 5.17 Sulfate/ammonium ratio (with dimensionless units) in aerosols. Data from the EMEP monitoring network over Europe (2004). Source: [39]

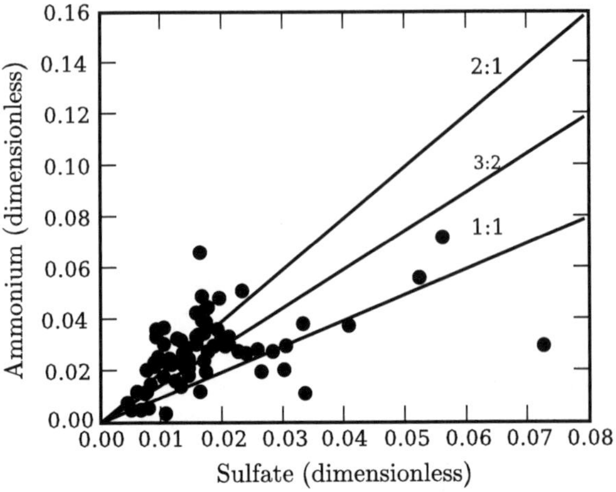

Fig. 5.18 Sulfate/ammonium ratio (with dimensionless units) in raindrops. Data from the EMEP monitoring network over Europe (2004). Source: [39]

Solution:
As $x = u \times t$, the evolution of the concentrations along the plume is given for SO_2 and for H_2SO_4 by

$$\frac{d[SO_2]}{dx} = -\frac{[SO_2]}{\tau_{SO_2} U},$$

$$\frac{d[H_2SO_4]}{dx} = \frac{[SO_2]}{\tau_{SO_2} U} - \frac{[H_2SO_4]}{\tau_{H_2SO_4} U}.$$

Upon integration,

$$[H_2SO_4](x) = \frac{\tau_{H_2SO_4}}{\tau_{H_2SO_4} + \tau_{SO_2}} \left[\exp\left(-\frac{x}{\tau_{SO_2} U}\right) - \exp\left(-\frac{x}{\tau_{H_2SO_4} U}\right) \right].$$

The maximal value of H_2SO_4 concentration is then obtained at

$$x^\star = U \frac{\tau_{SO_2} \tau_{H_2SO_4}}{\tau_{H_2SO_4} - \tau_{SO_2}} \ln \frac{\tau_{H_2SO_4}}{\tau_{SO_2}}.$$

This gives a distance of about 1300 km downwind the source. This illustrates the continental nature of acid rains.

5.3.2 *Wet Scavenging*

Wet scavenging processes (wet removal) play a major role for gas-phase soluble species and aerosols, leading to the low atmospheric residence time of aerosols (Sect. 5.1). In comparison, dry deposition (Problem 3.1, Chap. 3) does not have such a decisive role.

Wet scavenging is described by a *scavenging coefficient*, usually written as Λ, expressed as the inverse of a timescale. The time evolution of the scavenged concentrations, let us say c, is then given by

$$\frac{\mathrm{d}c}{\mathrm{d}t} = -\Lambda \times c. \tag{5.55}$$

The coefficient Λ parameterizes all the processes that result in scavenging. It is a function that depends on the rain properties (rain intensity, size distribution of the raindrops) and on the scavenged matter (solubility of a gas-phase species, size distribution of an aerosol population).

5.3.2.1 Scavenging of Gases

A soluble gas can be scavenged by rain, which results in a decrease in the gas-phase concentration. Two cases are usually distinguished:

- the case of irreversible soluble species (there is no evaporation flux from the aqueous phase): acid nitric HNO_3 is a good example (Table 5.10).
- the case of reversible soluble gases (condensation and evaporation fluxes have to be taken into account).

In the first case, the scavenging coefficient can be expressed as

$$\Lambda = \frac{6 \times 10^{-3} p_0\, K_c}{U_r\, D_r}, \tag{5.56}$$

with p_0 the rain intensity (a customary unit is mm h^{-1}), K_c a mass transfer coefficient (mainly depending on the raindrop size), U_r the falling velocity and D_r a representative diameter of the raindrop. In the general case, the evaporation flux has also to be taken into account. Note that, once at equilibrium, there is no mass transfer between the aqueous and the gaseous phases. A parameterization for Λ is then

$$\Lambda = \frac{6 \times 10^{-3} p_0\, K_c}{U_r\, D_r} \exp\left(-\frac{6K_c\, z}{D_r\, U_r\, H}\right), \tag{5.57}$$

with H the Henry's constant of the gas-phase species and z the distance from the cloud bottom. During the fall of the raindrop, the scavenging coefficient decreases because the concentrations approach mass transfer equilibrium. Note that $H \to \infty$ corresponds to the irreversible soluble case. For weakly soluble gases (low values of H), scavenging can be neglected.

5.3.2.2 Scavenging of Aerosols

For aerosols the scavenging processes depend both on the particle size and on the raindrop size. These (nonlinear) effects are usually parameterized by the following expression of the scavenging coefficient (Exercise 5.11),

$$\Lambda(d_p) = \frac{3}{2}\frac{E(D_r, d_p)p_0}{D_r}, \tag{5.58}$$

with d_p the particle diameter, D_r the representative diameter of the raindrops (in $[0.1, 1]$ mm), p_0 the rain intensity (in mm h^{-1}: typically from 1 to 10 mm h^{-1}) and $E(.,.)$ the so-called *collision efficiency*. The collision efficiency depends on all the processes that take part in scavenging (Fig. 5.19).

Figure 5.20 shows the dependence of the collision efficiency on the raindrop representative diameter. Stronger precipitation events result, as expected, in stronger scavenging.

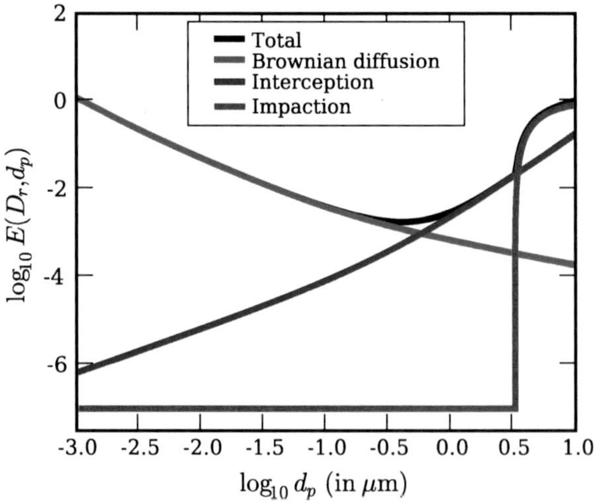

Fig. 5.19 Size distribution of the collision efficiency, $E(D_r, d_p)$, as a function of the particle size distribution (d_p): contributions of Brownian diffusion, of interception and of impaction. The representative raindrop diameter is $D_r = 0.1$ mm. Source: [135]

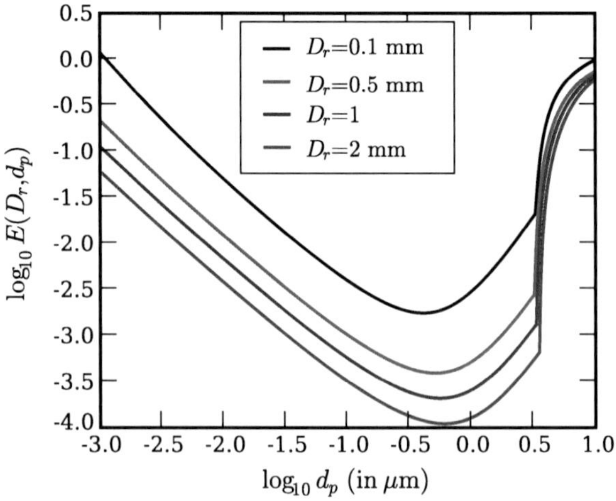

Fig. 5.20 Distribution of $E(D_r, d_p)$ as a function of d_p for different values of the representative raindrop diameter: $D_r = 0.1$ mm, 0.5 mm, 1 mm and 2 mm. Source: [135]

Note that the medium part of the aerosol distribution (corresponding to submicronic aerosols) is only weakly scavenged. As a consequence, these aerosols are the most present ones in the atmosphere. The decrease of the scavenging coefficient in the medium part of the aerosol distribution is usually referred to as the *Greenfield gap* (from the name of the scientist who first measured it in the 1950s; [45]).

Many observational data suggest that the scavenging coefficient for aerosols can be parameterized as ([135] and Exercise 5.11)

$$\Lambda = A p_0^B, \tag{5.59}$$

with p_0 in mm h^{-1}. For the coarse fraction of the aerosol distribution (for diameters above 1 μm), typical values of the coefficients are $A \in [10^{-5}, 10^{-3}]$ and $B \simeq 0.8$, while for the submicronic fraction $A \in [10^{-8}, 10^{-6}]$ and $B \simeq 0.6$.

For a rain intensity of 1 mm h^{-1}, the scavenging coefficient is about 10^{-4} s^{-1}, which corresponds to a decrease of scavenged concentrations by 30% in one hour ($\exp(-3600 \times 10^{-4}) \simeq 0.7$).

Exercise 5.11 (Scavenging of Aerosols by Rain) This exercise aims at explaining the parameterizations (5.58) and (5.59) for the scavenging coefficient.

The raindrop distribution is supposed to be monodispersed (only one raindrop size), with a representative diameter D_r. Rain is also characterized by the rain intensity (p_0), defined as the rain height per unit of ground surface and per unit of time.

1. Express p_0 as a function of the rain parameters. It is useful to introduce the falling velocity of raindrops, U_r, and the number distribution of raindrops, n_r (in number of raindrops per air volume).
2. Prove (5.58) after having introduced the collision efficiency, defined as the fraction of aerosols that actually collide with raindrops.
3. Observational data indicate that D_r (expressed in mm) is an increasing function of the rain intensity $D_r = \alpha p_0{}^{\beta}$ (with p_0 in mm h^{-1}), where $\alpha \in [0.25, 1]$ and $\beta \in [0.15, 0.25]$. Conclude by assuming a constant (realistic) value $E \simeq 0.1$.

Solution:

1. During a time interval of duration Δt, the volume of raindrops that reach a surface S at ground are $U_r \Delta t S$. There are $n_r U_r \Delta t S$ raindrops and the total volume of liquid water is then $n_r U_r \Delta t S \times \pi D_r^3/6$. The rain intensity is obtained by dividing by $S \Delta t$,

$$p_0 = n_r U_r \pi D_r^3/6.$$

2. The scavenged volume is defined, during the time interval of duration Δt, as the cylinder generated by the disk of surface $\pi D_r^2/4$ and with the height $U_r \Delta t$. This volume contains $n_r \times U_r \Delta t \pi D_r^2/4$ raindrops. The scavenging coefficient is obtained for the aerosols of diameter d_p by multiplying the number of raindrops by the collision efficiency $E(D_r, d_p)$ and by dividing by Δt,

$$\Lambda(d_p) = E(D_r, d_p) \times n_r U_r \pi D_r^2/4.$$

By eliminating $n_r U_r$, one gets

$$\Lambda(d_p) = E(D_r, d_p) p_0 \frac{\pi D_r^2/4}{\pi D_r^3/6} = \frac{3}{2} \frac{E(D_r, d_p) p_0}{D_r}.$$

3. It is easy to obtain

$$\Lambda = \frac{3}{2\alpha} E p_0^{1-\beta}.$$

As p_0 is expressed in $\mathrm{mm\,h^{-1}}$ the conversion of Λ to $\mathrm{s^{-1}}$ is carried out by dividing by 3600. Thus,

$$\Lambda = [4.2 \times 10^{-5},\ 1.7 \times 10^{-4}] \times p_0^{[0.75,\ 0.85]}.$$

To know more ([135]):
B. SPORTISSE, *A review of parameterizations for modeling dry deposition and scavenging of radionuclides*, Atmos. Env., **41** (2007), pp. 2683–2698

Problems Related to Chap. 5

Problem 5.1 (Marine Aerosols) The vertical profile of marine aerosols can be roughly approximated as the sum of a boundary layer contribution, $c_l(z)$, and a background contribution, $c_b(z)$. Marine aerosols are characterized by high concentrations in the first hundreds of meters above the sea surface, due to the wind-induced emissions of sea salts. In a first approximation we neglect the transport flux between sea and land.

The typical vertical distribution of marine aerosol concentrations is shown in Fig. 5.21. The boundary layer contribution is constant up to a height of 600 meters ($c_l(0) = 10\,\mathrm{\mu g\,m^{-3}}$) and decreases exponentially with a scale height of 1 km (h_l) above that altitude. The background contribution starts from $c_b(z) = 1\,\mathrm{\mu g\,m^{-3}}$ at ground and decreases exponentially with a scale height of 8 km (h_f).

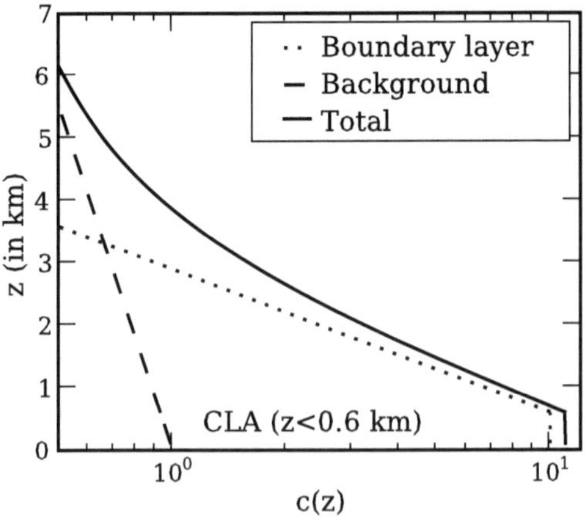

Fig. 5.21 Vertical profile of the marine aerosol concentration (in $\mathrm{\mu g\,m^{-3}}$). The concentration is plotted in log-scale. Problem 5.1

1. Express the vertical profile of the marine aerosol concentration $c(z)$.
2. Calculate the tropospheric marine aerosol column.
3. Calculate the residence time of the tropospheric marine aerosol.

Hints:
– seas cover about 70% of the Earth's surface (Earth's radius $R_t = 6400\,\text{km}$);
– the mass flux of marine aerosols is estimated to be $E = 3340\,\text{Tg}\,\text{yr}^{-1}$(Table 5.2);
– the top of the troposphere is supposed to be $z_t = 12\,\text{km}$.

Solution:

1. The marine aerosol concentration is given in the lower troposphere ($z \le h$ with $h = 600\,\text{m}$) by $c_m(z) = c_l(0) + c_b(0)\exp(-z/h_f)$, where $c_l(0) = 10\,\mu\text{g}\,\text{m}^{-3}$ and $c_b(0) = 1\,\mu\text{g}\,\text{m}^{-3}$. For $z \ge h$, the concentration is $c_m(z) = c_l(0)\exp(-(z-h)/h_l) + c_b(0)\exp(-z/h_f)$.
2. The tropospheric marine aerosol column is

$$\begin{aligned} C &= \int_0^{z_t} c_m(z)\,\mathrm{d}z \\ &= hc_l(0) + h_l c_l(0)\left[1 - \exp\left(-\frac{z_t - h}{h_l}\right)\right] + h_f c_b(0)\left[1 - \exp\left(-\frac{z_t}{h_f}\right)\right] \\ &\simeq (h + h_l)c_l(0) + h_f c_b(0)\left[1 - \exp\left(-\frac{z_t}{h_f}\right)\right] \\ &\simeq 2.2 \times 10^{-5}\,\text{kg}\,\text{m}^{-2}. \end{aligned}$$

As seas cover more than 70% of the Earth's surface, the total mass of tropospheric marine aerosol is

$$M = C \times 0.7 \times 4\pi R_t^2 \simeq 7.9 \times 10^9\,\text{kg}.$$

3. The residence time is $\tau = M/E$, namely about one day (0.88 day).

Problem 5.2 (Parameterization of Mass Transfer in a Cloud) This exercise aims at explaining the parameterization (5.52)–(5.53).

The evolution of concentrations in the aqueous (c_a) and gaseous (c_g) phases is given by the reaction-diffusion system

$$\begin{cases} \frac{\partial c_g}{\partial t} + D_g \Delta c_g = \chi_g(c_g) & \text{in } \Omega_g, \\ \frac{\partial c_a}{\partial t} + D_a \Delta c_a = \chi_a(c_a) & \text{in } \Omega_a, \end{cases}$$

with Ω_g the volume occupied by interstitial air and Ω_a the volume occupied by cloud droplets, supposed to be spheres of radius a. The source term χ_g (respectively χ_a) describes the gas-phase chemical reactions (respectively the aqueous-phase chemical reactions). D_g (respectively D_a) is the gas-phase molecular diffusion flux (respectively the aqueous-phase flux). The mass transfer flux at the drop surface Γ is given by

$$D_g \nabla c_g \cdot \mathbf{n} = -D_a \nabla c_a \cdot \mathbf{n} - \frac{1}{4}\alpha\bar{v}\left(\frac{c_a}{HRT} - c_g\right),$$

where $\mathbf{n}$ is the unitary vector normal to Γ (outwards oriented); α is the accommodation coefficient , $\bar{v}$ is the thermal velocity of the chemical species, H is Henry's coefficient (which describes the partitioning between the gaseous and aqueous phases for a given species), R is the universal gas constant and T the temperature.

We define the averaged concentrations in the gas and aqueous phases by $\bar{c}_g = \int_{\Omega_g} c_g \, d\omega$ and $\bar{c}_a = \int_{\Omega_a} c_a \, d\omega$. Let L be the liquid water content, expressed in unit of water volume per unit of air volume (it is the ratio of the liquid water volume, Ω_a, to the interstitial air volume Ω_g).

The diphasic model is parameterized by the homogeneous system

$$\begin{cases} \frac{d\bar{c}_g}{dt} = \chi_g(\bar{c}_g) - L\, I(\bar{c}_a, \bar{c}_g), \\ \frac{d\bar{c}_a}{dt} = \chi_a(\bar{c}_a) + I(\bar{c}_a, \bar{c}_g). \end{cases}$$

The mass transfer flux I is defined as $I(\bar{c}_a, \bar{c}_g) = k_{mt}(\bar{c}_g - \frac{\bar{c}_a}{HRT})$, with k_{mt} the mass transfer coefficient. A classical expression of k_{mt} is

$$k_{mt} = \left(\frac{a^2}{3D_g} + \frac{4a}{3\bar{v}\alpha}\right)^{-1}.$$

1. Calculate Γ/Ω_g as a function of the liquid water content L and of the droplet radius a. It is useful to consider n, the number of water droplets.
2. Assume homogeneity of the gas and aqueous phases. Formulate an equation for the time evolution of $\bar{c}_g(t)$. The concentration at the droplet surface will be written as c_g^s.
3. The assumption of homogeneity cannot be valid in the vicinity of the droplet surface because there is a strong gradient due to the mass transfer (Fig. 5.22). Assume that molecular diffusion is a fast process in the vicinity of the water droplet. Assume spherical symmetry: let r be the distance to the droplet center. The gas-phase concentration is then given in this boundary layer by

$$\Delta G = 0, \qquad \lim_{r\to\infty} G = \bar{c}_g, \qquad \lim_{r\to a} G = c_g^s.$$

   Solve this equation and calculate c_g^s. Conclude for the parameterization.
   Hint: the Laplacian operator can be computed as

$$\Delta G(r) = \frac{1}{r^2} \times \frac{d(r^2 \frac{dG}{dr})}{dr}.$$

Solution:

1. L is defined by $n \times 4\pi a^3/3\Omega_g$. The total surface is $\Gamma = n \times 4\pi a^2$. Thus we obtain $\Gamma/\Omega_g = 3L/a$.
2. After integration over the droplets, one gets, with the assumption of homogeneity,

$$\frac{d\bar{c}_g}{dt} = \chi_g(\bar{c}_g) - \frac{\Gamma}{\Omega_g}\frac{\alpha\bar{v}}{4}\left(c_g^s - \frac{c_a}{HRT}\right) = \chi_g(\bar{c}_g) - \frac{3L}{a}\frac{\alpha\bar{v}}{4}\left(c_g^s - \frac{c_a}{HRT}\right).$$

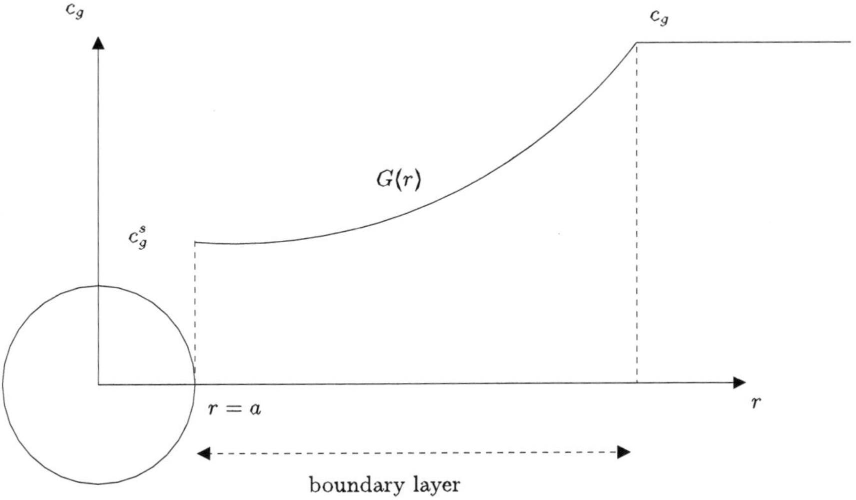

Fig. 5.22 Mass transfer and diffusion boundary layer for a cloud droplet. Problem 5.2

3. In the boundary layer, express the Laplacian operator as $G(r) = A/r + B$, where A and B are two unknown constants. The condition at $r = \infty$ gives $B = \bar{c}_g$ while the condition at the drop surface leads to

$$D_g \left(\frac{dG}{dr} \right)_{r=a} = \frac{\alpha \bar{v}}{4} \left(c_g^s - \frac{c_a}{HRT} \right) = -D_g \frac{A}{a^2}.$$

Thus $G(r) = \bar{c}_g - \frac{\alpha \bar{v} a^2}{4 D_g r} (c_g^s - \frac{c_a}{HRT})$. When evaluated at $r = a$, this expression yields

$$c_g^s - \frac{c_a}{HRT} = \frac{1}{1 + \frac{\alpha \bar{v} a}{4 D_g}} \left(\bar{c}_g - \frac{c_a}{HRT} \right).$$

The time evolution of $\bar{c}_g$ becomes

$$\frac{d\bar{c}_g}{dt} = \chi_g(\bar{c}_g) - L \times k_{mt} \left(\bar{c}_g - \frac{c_a}{HRT} \right),$$

with

$$\frac{1}{k_{mt}} = \frac{a^2}{3 D_g} + \frac{4a}{3 \bar{v} \alpha}.$$

To know more ([137]):

B. SPORTISSE AND R. DJOUAD, *Mathematical investigation of mass transfer for atmospheric pollutants into a fixed droplet with aqueous chemistry*, J. Geophys. Res., **108** (2003), p. 4073

Problem 5.3 (Vertical Distribution of Aerosol Concentrations) The vertical distribution of the coarsest aerosols (with a radius above 10 µm) is governed by the competition between atmospheric turbulence and gravitational settling. Let K_z be the vertical eddy diffusion coefficient and $v_s(r)$ be the settling velocity for particles of radius r.

1. Assume a stationary equilibrium between turbulence and sedimentation. Write the equation satisfied by the concentration c of particles of radius r.
   It is useful to introduce the air density ρ, given as a function of the altitude z with a scale height H. At ground ($z=0$), the emissions of particles are defined by an emission rate (E) and dry deposition is described by a dry deposition velocity $v_{dep}(r)$. The ground turbulent flux is (Chap. 6)
$$-\rho K_z \frac{\partial}{\partial z}\left(\frac{c}{\rho}\right)_{z=0} = E - v_{dep}(r) c_0,$$
   with c_0 the ground concentration. Moreover, the boundary conditions at the domain top are $\lim_{z\to\infty} c(z) = 0$.
2. Calculate the scale height $H(r)$ for the particles of radius r.

Hint: the vertical turbulent flux for aerosols (Chap. 3 and Chap. 6) will be parameterized by
$$\langle w'c'\rangle = -\rho K_z \frac{\partial}{\partial z}\left(\frac{c}{\rho}\right),$$
with w the vertical component of the wind velocity.

Solution:

1. Using the parameterization for the vertical turbulent flux we obtain
$$\frac{\partial c}{\partial t} + \frac{\partial}{\partial z}(-v_s(r)c) = \frac{\partial}{\partial z}\left[\rho K_z \frac{\partial}{\partial z}\left(\frac{c}{\rho}\right)\right].$$
   The assumption of stationary equilibrium leads to
$$-v_s(r)c - \rho K_z \frac{\partial}{\partial z}\left(\frac{c}{\rho}\right) = A,$$
   with A a constant. The boundary condition at ground ($z=0$) gives $A = E - v(r)c_0$, after having written $v(r) = v_s(r) + v_{dep}(r)$.
2. We solve this equation by taking as variable the mixing ratio $q(r) = c(r)/\rho$. Using $\rho(z) = \rho_0 \exp(-z/H)$ we obtain
$$\frac{\partial q}{\partial z} + \frac{v(r)}{K_z} q = \frac{v(r)c_0 - E}{\rho_0 K_z} \exp\left(\frac{z}{H}\right).$$
   Upon integration one gets
$$q(z) = B \exp\left(-\frac{v(r)}{K_z} z\right) + \frac{v(r)c_0 - E}{\rho_0 K_z} \frac{1}{\frac{1}{H} + \frac{v(r)}{K_z}} \exp\left(\frac{z}{H}\right),$$
   where B is a constant. Thus the aerosol concentration is
$$c(z) = \rho_0 B \exp\left[-\left(\frac{v(r)}{K_z} + \frac{1}{H}\right) z\right] + \frac{v(r)c_0 - E}{K_z} \frac{1}{\frac{1}{H} + \frac{v(r)}{K_z}}.$$

The boundary condition $\lim_{z\to\infty} c(z) = 0$ gives $c_0 = E/v(r)$ but this is also $\rho_0 B$. Thus we obtain

$$c(z) = \frac{E}{v(r)} \exp\left[-\left(\frac{v(r)}{K_z} + \frac{1}{H}\right)z\right].$$

3. The scale height for a particle of radius r is then

$$H(r) = \frac{1}{\frac{v(r)}{K_z} + \frac{1}{H}} = \frac{H}{1 + \frac{Hv(r)}{K_z}}.$$

This is, as expected, a decreasing function of the settling velocity.

Problem 5.4 (Evolution of Traffic-Induced Fine Particles) Regulations of traffic-induced emissions have resulted in a strong decrease in soot emissions (with a diameter of a few tens of nanometers). The use of catalytic oxidation and of particle filters, and the decrease of the sulfur concentration in fuels have contributed to this evolution. The indirect effects can unfortunately be difficult to estimate.

1. Recall briefly the processes that govern the production and growth of atmospheric aerosols. Detail the processes in which sulfuric acid (H_2SO_4) plays a leading role. We recall that sulfuric acid is produced by the oxidation of SO_2 by OH. Moreover it has a saturation vapor pressure with a very low value. Sulfuric acid is then in particulate phase.

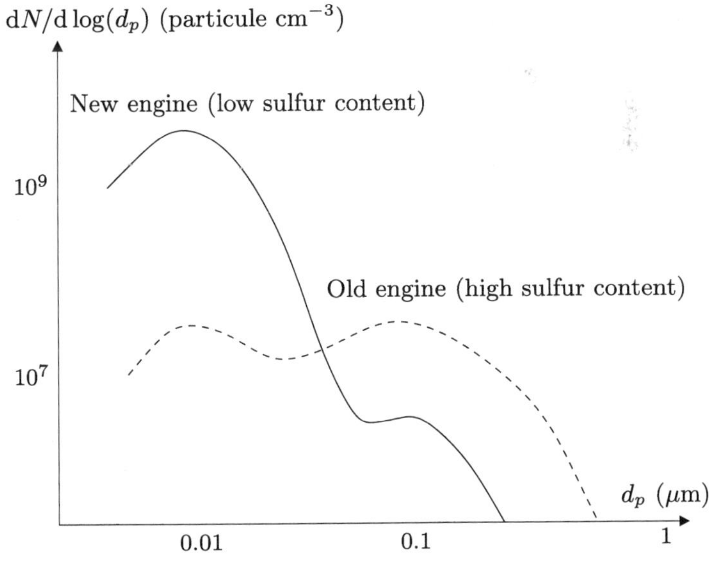

Fig. 5.23 Sketch of the evolution of the size distribution for particles in the vicinity of a car exhaust pipe: new engines (low concentration of sulfur) versus old engines (high concentration of sulfur). Problem 5.4. Source: [76]

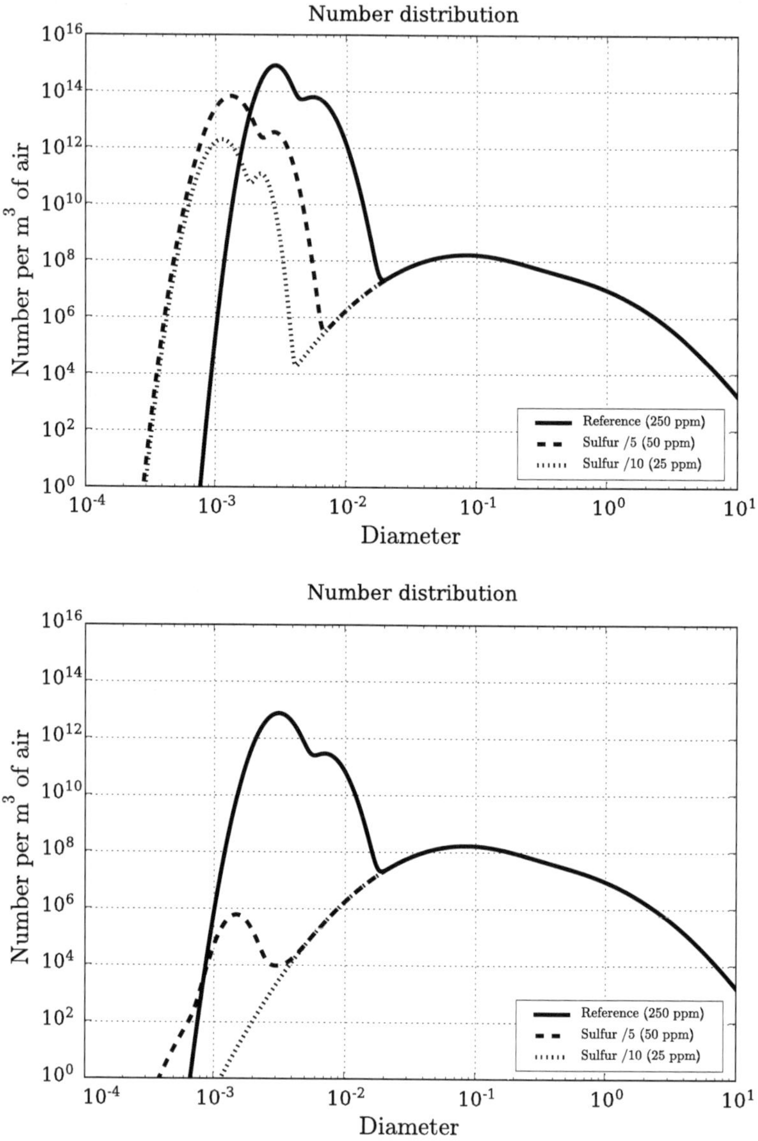

Fig. 5.24 Impact of a decrease in the sulfur concentration of fuels for the formation of fine particles in the vicinity of a car exhaust pipe. Simulation with the CFD model Mercure_Saturne and the aerosol model MAM. Case study: strong wind. *Upper panel*: winter (strong impact of nucleation); *lower panel*: summer (low impact of nucleation). Credit: Bastien Albriet, CEREA

2. We consider an aerosol distribution with two modes in Fig. 5.23: the first peak is centered at 10 nm while the second peak is centered at 100 nm. Compare the evolution of the particle number between the old and new engines. Comment on the evolution of the aerosol mass.

3. What process will favor a decrease of the concentration of the coarsest aerosols? Comment Fig. 5.23.

Solution:

1. The processes that contribute to the ageing of atmospheric aerosols are nucleation, especially for the binary mixture H_2O–H_2SO_4, and condensation. H_2SO_4 is implied both in nucleation and in condensation. There is therefore a competition between the two processes with respect to the available gas-phase sulfuric acid.
2. The ratio of the mass for spherical particles of diameter d_p to the mass for particles of diameter $d_p/10$ is equal to 1000 ($m = \pi d_p^3/6 \times \rho_p$ with ρ_p the aerosol density assumed to have a constant value). The mass of a particle with a diameter of 100 nm is therefore 1000 times larger than the mass of a particle of diameter 10 nm.
   The size distribution corresponding to old engines reveals high concentrations for the coarsest particles and therefore a higher mass. The size distribution corresponding to new engines has been shifted to the finest particles. The resulting mass is lower but the number of fine particles is much higher.
3. The emission of primary soots provides condensation surfaces for H_2SO_4. Decreasing the condensation surface leads, for new engines, to favor nucleation of H_2SO_4. This results in an increase in nucleated fine particles. This is an indirect effect of the reduced emissions of the coarsest aerosols (Fig. 5.24).

To know more ([76]):

D. KITTELSON, J. JOHNSON, W. WATTS, Q. WEI, M. DRAYTON, D. PAULSEN, AND N. BUKOWIECKI, *Diesel aerosol sampling in the atmosphere*, Technical paper series 2000-01-2212, SAE (Society of Automative Engineers), 2000

Chapter 6
Toward Numerical Simulation

This chapter aims at reviewing the fundamentals of numerical simulation for the physical processes described in the previous chapters.

The hypothesis of dilution states that the time evolution of concentrations can be split from the time evolution of the meteorological fields (wind velocity, air density, temperature, humidity). The time evolution of concentrations is then given by an advection-diffusion-reaction partial differential equation, the so-called equation of *reactive dispersion*. Advection corresponds to wind transport, diffusion to turbulent mixing, and reactions to physical and chemical processes. The three-dimensional numerical models that solve these equations are the so-called *chemistry-transport models* (CTM). Such models are the basis of any state-of-the-science modeling systems for air quality.

Numerical simulation of the dispersion equation is a challenging issue. For example, the dimension of the partial differential equations can be up to tens of chemical species (Table 6.1). The characteristic timescales also reveal a wide range: from fractions of seconds for the fastest chemical reactions or condensation onto the finest aerosols, to hours, days or years for the most stable species. This induces the so-called numerical *stiffness*, one of the most difficult issues for numerical simulation, with several consequences: computational burden, use of tailored numerical solvers for the time integration, and so on.

This chapter is organized as follows. The reactive dispersion equation is formulated for averaged variables in Sect. 6.1. The underlying model hierarchy (from Gaussian models to three-dimensional Eulerian chemistry-transport models) is then presented. Section 6.2 is devoted to the fundamentals of numerical analysis required for the chemistry-transport models. Attention is paid to *operator splitting* in Sect. 6.2.1, time integration of chemical kinetics in Sect. 6.2.2, and advection in Sect. 6.2.3. The numerical simulation of aerosol dynamics is investigated in Sect. 6.3. To conclude, high-level methods, such as the propagation of uncertainties or data assimilation (coupling between the model outputs and the measured data in order to constrain the simulation), are briefly introduced in Sect. 6.4.

B. Sportisse, *Fundamentals in Air Pollution*,

Table 6.1 Number of chemical variables per grid cell in a CTM for different applications (accidental release, photochemistry, multiphase chemistry). Note that photochemistry, cloud chemistry and aerosols have to be taken into account for a multiphase model

Application	Number of species
Passive tracer	1
Photochemistry	50–100
Cloud chemistry	10–50
Aerosols	$\geq$ 100 (20 times, *a minima*, 5 size sections)

6.1 Reactive Dispersion Equation

6.1.1 Dilution and Off-Line Coupling

It is desired to describe the evolution of trace species concentrations in the atmosphere: the trace species can be chemical species, biological compounds or radionuclides in the gas and particulate phase.

The evolution of the reactive fluid is given by the reactive Navier-Stokes equations, defined by adding species continuity equations to the Navier-Stokes equations (Sect. 3.5). When solved, this coupled model is usually referred to as an *on-line* model.

Evolution of meteorological fields and evolution of species concentrations are usually decoupled: this is the so-called *dilution* hypothesis. This can be motivated when the contributions of trace species to the evolution of temperature (or of internal energy) can be approximated by a constant value. A first (direct) contribution is related to the heat fluxes generated by chemical reactions. A second (indirect) contribution is related to interactions between atmospheric matter and solar and terrestrial radiations (Chap. 2).

In a first approximation, the first contribution can be neglected in the troposphere. The second contribution plays a leading role in atmospheric dynamics: it can be neglected for air quality studies at regional scales but not at a global scale for long-term climate studies.

In the sequel, we assume that the dilution hypothesis is satisfied. The meteorological fields (wind velocity, temperature, density, relative humidity, etc.) are supposed to be computed by a meteorological model in a preprocessing step. They are then given as input data for the dispersion equations of the trace species. This approach is usually referred to as *off-line coupling*.

6.1.2 Advection-Diffusion-Reaction Equations

Concentrations The processes that govern evolution of the atmospheric trace species are shown in Fig. 6.1. The species X_i is represented by a concentration,

c_i, expressed in a number of moles (or of molecules) per air volume, or by a mass concentration expressed in units of mass per air volume.

Let us consider a gas-phase species. The evolution of c_i is then given by a system of partial differential equations (PDE),

$$\frac{\partial c_i}{\partial t} + \mathrm{div}[V(x,t)c_i] = \mathrm{div}(K_{molec}\nabla c_i) + \chi_i[c, T(x,t), t] + S_i(x,t) - \Lambda_i c_i, \quad (6.1)$$

where x and t stand for the spatial and time coordinates, respectively, and c is the vector of concentrations.

Different data sets are required:

- *Meteorological fields.*
  $V(x,t)$ is the wind velocity, K_{molec} is the molecular-diffusion matrix (we take into account, *a priori*, intermolecular diffusion), $T(x,t)$ is the temperature.
- *Emission data.*
  $S_i(x,t)$ is the source term for species X_i: it describes the point emissions (typically a chimney).
- *Physical and chemical parameterizations.*
  χ_i stands for the chemical source term of X_i in the gaseous phase (Chap. 4). It depends on concentrations, temperature, radiative actinic flux and relative humidity (the last two variables are omitted for the sake of clarity).
  Scavenging terms (for wet removal by precipations and clouds) are usually parameterized by the term $-\Lambda_i c_i$. The scavenging coefficient, Λ_i, is a function of the meteorological fields (liquid water content or rain intensity) and of several microphysical variables (Sect. 5.3.2).

Mixing Ratio Another description is based on the mixing ratio, written as $C_i = c_i/\rho$ (expressed in mol mol^{-1}). The air density satisfies the continuity equation (Chap. 3)

$$\frac{\partial \rho}{\partial t} + \mathrm{div}(\rho V) = 0. \quad (6.2)$$

It is then easy to prove (Exercise 6.1) that the evolution of C_i is given by

$$\frac{\partial C_i}{\partial t} + V \cdot \nabla C_i = \frac{1}{\rho}\mathrm{div}(K_{molec}\nabla(\rho C_i)) + \frac{\chi_i(\rho C, T(x,t), t) + S_i(x,t)}{\rho} - \Lambda_i C_i. \quad (6.3)$$

Exercise 6.1 (Mixing Ratio and Mass Consistency) We consider the case of pure advection by the wind. Formulate the evolution equation of the mixing ratio. Prove the *mass consistency* property: a passive tracer that is initially well mixed (namely with a homogeneous mixing ratio) remains well mixed.

Solution:

It is straightforward to obtain:

$$\frac{\partial C_i}{\partial t} = \frac{1}{\rho}\frac{\partial c_i}{\partial t} - \frac{c_i}{\rho^2}\frac{\partial \rho}{\partial t} = \frac{-\mathrm{div}(Vc_i)}{\rho} + \frac{c_i}{\rho^2}\mathrm{div}(\rho V) - \frac{-\mathrm{div}(\rho V C_i) + C_i \mathrm{div}(\rho V)}{\rho}$$

$$= -\left(V\cdot\nabla C_i + \frac{C_i}{\rho}\mathrm{div}(\rho V)\right) + \frac{C_i}{\rho}\mathrm{div}(\rho V) = -V\cdot\nabla C_i.$$

It is then easy to check the mass consistency property: if $C_i(t=0)=k$, with k a constant, $C_i(t)=k$ is the unique solution for the advection equation.
To know more ([138]):
B. SPORTISSE, D. QUÉLO, AND V. MALLET, *Impact of mass consistency errors for atmospheric dispersion*, Atmos. Env., **41** (2007), pp. 6132–6142

6.1.3 Averaged Models and Closure Schemes

Equation (6.1) is only valid at the "microscopic" scale (at least in a continuum description) but cannot be used, as such, for turbulent flows. As it is illustrated in Chap. 3, it is not possible to simulate all scales in a three-dimensional case and averaging approaches are necessary.

We assume that the fields (meteorological fields, trace species concentrations) can be decomposed as $\Psi = \langle\Psi\rangle + \Psi'$ with $\langle\Psi\rangle$ an averaged value and Ψ' a fluctuation. We do not detail the averaging operator: note that it is supposed to satisfy commutation of derivation (with respect to time and space) and $\langle\Psi'\rangle = 0$.

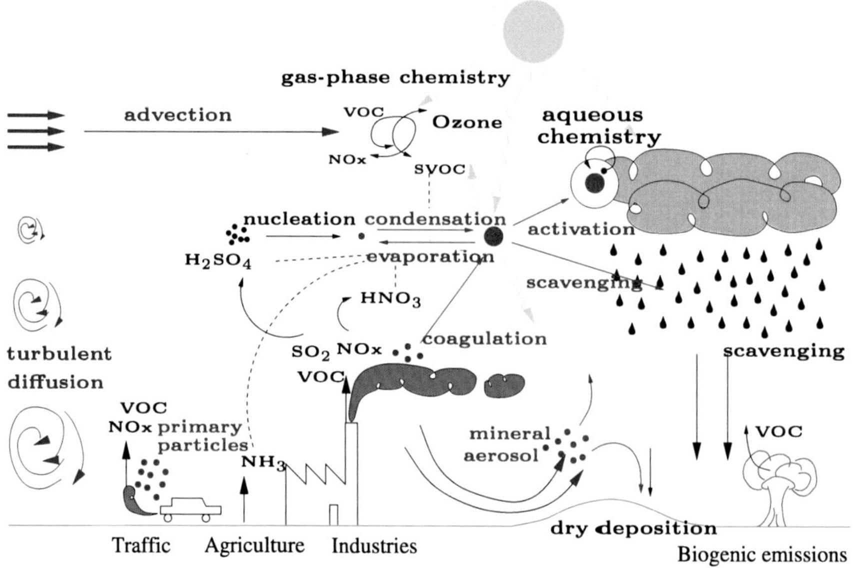

Fig. 6.1 Sketch of the processes described in a tropospheric chemistry-transport model

When applied to (6.1), this decomposition leads, after having averaged the resulting equation, to

$$\begin{aligned}\frac{\partial \langle c_i \rangle}{\partial t} &+ \mathrm{div}(\langle V(x,t)\rangle \langle c_i \rangle) \\ &= \mathrm{div}(K_{molec} \nabla \langle c_i \rangle) + \langle \chi_i(c, T(x,t), t)\rangle \\ &+ \langle S_i(x,t)\rangle - \langle \Lambda_i c_i \rangle - \mathrm{div}(\langle c_i' V' \rangle).\end{aligned} \tag{6.4}$$

The key intermediate step is $\langle c_i + c_i' \rangle \langle V + V' \rangle = \langle V \rangle \langle c_i \rangle + \langle c_i' V' \rangle$.

Note that the *linear* terms are not modified in the averaged equation. The *non-linear* terms (here in a quadratic form) generate *correlations* (averaged products of fluctuations). A *closure scheme* for the averaged equations consists in expressing these correlations as a function of the *resolved* variables (the averaged fields).

The two main processes requiring a closure scheme are wind advection (vertical turbulent flux) and chemical production (segregation effect).

6.1.3.1 Turbulent Flux

The averaging of the continuity equation for species X_i introduces a new term associated to wind advection, $\mathrm{div}\langle c_i' V' \rangle$. This term can be viewed as a turbulent flux.

The closure scheme is usually based on the so-called K-theory, by analogy with molecular diffusion (Chap. 3). The parameterization is then given for a field Ψ by

$$\langle \Psi' V' \rangle = -K_{turb}^{\Psi}(x,t) \nabla \langle \Psi \rangle, \tag{6.5}$$

with K_{turb}^{Ψ} the eddy diffusion matrix. In practice K_{turb}^{Ψ} is a function of the meteorological fields and their gradients, and it depends on space and time.

The application of this parameterization to the concentration c_i or to the mixing ratio C_i does not lead to the same result. Actually, it is more relevant to apply the parameterization to the mixing ratio,

$$\langle C_i' V' \rangle = -K_{turb} \nabla \langle C_i \rangle. \tag{6.6}$$

The fluctuation of density is usually neglected ($\rho' = 0$). As $c_i = \rho C_i$, one gets $c_i' V' \simeq \langle \rho \rangle C_i' V'$ and $\langle c_i \rangle \simeq \langle \rho \rangle \langle C_i \rangle$. The resulting parameterization for the concentration is then

$$\langle c_i' V' \rangle = -\langle \rho \rangle K_{turb} \nabla \left(\frac{\langle c_i \rangle}{\langle \rho \rangle} \right). \tag{6.7}$$

The advantage of this parameterization is that the mass consistency property is still satisfied (Exercise 6.1). If $\langle c_i \rangle / \langle \rho \rangle$ is initially homogeneous, the homogeneity is conserved for the averaged equation. This is easy to prove since the averaged equation is exactly the initial equation, as the turbulent flux is equal to zero. Such a result would not have been obtained by applying the K-theory to the concentration.

The turbulent diffusion matrix is supposed to be the same one for all species, and inter-molecular turbulent diffusion is neglected. The matrix K_{turb} is usually a *diagonal* matrix: we write K_x, K_y and K_z the eddy diffusion coefficients for the directions x, y and z, respectively.

In pratice, K_{turb} is much higher than K_{molec} (about $10^{-5}\,\mathrm{m^2\,s^{-1}}$): the diffusion flux is chiefly the turbulent diffusion flux except in a thin laminar layer just above the ground.

Moreover, the horizontal diffusion terms can be usually neglected since it dominates the horizontal advection transport. They are somehow not well parameterized and are exceeded by the *numerical diffusion* of advection schemes (Sect. 6.2.3 and Exercise 6.6). To conclude, only the eddy vertical diffusion is usually taken into account. It is described by K_z (typically about $10\,\mathrm{m^2\,s^{-1}}$).

6.1.3.2 Segregation Effect

The averaging procedure leads to a closure problem for nonlinear chemical kinetics. The rate of a bimolecular chemical reaction with reactants X_i and X_j (Chap. 4) is

$$\langle c_i c_j \rangle = \langle c_i \rangle \langle c_j \rangle + \langle c'_i c'_j \rangle, \tag{6.8}$$

where the correlation term $\langle c'_i c'_j \rangle$ is not resolved. This term is usually neglected. This assumption is sometimes referred to as the *well-stirred tank reactor* hypothesis. In this case we obtain

$$\langle \chi(c) \rangle \simeq \chi(\langle c \rangle). \tag{6.9}$$

This assumption is valid only if the characteristic timescales of chemical kinetics are much higher than the timescales of turbulent mixing (Exercise 6.2).

Note that the correlation term is sometimes called a *segregation* term. From this viewpoint, we can also write

$$\langle c_i c_j \rangle = \langle c_i \rangle \langle c_j \rangle (1 + I_s), \quad I_s = \frac{\langle c'_i c'_j \rangle}{\langle c_i \rangle \langle c_j \rangle}, \tag{6.10}$$

with I_s the *segregation intensity*. It is easy to check that $I_s \geq -1$ due to the positivity of concentrations.

Exercise 6.2 (Damköhler Number) We define a Damköhler number, Da, as the ratio of a characteristic timescale of chemical kinetics to a characteristic timescale of turbulent mixing. Comment the different regimes defined by the magnitude of Da. Classify a few chemical species.

Solution:

When $Da \ll 1$ (typically for long-lived VOCs, PAN, etc.), we can consider that the environment is well mixed: there is no segregation. When $Da \gg 1$, chemical kinetics is at equilibrium before turbulence mixing of pollutants: chemical kinetics first governs the evolution of species (for short-lived species, such as radicals, Exercise 6.5: typically OH and HO_2). When $Da \sim 1$, the characteristic timescales of chemical kinetics and of turbulent mixing have similar values and the processes are strongly coupled: the laminar assumption is no longer valid

and the segregation effect has to be taken into account (typically for O_3 and NO_2).
To know more ([147]):
J. VILA-GUERAU DE ARELLANO, A. DOSIO, J. VINUESA, A. HOLTSLAG, AND S. GALMARINI, *The dispersion of chemically reactive species in the atmospheric boundary layer*, Meteor. Atmos. Phys., **87** (2004)

If two species are not correlated, $I_s = 0$. Otherwise, the well-stirred tank reactor hypothesis can lead to a deviation to the effective chemical source term, defined at the *model resolution*. The classical illustration is the titration reaction for ozone and nitrogen monoxide, a decisive reaction for photochemistry (Chap. 4),

$$NO + O_3 \longrightarrow NO_2 + O_2. \quad \text{(R 148)}$$

Nitrogen monoxide (NO) is emitted at ground (about 90% of NO_x emissions are in this form), while ozone (O_3), that is transported and formed at regional scales, is dominant above the boundary layer. Henceforth NO and O_3 are not well mixed: NO is rather located in *updrafts* while O_3 is rather located in *downdrafts*. The segregation intensity is then negative ($I_s < 0$) and the homogeneity hypothesis leads to an overestimation of the reaction rate for the titration reaction.

Figure 6.2 illustrates the heterogeneity of the reaction rate in the atmospheric boundary layer. The computation is carried out with a DNS model (*Direct Navier Stokes*), to be viewed as an "exact" computation. The zones with high values of the reaction rate (just above the ground) correspond to the zones in which the reactants are in contact (and the reaction can take place).

Similarly, measurements of the segregation intensity are shown in Fig. 6.3. The time evolution reveals peaks (in absolute value) corresponding to emissions peaks of NO_x (in the early morning and in the afternoon). The values range from 0 to −0.65 and are very weak at night.

More generally, the homogeneity hypothesis is not satisfied for the fastest reactions in the vicinity of emissions. However, the parameterization of I_s is still an open question for current state-of-the-science models (in 2008).

6.1.3.3 Averaged Dispersion Equation

Let us assume that the segregation intensity is zero. If we neglect the correlations for the scavenging term, the averaged dispersion equation is then

$$\begin{aligned} \frac{\partial \langle c_i \rangle}{\partial t} &+ \operatorname{div}(\langle V(x,t)\rangle \langle c_i \rangle) \\ &= \operatorname{div}\left[\langle \rho \rangle K_{turb} \nabla \left(\frac{\langle c_i \rangle}{\langle \rho \rangle} \right)\right] + \chi_i(\langle c \rangle, \langle T(x,t)\rangle, t) \\ &+ \langle S_i(x,t)\rangle - \langle \Lambda_i \rangle \langle c_i \rangle. \end{aligned} \quad (6.11)$$

In the sequel, for the sake of clarity, the notation Ψ will stand for the averaged field.

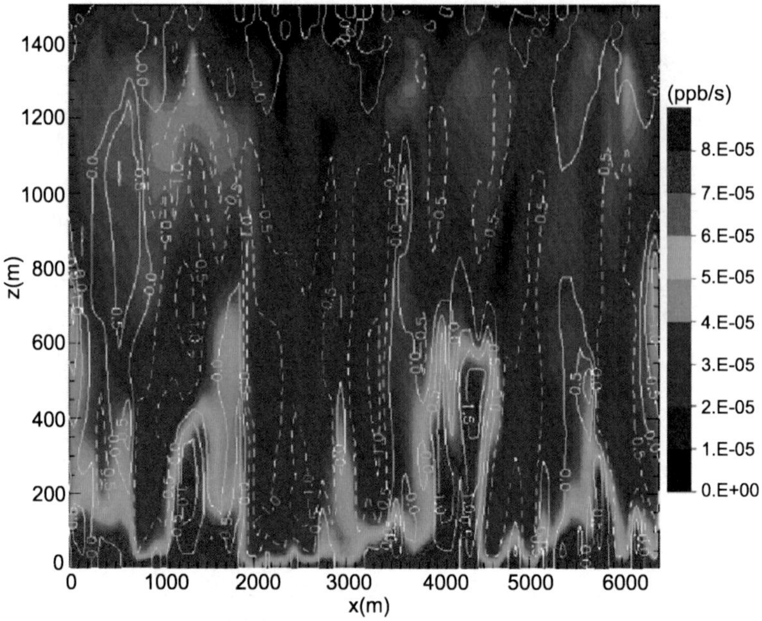

Fig. 6.2 Reaction rate of the titration reaction $NO + O_3$ in the atmospheric boundary layer. Computed with a DNS approach (*Direct Navier Stokes*). Credit: Jordi Vila-Guerau de Arellano ([97])

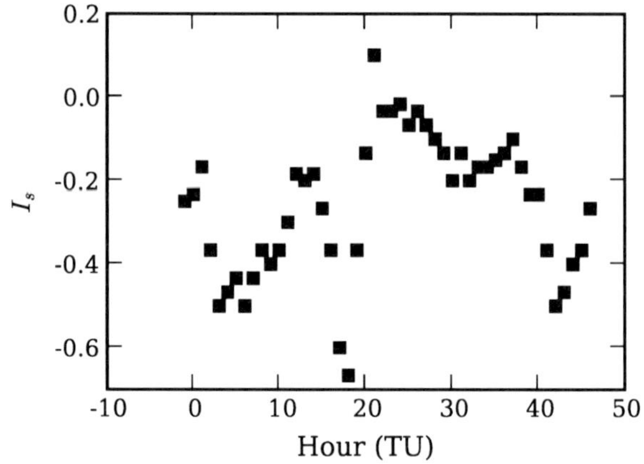

Fig. 6.3 Time evolution of the segregation intensity for the titration reaction, at ground. The segregation intensity is estimated with a weighting approach applied to measurement stations in the Netherlands (15–16 September 1989). Source: [148]

6.1.4 Boundary Conditions

Initial conditions and boundary conditions have to be specified for these advection-diffusion-reaction equations.

The horizontal transport is dominated by wind advection while the vertical transport is dominated by eddy diffusion (Chap. 3). The horizontal boundary conditions are then associated to inward-pointed wind velocities while the vertical boundary conditions are defined by diffusion fluxes.

For example, the emissions and the dry deposition define the ground boundary conditions. If z stands for the vertical coordinates, the boundary condition at $z = 0$ reads

$$-K_z \frac{\partial c_i}{\partial z} = E_i(x,t) - v^i_{dep}(x,t) c_i. \tag{6.12}$$

The surface emission of species X_i, E_i, depends on emission conditions (rural, urban, regional) and can be split into an anthropogenic part (e.g. related to traffic-induced emissions) and a biogenic part. The dry deposition velocity v^i_{dep} is parameterized for each chemical species according to the meteorological conditions in the surface boundary layer (Chap. 3) and to the ground type. The ground type is usually characterized by its *land use cover* (LUC) and by a specific roughness parameter (Table 3.5).

At the top of the domain ($z = z_H$), namely at the interface with the so-called *free atmosphere*, the boundary condition is a null diffusion flux,

$$-K_z(x,t) \frac{\partial c_i}{\partial z} = 0. \tag{6.13}$$

6.1.5 *Model Hierarchy*

Several approaches can be used to solve (6.11). A selection of available model types is summarized in Table 6.2.

6.1.5.1 Gaussian Models

Academic Case The most simple models are the Gaussian models ([10] for a deeper understanding). They can be introduced by an academic case. We consider a passive tracer that is emitted by a point source. The environment is supposed to have

Table 6.2 A selection of models. A few *on-line* coupled models are used for research applications

Model type	applications
Gaussian and plume	accidental release
Particle Lagrangian	accidental release
Eulerian box	forecast (1980–2000)
Lagrangian box	impact study (1980–2000)
3D-Eulerian gas-phase CTM	forecast and impact study (1990–.)
3D Eulerian multiphase CTM	forecast and impact study (2000–.)
On-line coupled model	research (2000/2005–)

a constant wind velocity (u), aligned along the x-axis, and a constant air density. The dispersion equation for the mixing ratio is therefore

$$\frac{\partial C}{\partial t} + u\frac{\partial C}{\partial x} = K_x\frac{\partial^2 C}{\partial x^2} + K_y\frac{\partial^2 C}{\partial y^2} + K_z\frac{\partial^2 C}{\partial z^2}. \tag{6.14}$$

We investigate a point emission, emitted at $(x, y, z) = (0, 0, 0)$ and at time $t = 0$. The initial condition is supposed to be $C(x, y, z, 0) = S\delta(x)\delta(y)\delta(z)$, with $\delta(.)$ the Dirac function at 0. In the academic case of *infinite* dimension (no boundary conditions), the exact solution is given by the Gaussian function,

$$C(x, y, z, t) = \frac{S}{(2\pi\, t)^{3/2}\sqrt{K_x K_y K_z}} \exp\left[-\frac{(x-ut)^2}{4K_x t} - \frac{y^2}{4K_y t} - \frac{z^2}{4K_z t}\right]. \tag{6.15}$$

The plume extension, for instance in the x direction, is then determined by the variance $2K_x t$ (often written as σ_x^2).

This example provides the basis for Gaussian models. In practice, there exist many situations, defined by the characteristics of sources and by ground boundary conditions. For each case, a specific analytical formula can be derived. The domain of validity is the vicinity of the sources (a few kilometers).

Gaussian Plume Model A *Gaussian plume model* (Fig. 6.4) is a stationary model that can be used in the case of a continuous source (continuous with respect to time), emitted at $x = y = 0$ and at the height h, with a ground boundary condition of total reflection. The formula is

$$C(x, y, z) = \frac{S}{2\pi u \sigma_y \sigma_z} \exp\left(-\frac{y^2}{2\sigma_y^2}\right)\left[\exp\left(-\frac{(z-h)^2}{2\sigma_z^2}\right) + \exp\left(-\frac{(z+h)^2}{2\sigma_z^2}\right)\right]. \tag{6.16}$$

Taking into account dry deposition can be a challenging issue and several approaches exist. The variances, $\sigma_\star^2$, are given by parameterizations as a function of

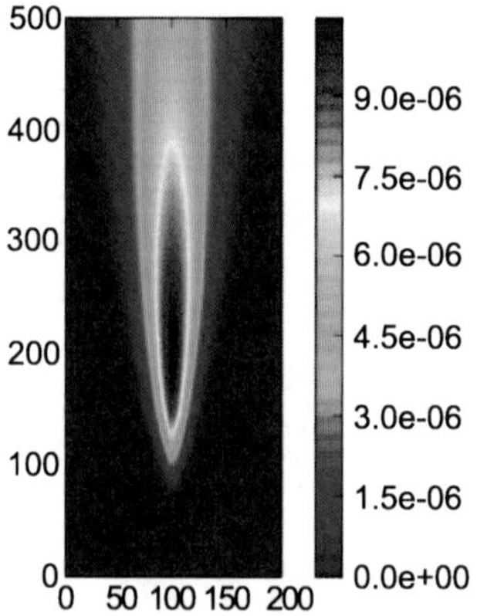

Fig. 6.4 Ground activity of a radionuclide (in Bq), computed by a Gaussian model (with the Briggs parameterization for $\sigma_\star^2$). Simulation with the POLYPHEMUS system. Credit: Irène Korsakissok, CEREA

the distance to the point source and of the meteorological fields. These (rather) empirical parameterizations can be functions of the atmospheric stability or computed from the meteorological fields (such as the Monin-Obukhov length or the mixing height, Sect. 3.5.2.4).

Gaussian Puff Model A natural extension of the Gaussian plume model is the *Gaussian puff model*, that takes into account the time variation of emissions and of meteorological conditions. We suppose that a *puff* is emitted during N successive time intervals of duration Δt. Each puff has its own evolution, given by a Gaussian model, independently from the others. The resulting concentration is then calculated by summing over all puffs (Fig. 6.5),

$\forall t \geq N\Delta t$,

$$C(x,y,z,t) = \frac{1}{(2\pi)^{3/2}} \sum_{i=1}^{N} \frac{S_i \Delta t}{\sigma_x \sigma_y \sigma_z} \exp\left[-\frac{(x-x_i(t))^2}{2\sigma_x^2} - \frac{(y-y_i(t))^2}{2\sigma_y^2}\right] \times \left[\exp\left(-\frac{(z-z_i(t)-h)^2}{2\sigma_z^2}\right) + \exp\left(-\frac{(z-z_i(t)+h)^2}{2\sigma_z^2}\right)\right], \quad (6.17)$$

with $S_i \Delta t$ the emission at time $t_i = i\Delta t$. The location of the puff center i is given, for instance along the x-axis, by

$$x_i(t) = x_i(t-\Delta t) + u[x_i(t-\Delta t), t-\Delta t]\Delta t, \quad (6.18)$$

with $u(x,t)$ the component of the wind velocity along the x-axis. Modeling the evolution of the puff size is a difficult point for such models: many parameterizations describe the time and space evolutions for $\sigma_\star^2$ ([10]).

Even if these models have many limitations, they are easy to use, which explains their popularity, especially for applications related to accidental releases of *passive* or *linear* tracers, and to short-range dispersion (Fig. 6.4). Unfortunately, the extension of these models to the reactive case is a challenging issue and lacks rigorous justification.

6.1.5.2 Lagrangian Particle Model

A *Lagrangian particle model* is based on the tracking of *numerical particles*. The theoretical background is the stochastic interpretation of the passive dispersion equation (Fokker-Planck equation). For a wind velocity $\mathbf{V} = (u, v, w)$ and a diagonal diffusion matrix $K = (K_x, K_y, K_z)$, the trajectory of a numerical particle is $(x(t), y(t), z(t))$. The iteration from time t_n to time $t_{n+1} = t_n + \Delta t$ is then

$$\begin{aligned} x(t_{n+1}) &= x(t_n) + \left(u + \frac{\partial K_x}{\partial x}\right)\Delta t + \sqrt{2K_x}\Delta W_x, \\ y(t_{n+1}) &= y(t_n) + \left(v + \frac{\partial K_y}{\partial y}\right)\Delta t + \sqrt{2K_y}\Delta W_y, \\ z(t_{n+1}) &= z(t_n) + \left(w + \frac{\partial K_z}{\partial z}\right)\Delta t + \sqrt{2K_z}\Delta W_z. \end{aligned} \quad (6.19)$$

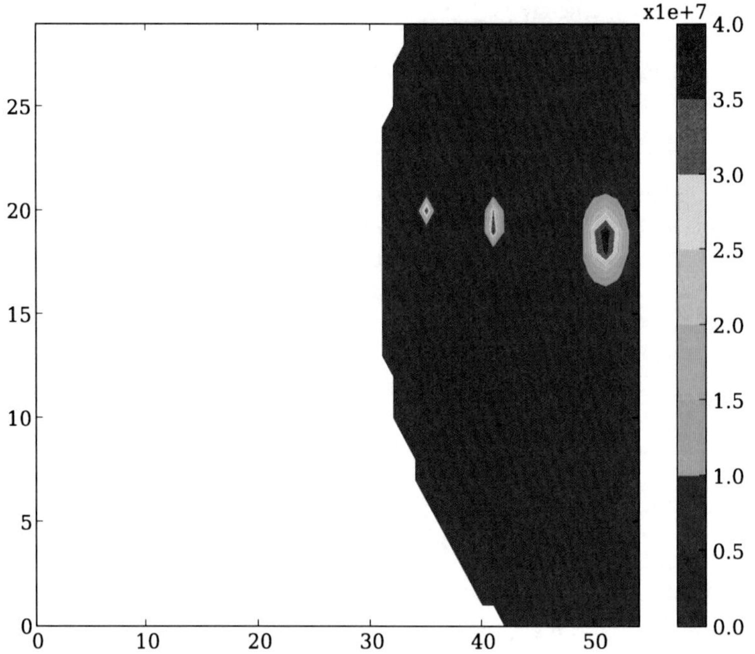

Fig. 6.5 Ground concentration of an aerosol of diameter 1 μm, simulated with a puff model. The concentration is dimensionless. Simulation with the POLYPHEMUS system. Credit: Irène Korsakissok, CEREA

$\Delta W_\star$ is a "white noise", that is to say a stochastic process following a normal law $\mathcal{N}(0, \Delta t)$ (with a null mean and a variance Δt). The resulting concentration c is calculated by summing over all particles in a given grid cell. It satisfies at the continuous limit (with an infinite number of particles),

$$\frac{\partial c}{\partial t} + \operatorname{div}(Vc) = \operatorname{div}(K\nabla c). \tag{6.20}$$

The advantage of a Lagrangian particle model is a low numerical diffusion, when applied to the dispersion of point sources. The drawback is that the convergence requires a large number of numerical particles (up to tens of thousands). A powerful technique is to use *kernels* by assuming a density to the numerical particles. The extension to the reactive case is however difficult (the correlations between the numerical particles have to be tracked).

6.1.5.3 Box Models

The *box models* are defined by considering large domains ("boxes"), as compared to a grid cell, in which the concentrations are supposed to be homogeneous. For an *Eulerian box model*, one gets upon integration of (6.11) and with the boundary conditions,

$$\frac{d\bar{c_i}}{dt} = \chi_i(\bar{c}) + \frac{1}{z_H}(E_i - v^i_{dep}\bar{c}_i). \tag{6.21}$$

Here, the resolved variable is $\bar{c} = (\int_0^{z_H} c(z)dz)/z_H$. We have also assumed that the chemical source term is homogeneous, that is $(\int_0^{z_H} \chi_i(c(z))dz)/z_H \simeq \chi_i(\bar{c})$.

Many boxes can be connected by introducing flux terms between boxes. For example, one box can represent the mixing layer (z_H is then the mixing height, a function of time) while another box can represent the residual layer (Exercise 3.3). These models were very popular in the 1990s but are no longer used.

The *Lagrangian box models* omit turbulent mixing and solve the equations of chemical kinetics (with dry deposition, emissions and wet scavenging) along the characteristic curves of the flow (Exercise 6.4). The computational cost is low, which motivates the use of these models for impact studies in the 1990s (especially for the estimation of source/receptor matrices for transboundary pollution).

6.1.5.4 Chemistry-Transport Models (CTM)

The chemistry-transport models solve (6.11) with appropriate numerical algorithms. The physical parameterizations associated to the different processes are the key components of a CTM. A flow chart is shown in Fig. 6.6.

Emissions (E_i and S_i) They are usually given by *emission inventories*. The emission inventories are usually poorly accurate. They are based on the so-called *Selected Nomenclature for Air Pollution* (SNAP, Table 6.3 for Greater Paris), defined

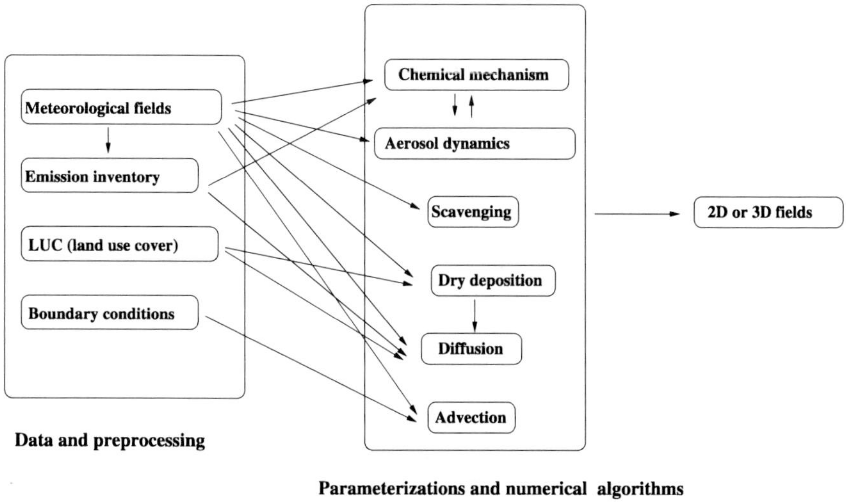

Fig. 6.6 Flow chart of a CTM (*off-line* coupling)

Table 6.3 Emission inventory over Greater Paris (Île-de-France, 2006). The total is expressed in kt year^{-1} (kt for kiloton); the contribution is in % for each SNAP class. Source: Airparif

	SNAP	NO_x	CO	SO_2	VOCs	PM_{10}	CO_2
1	Energy and urban heating	10	1	40.5	0.5	9	12
2	Residential	12	17	42	7.5	28.5	46
3	Industries	4.5	0	10	0	2	5.5
4	Oil industries	0	0	1	1	18	0
5	Oil distribution	0	0	0	3.5	0	0
6	Solvents	0	0	0	41	0	0
7	Road transport	52	77	3.5	29.5	36.5	29
8	Other mobile sources	9	4	1	2.5	3	3
9	Waste incineration	4.5	1	2.	1	2.5	4.5
10	Agriculture	8	0	0	0	0.5	0
11	Biogenic sources	0	0	0	13.5	0	0
	Total (kT year^{-1})	161	398	68	178	22	49567

as a classification of anthropogenic emissions according to the economic sectors. The time distribution of emissions is often much coarser than the distribution required for a CTM (typically with a resolution of one hour). A challenging issue is to determine the chemical composition of emissions (and the size distribution for primary aerosols). For example, the VOCs are given as a whole. One has first to specify the emissions for the true species (*speciation*) and second to map them for the *model species*, the species actually described in the chemical mechanisms (*aggregation*).

Meteorological Fields (V, T, ρ) The meteorological fields are outputs of the meteorological models. They are mapped onto the CTM meshes. The projection and interpolation steps may be prevailing issues in the preprocessing of a CTM. A specific issue is the so-called *mass consistency* (Exercise 6.7).

Eddy Diffusion (K_z) The parameterization of K_z is a decisive component of a CTM. Several parameterizations are available: they are functions of the gradients of the potential temperature and of the wind velocity (Chap. 3). A difficult point is the choice of thresholds for K_z (e.g. its minimal value, typically 0.5 m^2 s^{-1}), to be used for the nocturnal boundary layer, characterized by a strong stability (Chap. 3).

Chemical Mechanisms ($\chi_i(.)$) The chemical mechanism is the second key component of a CTM. For passive tracers or radionuclides, it is replaced by a parameterization of the atmospheric ageing (in practice a lifetime) or by a filiation lifetime. For reactive species the chemical mechanism depends on the targets and on the altitude (see Chap. 4 for a description of tropospheric and stratospheric chemistry of ozone). To date there does not exist any "universal" atmospheric chemical mechanism. A highly detailed mechanism is sometimes referred to as a *master* chemical mechanism. For example, in the case of tropospheric photochemistry, a standard gas-phase chemical mechanism may describe up to one hundred gas-phase chemical species and a few hundreds of chemical reactions.

Wet Scavenging (Λ_i) Scavenging by clouds and precipitations plays a leading role for the soluble gas-phase species and the aerosols (Chap. 5) as it governs their atmospheric residence time. These processes are still poorly known and their parameterization inside three-dimensional models is still a challenging issue. It is all the more difficult that there are large uncertainties in the microphysical parameters that describe rain or clouds.

Dry Deposition (v_i^{dep}) Dry deposition depends on the deposited species, on the meteorological conditions in the surface boundary layer and on the land use cover. It is usually described by a dry deposition velocity (see (6.12)) that is classically parameterized by resistance models (Exercise 3.1). Due to the uncertainties (for instance related to the seasonal evolution of land use coverage), the values of dry deposition velocities are not accurate. Moreover, a few processes, such as codeposition of species, are poorly known.

6.2 Fundamentals of Numerical Analysis for Chemistry-Transport Models

This section and the next one present the fundamentals of numerical analysis for chemistry-transport models. The next section investigates the general dynamic equation for aerosols. The content of both sections is quite technical and demands bases in numerical analysis and applied mathematics. A reader interested in the high-level uses of modern simulation systems can omit these sections and should focus on Sect. 6.4 where the numerical model is viewed as a "black box".

6.2.1 Operator Splitting Methods

6.2.1.1 Principle

The reactive dispersion equation can be rewritten in the form

$$\frac{\mathrm{d}c}{\mathrm{d}t} = \sum_{j=1}^{n_p} f_j(c), \qquad (6.22)$$

where n_p is the number of processes that are taken into account, and $f_j(c)$ is the source term associated to the process j (among advection, diffusion, chemical kinetics, etc.). All processes are coupled *a priori* and this equation should be solved by taking into account the n_p processes simultaneously. In practice, the numerical methods for solving this equation are usually based on the so-called *operator splitting* method (*operator* stands for process). The principle is to integrate the processes one by one, by solving equations in the form

$$\frac{\mathrm{d}c}{\mathrm{d}t} = f_j(c). \tag{6.23}$$

There are at least three motivations for splitting:

- appropriate numerical schemes can be applied for each process with tailored properties (see below);
- the computational burden of a coupled approach can be too high (for instance with implicit time integration, see Sect. 6.2.2);
- from a pragmatic viewpoint, such an approach makes it possible to use off-the-shelf submodels, to be plugged into the chemistry-transport model: hence it is easy to modify a parameterization or to add a new process.

Splitting processes that are physically coupled results in errors. The different operator splitting methods aim at reducing the so-called *splitting error*. In case of linear processes ($f_j(c) = A_j c$ with A_j a matrix), analytical formulas are available. The splitting error is then easy to estimate in this case.

6.2.1.2 A Few Methods

Operator splitting methods are usually presented in the case of two linear processes,

$$\frac{\mathrm{d}c}{\mathrm{d}t} = A_1 c + A_2 c. \tag{6.24}$$

Let c_n be the numerical solution computed at time t_n. The operator splitting method provides a numerical solution c_{n+1} at time $t_{n+1} = t_n + \Delta t$ with Δt the so-called *splitting time step*.

First-Order Splitting The simplest method is based on a time-integration sequence. For instance, we first integrate A_1 and then A_2 (the sequence can be reversed). The first step is time integration over $[0, \Delta t]$ of

$$\frac{\mathrm{d}c^\star}{\mathrm{d}t} = A_1 c^\star, \qquad c(0) = c_n. \tag{6.25}$$

We obtain a solution $c^\star(\Delta t)$, to be used as an initial condition for the second step. The second step is time integration over $[0, \Delta t]$ of

$$\frac{\mathrm{d}c^{\star\star}}{\mathrm{d}t} = A_2 c^{\star\star}, \qquad c^{\star\star}(0) = c^\star(\Delta t). \tag{6.26}$$

The final solution at Δt provides $c_{n+1} = c^{\star\star}(\Delta t)$. It is easy to prove that the *local splitting error* (the error made during one time step) is of second order in Δt

(Exercise 6.3). By summing over the global time interval, one gets a *global error* in $O(\Delta t)$. Hence this method is said to be a first-order method. We will detail the notions of local and global errors in Sect. 6.2.2.

Exercise 6.3 (Splitting Error) Calculate the local splitting error for the first-order method.
Solution:
We compare the exact solution $c(t_{n+1})$ and the numerical solution c_{n+1},

$$c(t_{n+1}) = \exp[(A_1 + A_2)\Delta t]c_n, \qquad c_{n+1} = \exp(A_2\Delta t)\exp(A_1\Delta t)c_n,$$

where exp stands for the exponential of a matrix. The error is due to the commutation default of matrices:

$$\exp[(A_1 + A_2)\Delta t] = \exp(A_2\Delta t)\exp(A_1\Delta t)$$

is satisfied if and only if $A_1 A_2 = A_2 A_1$. We recall the asymptotic expansion $\exp(A\Delta t) = I + A\Delta t + A^2\Delta t^2/2 + O(\Delta t^3)$ where I is the identity matrix. It is easy to check that the leading term of the splitting error, $c(t_{n+1}) - c_{n+1}$, is $(A_1 A_2 - A_2 A_1)\Delta t^2/2$.

Another method is the so-called *source splitting* method. The integration of the first process does not result in a modified initial condition: a source term is added to the contribution of the second process in the second step,

$$\frac{\mathrm{d}c^{\star\star}}{\mathrm{d}t} = A_2 c^{\star\star} + \frac{c^\star(\Delta t) - c_n}{\Delta t}, \qquad c^{\star\star}(0) = c_n. \tag{6.27}$$

This method is often advocated when A_1 is a slow process while A_2 includes fast dynamics (Sect. 6.2.1.3).

Second-Order Splitting A simple way for reducing the splitting error is to remove the term in $O(\Delta t^2)$ from the local error splitting. It can be carried out by computing a solution with the sequence A_1–A_2 (let say $c_{A_1-A_2}$) and a second solution with the reversed sequence (let say $c_{A_2-A_1}$). The local error associated to the solution $(c_{A_1-A_2} + c_{A_2-A_1})/2$ is then of fourth order. This approach requires integration of *four* processes. A more efficient approach is the so-called *Strang splitting*.[1] The symmetry default of the first-order method can be avoided by integrating A_1 over $[0, \Delta t/2]$, A_2 over $[0, \Delta t]$ and then A_1 over $[0, \Delta t/2]$. Each process is thus integrated over a time interval $[0, \Delta t]$. The resulting solution is $c_{n+1} = \exp(A_1\Delta t/2)\exp(A_2\Delta t)\exp(A_1\Delta t/2)c_n$. It is easy to prove, with an asymptotic expansion, that the local error is of third order with respect to Δt. The method is therefore a second-order method. Note that we have to solve only *three* processes with this algorithm.

[1] Named after the American mathematician Gilbert Strang.

6.2.1.3 Additional Properties

Application to the Reactive Dispersion Equation The error analysis is usually carried out for linear problems. The extension to nonlinear problems is technical and is beyond the scope of this presentation ([81]). Nevertheless, the calculations are easy in the case of an advection-diffusion-reaction equation. We refer to Exercise 6.4 for the case of advection and chemical kinetics. The following theorem summarizes the main results.

Theorem 6.2.1 *The following pairs of processes have no splitting error:*

- *advection and chemical kinetics if the wind velocity is divergence free and if chemical kinetics does not depend on the spatial location;*
- *diffusion and advection if the wind velocity and the diffusion matrix do not depend on the spatial location;*
- *diffusion and chemical kinetics if the chemical reactions are monomolecular and do not depend on the spatial location.*

Assuming that a process has a weak dependence on the spatial location is *locally* realistic. On the other hand, assuming that the chemical reactions are monomolecular is not possible, except for passive or linear tracers (e.g. radionuclides). The splitting error is then mainly related to the splitting between chemical kinetics and vertical eddy diffusion. Note that the boundary conditions may also induce splitting errors (which is out of the scope of this presentation).

Exercise 6.4 (Advection-Reaction Model) We neglect diffusion and we assume that chemical kinetics does not depend on the spatial location. The dispersion model is then

$$\frac{\partial c}{\partial t} + \operatorname{div}(Vc) = \chi(c).$$

We suppose that the wind field is divergence free, namely $\operatorname{div} V = 0$. Integrate along the characteristic curves defined by $\mathrm{d}X/\mathrm{d}t = V$. Show that there is no splitting error.

Solution:

Let $\bar{c}(t) = c(X(t), t)$ be the solution along the characteristic curve $X(t)$. It is straightforward to obtain

$$\frac{\mathrm{d}\bar{c}}{\mathrm{d}t} = \frac{\partial c}{\partial t} + \frac{\mathrm{d}X}{\mathrm{d}t}\frac{\partial c}{\partial x} = \frac{\partial c}{\partial t} + V\frac{\partial c}{\partial x} = \chi(\bar{c}),$$

since $\operatorname{div}(Vc) = V \cdot \nabla c$ from $\operatorname{div} V = 0$. The exact solution can then be computed by splitting: we integrate chemical kinetics along the characteristic curves. This can be interpreted in terms of splitting (the splitting error is null). This result provides the basis for the Lagrangian box models.

Order Reduction The error analysis is based on asymptotic expansions with respect to small Δt. In practice, the time step has not a vanishing value: it is typically a few hundreds of seconds for a CTM. A value is "small" only in comparison to a reference value given by the characteristic timescales of the physical processes. We will see in Sect. 6.2.2 that chemical kinetics is characterized by a wide range of timescales. If τ_f is the characteristic timescale of the fastest process, the numerical methods to be used for the time integration of chemical kinetics will be designed so that the numerical timestep is much larger than τ_f (otherwise all timescales have to be resolved). Note that the splitting timestep, Δt, is larger than the subcycling timestep used for each process. Hence, we have always $\Delta t/\tau_f \gg 1$! The asymptotic analysis with vanishing values of Δt is then no longer valid. Consequently, the error analysis has to be carried out with another approach. For example, the fastest processes can be filtered out before calculating the local splitting error ([134]). Two results may be retained from this analysis:

- the method order may be reduced: this property is usually referred to as *order reduction* and depends on the characteristic timescales of the species (the fastest species are the most affected ones, as expected);
- the sequence order has an impact when the processes do not have similar characteristic timescales. In practice, advection and diffusion can be viewed as slow processes while chemical kinetics comprises both slow and fast chemical reactions. It is advocated to end the splitting sequence with the fastest processes, so that the system is stabilized.

6.2.2 *Time Integration of Chemical Kinetics*

A key component of a CTM is the time integration of chemical kinetics,

$$\frac{\mathrm{d}c}{\mathrm{d}t} = \chi(c, T, h\nu), \tag{6.28}$$

where the chemical source term depends on temperature through thermal chemical reactions and on radiation through photolysis reactions. This system of ordinary differential equations (ODEs) is usually characterized by its large dimension (up to one hundred species), by its nonlinear nature (the equations are coupled) and by the wide range of timescales. It is desired to integrate this system of ODE, written in the generic form

$$\frac{\mathrm{d}c}{\mathrm{d}t} = f(c), \qquad c(0) = c_0. \tag{6.29}$$

The time integration algorithms are usually based on an iterative scheme. We suppose that a numerical solution $(c_k)_{k=1,\ldots,n}$ has been computed at times $(t_k)_{k=1,\ldots,n}$ (the subscript k stands here for time index, not for species). The objective is then to calculate the solution c_{n+1} at time $t_{n+1} = t_n + \Delta t$ as a function of the solutions obtained at previous timesteps (possibly only as a function of c_n). In this section, Δt stands for the integration timestep of chemical kinetics: it is often referred to as the *subcycling* timestep, with a value lower or equal to the splitting timestep.

6.2.2.1 Timescales

For a scalar system $\mathrm{d}c/\mathrm{d}t = \lambda c$ (here $c \in \mathbb{R}$ corresponds to a single species), the solution is $c(t) = \exp(\lambda t)c(0)$ and the characteristic timescale is $1/|\lambda|$. For a linear system, $\mathrm{d}c/\mathrm{d}t = Ac$, with $c \in \mathbb{R}^{n_s}$ (concentration vector), the characteristic timescales are defined by the inverse values of the Jacobian matrix (A) eigenvalues, $\{\lambda_i\}_i$. For the sake of clarity, we assume that the Jacobian matrix can be diagonalized in $\mathbb{R}$ and is nonsingular ($\lambda_i \neq 0$). For a nonlinear system, $\mathrm{d}c/\mathrm{d}t = f(c)$, the linearized system for small perturbations is given by

$$\frac{\mathrm{d}(\delta c)}{\mathrm{d}t} = \frac{\partial f}{\partial c}\delta c, \tag{6.30}$$

with $\partial f/\partial c$ the *Jacobian matrix*, often denoted by J. By extension, the characteristic timescales are *locally* defined by the inverse values of the real parts of the eigenvalues of J.

6.2.2.2 Implicit Schemes Versus Explicit Schemes

Explicit Schemes The simplest algorithm is the so-called forward Euler algorithm (or explicit Euler algorithm). The time derivative is approached by a finite difference and the source term is evaluated at time t_n,

$$\frac{c_{n+1} - c_n}{\Delta t} = f(c_n), \tag{6.31}$$

which leads to

$$c_{n+1} = c_n + \Delta t\, f(c_n). \tag{6.32}$$

Implicit Schemes In the backward Euler algorithm (also referred to as implicit Euler algorithm), the source term is evaluated at time t_{n+1},

$$\frac{c_{n+1} - c_n}{\Delta t} = f(c_{n+1}). \tag{6.33}$$

It is therefore not possible to compute explicitly c_{n+1}, given as the implicit solution of an algebraic equation. In case of a linear system, with $f(c) = Jc$, it is straightforward to obtain

$$(I - J\Delta t)c_{n+1} = c_n, \tag{6.34}$$

with I the identity matrix. The solution can be computed once the inversion of the matrix $I - J\Delta t$ is carried out. In the case of a nonlinear system, an iterative algorithm has to be used in order to solve the algebraic system. For example, c_{n+1} can be defined as the root of a fixed-point problem,

$$c_{n+1} = g(c_{n+1}), \tag{6.35}$$

with $g(c) = c_n + \Delta t\, f(c)$. The fixed-point algorithm leads to a sequence of iterations labelled by k, with

$$c_{n+1}^{k+1} = g(c_{n+1}^k) = c_n + \Delta t\, f(c_{n+1}^k). \tag{6.36}$$

The convergence is reached is k is large enough, which provides an estimation of c_{n+1}. When the algorithm is stopped after one iteration, we recover the explicit Euler scheme. An alternative is to use a Newton-like algorithm by rearranging the algebraic equation. The problem is then rewritten in the form

$$g(c_{n+1}) = 0, \tag{6.37}$$

with $g(c) = c - \Delta t\, f(c) - c_n$ (we have kept the notation g). The Newton algorithm is obtained by linearization, which results in the iterative sequence

$$0 = g(c_{n+1}^{k+1}) \simeq g(c_{n+1}^k) + \frac{\partial g}{\partial c}(c_{n+1}^k)(c_{n+1}^{k+1} - c_{n+1}^k). \tag{6.38}$$

As $\partial g/\partial c = I - \Delta t\, \partial f/\partial c$, we obtain

$$(I - J(c_{n+1}^k)\Delta t)(c_{n+1}^{k+1} - c_{n+1}^k) = -(c_{n+1}^k - \Delta t\, f(c_{n+1}^k) - c_n), \tag{6.39}$$

with $J(c_{n+1}^k)$ the Jacobian matrix of f at point c_{n+1}^k. Note that this matrix can be approached by $J(c_n)$ so that the computational cost can be reduced (only the inversion of $I - J(c_n)\Delta t$ is required). For a fixed time step, an explicit method is always easier to use than an implicit method. The choice of an implicit method is forced by stability constraints.

6.2.2.3 Accuracy and Stability

The timestep choice is determined by the error analysis. For a linear system, we can write the previous algorithms in the generic form

$$c_{n+1} = c_n + (1-\theta)\Delta t f(c_n, t_n) + \theta \Delta t\, f(c_{n+1}, t_{n+1}), \tag{6.40}$$

with θ a numerical parameter in [0, 1]. If $\theta = 0$ (1, respectively), we recover the explicit Euler method (the *implicit* Euler method, respectively). The application of this numerical algorithm to the exact solution, $c(t_n)$, defines the residual η_n (also referred to as the local truncation error),

$$c(t_{n+1}) = c(t_n) + \Delta t(1-\theta) f(c(t_n), t_n) + \Delta t \theta\, f(c(t_{n+1}), t_{n+1}) + \eta_n. \tag{6.41}$$

With a Taylor expansion,

$$\eta_n = \left[\frac{1}{2}(1-2\theta)\frac{\mathrm{d}^2 c}{\mathrm{d}t^2}\right]\Delta t^2 + O(\Delta t^3), \tag{6.42}$$

where $d^2c/dt^2 = (\partial f/\partial c)\ f + \partial f/\partial t$. A key point is the propagation of this error. Let $\varepsilon_n = c(t_n) - c_n$ be the *global error* that results from the previous local errors and from their propagations. In the linear case ($f(c) = \lambda c$), we obtain by subtracting (6.40) to (6.41)

$$\varepsilon_{n+1} = \varepsilon_n + (1-\theta)\lambda\Delta t\varepsilon_n + \theta\lambda\Delta t\varepsilon_{n+1} + \eta_n. \tag{6.43}$$

Let us write $R(z) = (1+(1-\theta)z)/(1-\theta z)$. The error propagation is then given by

$$\varepsilon_{n+1} = R(\lambda\Delta t)\varepsilon_n + \delta_n, \tag{6.44}$$

with $\delta_n = (1-\theta\lambda\Delta t)^{-1}\eta_n$. As expected, there are two contributions:

- a *local* contribution (related to the truncation error), δ_n, corresponding to the error made in one timestep when the solution at time t_n is supposed to be exact;
- a contribution corresponding to the error propagation, given by $R(\lambda\Delta t)$, which defines the stability property of the algorithm.

With an obvious notation, $R(z)$ is named the *stability function*. It is a function of the complex variable $z \in \mathbb{C}$. Note that the eigenvalues of J do not have necessarily real values. It is straightforward to obtain

$$\varepsilon_n = [R(\lambda\Delta t)]^n\varepsilon_0 + \sum_{i=0}^{n-1}[R(\lambda\Delta t)]^i\delta_{n-1-i}. \tag{6.45}$$

The stability constraint is supposed to guarantee that the error is bounded. With $|R(\lambda\Delta t)| < 1$, we get

$$|\varepsilon_n| \le |\varepsilon_0| + \sum_{i=0}^{n-1}|\delta_i|. \tag{6.46}$$

Let $T = n\Delta t$ be the final time. For a local error satisfying $\delta_i = O(\Delta t^{p+1})$ (here $p = 1$ or $p = 2$), the global error is controlled by $n \times O(\Delta t^{p+1}) = T \times O(\Delta t^p)$. This illustrates the order reduction from the local error to the global error.

6.2.2.4 Stiff Systems

Let us consider dynamical systems that are physically stable (this is the case of atmospheric chemical kinetics): the real part of the eigenvalues of J are then negative and the linearized system can be written in the form $dc/dt = -|\lambda|c$. The stability condition is $1/(1+|\lambda\Delta t|) < 1$ for the implicit Euler scheme. This condition is always satisfied and the scheme is said to be *unconditionally stable*. The stability condition is $|(1-|\lambda\Delta t|)| < 1$ for the explicit Euler scheme, namely $|\lambda\Delta t| \le 2$. The choice of the timestep Δt is then constrained by the characteristic timestep so that $\Delta t \le 2/|\lambda|$.

A dynamical system is said to be *stiff* if there is a wide range of characteristic timescales: this means that it comprises both fast processes ($|\lambda| \gg 1$) and slow processes ($|\lambda| \sim 1$). The stability condition is more stringent for the fastest processes: for an explicit scheme, the integration timesteps are then forced to be about the smallest characteristic timescales. In practice, this is not affordable (remember that this corresponds to fractions of seconds for chemical kinetics) and implicit schemes have to be used.

Easy-to-use tailored numerical schemes are still very popular for the time integration of chemical kinetics. Their advantage is that they are simple to implement. Moreover, they are apparently "quicker" than implicit methods since they do not require the inversion of large-dimensional matrices. Note that rigorous analyses, namely by comparing the computational cost against the numerical error (see for instance [125]), have always indicated that these schemes are actually not efficient. We present them for the sake of completeness. Let us consider the loss-consumption form for the time evolution of the concentration of species X_i (Sect. 4.1.4). The chemical source term is then $f_i(c) = P_i(c) - L_i(c)c_i$. The *asymptotic schemes* are based on the linearization of the source term. If P_i and L_i are supposed to have constant values over a time interval of length Δt, the solution is

$$c_i^{n+1} = \exp(-L_i^n \Delta t)c_i^n + (1 - \exp(-L_i^n \Delta t))\frac{P_i^n}{L_i^n}, \tag{6.47}$$

where $c_i^n = c_i(t_n)$, $c^n = c(t_n)$, $L_i^n = L_i(c^n)$ and $P_i^n = P_i(c^n)$. The QSSA scheme (QSSA stands for *Quasi-Steady-State Approximation*, Sect. 4.1.4) is based on the partitioning of species according to their lifetime $\tau_i^n = (L_i^n)^{-1}$. For example, for the so-called fast species defined by $\tau_i^n/\Delta t \ll 1$, the formula yields $c_i^{n+1} = P_i^n/L_i^n$. In practice, there are numerous implementations that may strongly differ. Exercise 6.5 details the theoretical background.

Exercise 6.5 (Slow/Fast Models) The wide range of characteristic timescales results in the numerical stiffness. It is deeply associated to the existence of underlying

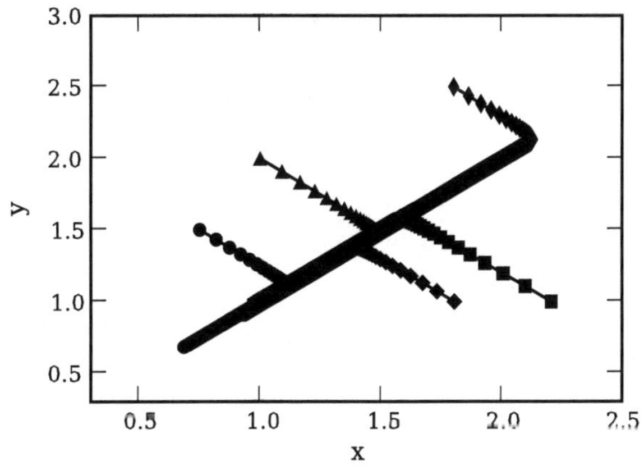

Fig. 6.7 Exercise 6.5: convergence of the exact model to the reduced model ($x = y$). The motion along each trajectory is given by the time evolution; each trajectory is defined by an initial condition. The time step between two symbols is constant (10^{-3}) along a given trajectory. The timescale ratio is $\varepsilon = 10^{-2}$

reduced models. The simplest example is the chemical mechanism

$$\mathrm{X} \overset{1/\varepsilon}{\rightarrow} \mathrm{Y}, \qquad \mathrm{Y} \overset{1/\varepsilon}{\rightarrow} \mathrm{X}, \qquad \mathrm{Y} \overset{1}{\rightarrow} \mathrm{Z}.$$

The values above the reaction arrows correspond to the magnitude of the kinetic rates. The equilibrium between X and Y is characterized by large kinetic rates (of magnitude $1/\varepsilon$ with $\varepsilon \simeq 0$) while the production of Z is a slow process (with a kinetic rate of magnitude 1). Write the equations for the time evolution of concentrations x and y for X and Y. What is the reduced model defined by $\varepsilon \to 0$? Hint: change the species basis by using (u, y) with $u = x + y$ (species *lumping*).

Solution:

The time evolution of concentrations is governed by

$$\varepsilon \frac{\mathrm{d}x}{\mathrm{d}t} = y - x, \qquad \varepsilon \frac{\mathrm{d}y}{\mathrm{d}t} = x - y - \varepsilon y.$$

With the species *lumping* $u = x + y$, one gets in the basis (u, y),

$$\frac{\mathrm{d}u}{\mathrm{d}t} = -y, \qquad \varepsilon \frac{\mathrm{d}y}{\mathrm{d}t} = u - 2y - \varepsilon y.$$

For $\varepsilon \to 0$, the evolution equation of y can be replaced by the algebraic constraint $y = u/2$. Hence the system becomes

$$\frac{\mathrm{d}u}{\mathrm{d}t} = -\frac{u}{2}, \qquad y = \frac{u}{2}.$$

The underlying model is said to be a *reduced model*. It describes only the time evolution of the slow species (here u) while the fast species (here y) are supposed to be at equilibrium with the slow species. Note that there is still a time evolution for the fast species but it is not given as such by an evolution equation. All the trajectories, defined by different initial conditions, quickly converge to a reduced model defined by the algebraic constraint $x = y = u/2$ (Fig. 6.5). A key point is that a change of basis (species lumping) may be required. The asymptotic analysis applied to the initial system would have led to the same algebraic constraint for both species ($x = y$, that is $y = u/2$). The reduced system would then have been underdetermined (one equation for two variables).

To know more ([136]):

B. Sportisse and R. Djouad, *Reduction of chemical kinetics in air pollution modelling*, J. Comp. Phys., **164** (2000), pp. 354–376

6.2.3 Advection Schemes

Advection is a key component of a CTM, especially for a passive or linear tracer that is not implied in chemical reactions. The advection equation is

$$\frac{\partial c}{\partial t} + \mathrm{div}(V\, c) = 0. \tag{6.48}$$

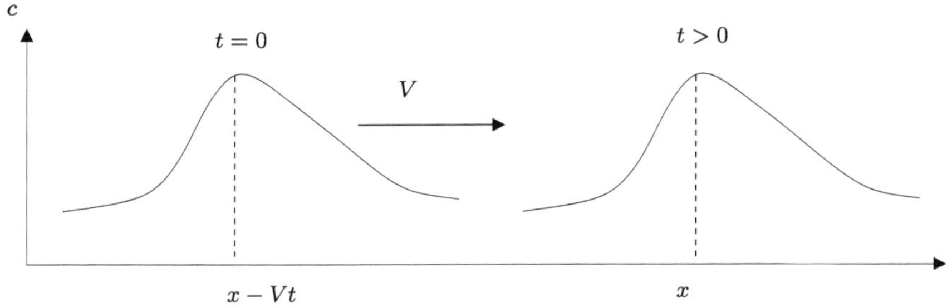

Fig. 6.8 Tracer advection for a one-dimensional constant wind velocity. The initial concentration distribution is conserved along time

Advection does not couple species and, as a result, we consider only one species. Simulating advection is a challenging issue because a few properties have to be satisfied at the numerical level: mass conservation, positivity of concentrations and monotony (no creation of purely numerical maxima—equivalent to positivity in the linear case).

6.2.3.1 Lagrangian and Eulerian Methods

With the *method of characteristics* applied to (6.48), it is straightforward to get the exact solution. Let $c_0(x)$ be the initial condition. We consider a one-dimensional wind field. It is easy to check that $c(x,t) = c_0(x - Vt)$ satisfies the advection equation (Fig. 6.8). Similarly, the solution is constant along the so-called *characteristic curves* defined by $\mathrm{d}X/\mathrm{d}t = V$. As $\bar{c}(t) = c(X(t),t)$,

$$\frac{\mathrm{d}\bar{c}}{\mathrm{d}t} = \frac{\partial c}{\partial t} + \frac{\partial c}{\partial x}\frac{\mathrm{d}X}{\mathrm{d}t} = 0. \tag{6.49}$$

In the general case (V is not constant), the most appropriate variable is the mixing ratio C, whose evolution is governed by $\partial C/\partial t + V \cdot \nabla C = 0$. We usually separate the *Lagrangian methods* (and also the *semi-Lagrangian methods*) that are based on this approach, from the *Eulerian methods* that directly solve the PDE by discretization. The drawback of Lagrangian methods is often that they do not satisfy the mass conservation law. We have chosen to focus on Eulerian methods in the following.

6.2.3.2 Finite Difference Schemes for Advection

We consider a discretization of the x axis: let (x_i) be the resulting nodes with a fixed grid step Δx (Fig. 6.9). We also discretize the time period with the sequence (t_n), where the timestep $\Delta t = t_{n+1} - t_n$ is supposed to have a constant value. For the sake of clarity, the grid and time steps are supposed to be constant. The extension to variable steps is straightforward.

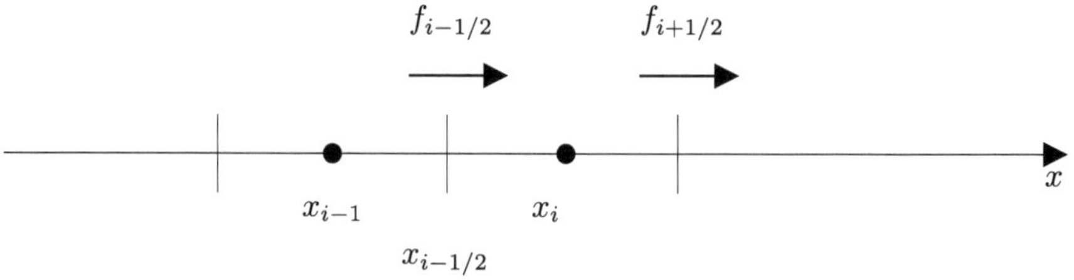

Fig. 6.9 Finite difference scheme for advection. The flux $f_{i-1/2}$ is defined at the interface $x_{i-1/2}$ between the grid cells centered at x_{i-1} and at x_i

Spatial Discretization Let us first introduce the *method of lines*. The principle is first to discretize with respect to space and then to integrate in time the resulting ODE. We use the misleading notation $c_i(t) \simeq c(x_i, t)$, where i corresponds here to the grid cell and not to species (remember that the species are advected one by one).

A *conservative form* of the numerical scheme is

$$\frac{\mathrm{d}c_i}{\mathrm{d}t} = \frac{f_{i-1/2} - f_{i+1/2}}{\Delta x}, \tag{6.50}$$

where $f_{i-1/2}$ is a numerical approximation of the flux at $x_{i-1/2}$, that is, $Vc(x - \Delta x/2, t)$. This form guarantees mass conservation (easy to prove by summing over all cells).

We suppose that $V > 0$. The simplest scheme is to approximate the flux by $f_{i-1/2} = Vc_{i-1}$. Note that this is equivalent to approximate the spatial gradient by the finite difference $\mathrm{d}c/\mathrm{d}x(x) \simeq [c(x) - c(x - \Delta x)]/\Delta x$. This approach defines the so-called *upwind scheme* (also referred to as the *donor-cell scheme*, with an obvious terminology),

$$\frac{\mathrm{d}c_i}{\mathrm{d}t} = V\frac{c_{i-1} - c_i}{\Delta x}. \tag{6.51}$$

A second approach is to approximate the flux at $x_{i-1/2}$ by $f_{i-1/2} = V(c_{i-1} + c_i)/2$ or, similarly, to use a Taylor expansion for the approximation $\mathrm{d}c/\mathrm{d}x(x) \simeq [c(x + \Delta x) - c(x - \Delta x)]/(2\Delta x)$. The resulting scheme is

$$\frac{\mathrm{d}c_i}{\mathrm{d}t} = V\frac{c_{i-1} - c_{i+1}}{2\Delta x}. \tag{6.52}$$

Time Integration Let c_i^n be the numerical solution at time t_n and at node x_i. For example, for the upwind scheme, several time integration schemes can be used. The simplest one is the explicit Euler scheme, namely

$$\frac{c_i^{n+1} - c_i^n}{\Delta t} = V\frac{c_{i-1}^n - c_i^n}{\Delta x}. \tag{6.53}$$

Note that this scheme could have been introduced by discretizing simultaneously space and time. This approach is sometimes referred to as the *direct space time*

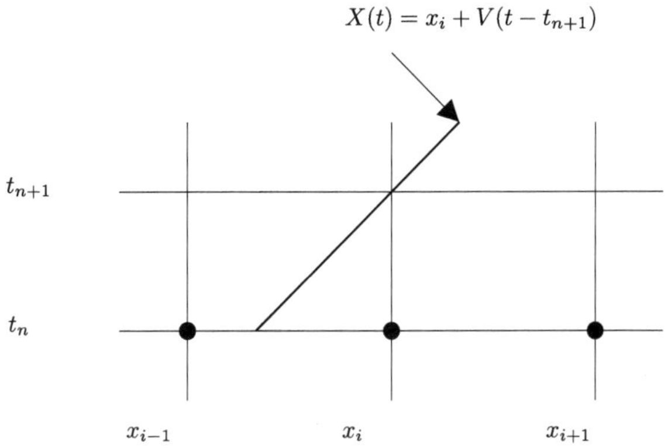

Fig. 6.10 Advection schemes and interpolation. Along the characteristic curve defined by $\mathrm{d}X/\mathrm{d}t = V$, $c(X(t), t)$ is constant. The *black circles* indicate the known numerical values at t_n. Note that $V\Delta t/\Delta x < 1$ in this case

method. We start from the exact equation satisfied along the characteristic curves

$$c(x_i, t_{n+1}) = c(x_i - V\Delta t, t_n). \tag{6.54}$$

The concentration $c(x_i - V\Delta t, t_n)$ has then to be interpolated as a function of the computed values (c_j^n) at discretization points (x_j). For the *upwind* scheme, if $V > 0$ and $V\Delta t \leq \Delta x$ (we will comment on this condition below), it is relevant to interpolate the concentration at $x_i - V\Delta t$ from the values at x_{i-1} and x_i (Fig. 6.10). A linear interpolation gives straightforward the upwind scheme written in the form

$$c_i^{n+1} = (1 - a)c_i^n + ac_{i-1}^n, \tag{6.55}$$

with $a = |V|\Delta t/\Delta x$ the so-called Courant-Friedrichs-Lewy number (CFL).

Note that (6.55) has been obtained with the constraint $a < 1$.

6.2.3.3 Challenging Issues: CFL Condition, Numerical Diffusion and Mass Consistency

The implementation of the advection numerical schemes faces at least three difficulties:

- the so-called CFL condition, $a < 1$, has to be satisfied so that stability is guaranteed;
- the spatial discretization generates numerical diffusion, especially for coarse meshes;
- due to the preprocessed interpolation of the meteorological fields, the mass consistency property may not be met by a CTM.

Stability and CFL Condition In a way similar to chemical kinetics (Sect. 6.2.2), the numerical discretization error has two components:

- the local error related to the discretization error for one time step (function of the scheme *order*);

- the propagation of local errors (to be investigated with a *stability* analysis).

The local error is estimated by replacing the numerical solution c_i^n by the exact solution $c(x_i, t_n)$ in the iterative scheme. For example, the residual η_i^n of the upwind scheme is

$$\frac{c(x_i, t_{n+1}) - c(x_i, t_n)}{\Delta t} + V\frac{c(x_i, t_n) - c(x_{i-1}, t_n)}{\Delta x} = \eta_i^n. \tag{6.56}$$

Using a Taylor expansion, we obtain $\eta_i^n = O(\Delta t) + O(\Delta x)$. Upon subtraction, the global error $\varepsilon_i^n = c(x_i, t_n) - c_i^n$ satisfies

$$\frac{\varepsilon_i^{n+1} - \varepsilon_i^n}{\Delta t} + V\frac{\varepsilon_i^n - \varepsilon_{i-1}^n}{\Delta x} = \eta_i^n, \tag{6.57}$$

which clearly indicates two contributions. The stability analysis (see also Exercise 6.6) leads to the so-called *Courant-Friedrich-Lewy condition* (CFL condition),

$$a = \frac{|V|\Delta t}{\Delta x} \leq 1. \tag{6.58}$$

Note that this condition is equivalent to the positivity of the solution written in the form (6.55).

The CFL condition can be viewed as a maximal bound for the timestep ($\Delta t < \Delta x/|V|$). Note that the constraint is stringent for high values of wind velocity and for fine meshes. In practice, the CFL condition plays a leading role at the local scale ($\Delta x \simeq 1$ km in this case, against $\Delta x \simeq 10$ km at the continental scale).

Exercise 6.6 (Numerical Diffusion and CFL Condition) This exercise aims at investigating stability and numerical diffusion.

1. We consider a one-dimensional case with a constant wind velocity V. Using a Taylor expansion of the residual η, prove that the numerical solution computed with the upwind scheme is an approximation of the PDE
$$\frac{\partial c}{\partial t} + V\frac{\partial c}{\partial x} = \frac{V\Delta x}{2}(1-a)\frac{\partial^2 c}{\partial x^2}.$$
2. This PDE is named the *equivalent PDE* of the upwind scheme: the behavior of the numerical scheme is actually closer to the behavior of the solution to this PDE than to the one of the solution to the advection equation (because the omitted terms are of higher order). Define the *numerical diffusion*. Calculate the numerical diffusion at the continental scale.

Data: for horizontal advection, $V \simeq 5\,\mathrm{m\,s^{-1}}$, $\Delta x \simeq 10$ km; for vertical advection, $V \simeq 0.1\,\mathrm{cm\,s^{-1}}$, $\Delta x \simeq 50$ m; in both case, $\Delta t \simeq 900$ s.

Solution:

1. Let us replace c_i^n by the exact solution $c(x_i, t_n)$ in the scheme iteration. Using a Taylor expansion up to second order, we obtain
$$\eta_i^n \simeq \left(\frac{\partial c}{\partial t} + V\frac{\partial c}{\partial x}\right) + \frac{1}{2}\left(\Delta t\frac{\partial^2 c}{\partial t^2} - V\Delta x\frac{\partial^2 c}{\partial x^2}\right).$$

The PDE associated with a null residual is the PDE that is the best approximation of the numerical scheme. Up to first order we get the advection equation. There are complementary terms for the second order. If $\partial c/\partial t + V\partial c/\partial x = 0$, it is straightforward to check that $\partial^2 c/\partial t^2 = V^2\partial^2 c/\partial x^2$ (differentiate c with respect to x and then with respect to t; reverse the differentiation sequence; equalize the results). If $\partial c/\partial t + V\partial c/\partial x = \mathcal{O}(\Delta t, \Delta x)$, this is still valid up to $\mathcal{O}(\Delta t, \Delta x)$. Thus, the residual is up to second order

$$\eta_i^n = \left(\frac{\partial c}{\partial t} + V\frac{\partial c}{\partial x}\right) - \underbrace{\frac{V\Delta x}{2}(1-a)}_{K_{num}}\frac{\partial^2 c}{\partial x^2},$$

with $a = V\Delta t/\Delta x$. This defines the equivalent PDE.

2. This PDE comprises a diffusion term: let K_{num} be the numerical diffusion. Note that the numerical diffusion is larger for coarse meshes (large values of Δx), as expected. Moreover, the stability of a diffusion equation requires the positivity of the diffusion coefficient (otherwise the gradients would grow): hence $a < 1$ (CFL condition).
At the continental scale, one has for the horizontal advection $K_{num} \sim 10^4\,\text{m}^2\,\text{s}^{-1}$ and for the vertical advection $K_{num} \sim 10^{-1}\,\text{m}^2\,\text{s}^{-1}$ (to be compared to the values of K_z, typically ranging from 1 to $10\,\text{m}^2\,\text{s}^{-1}$). The numerical diffusion is then dominant with respect to horizontal diffusion.

Numerical Diffusion One of the prevailing issues associated to advection is *numerical diffusion*. Numerical diffusion is actually much larger than horizontal physical diffusion, that is therefore usually neglected. A direct consequence of the interpolation of numerical values along the characteristic curve is to diffuse peaked concentrations over neighboring cells.

Minimizing numerical diffusion is a decisive requirement for applications related to accidental releases of point sources. Indeed, the resulting concentration fields are characterized by strong gradients: there is a polluted plume in a "clean" background (see Fig. 0.7 in Introduction). For photochemistry, the spatial fields are much smoother. Exercise 6.6 presents a powerful approach for the study of numerical diffusion (with the notion of *equivalent PDE*).

Exercise 6.7 (Mass Consistency) The objective of this exercise is to introduce a few methods that ensure mass consistency. The simplest numerical approach is the so-called "renormalization technique".

Let $c^{n+1} = \mathcal{S}(c^n, V^n)$ be the solution defined by the advection scheme. We assume that the scheme is explicit (this is the case in practice). Let $\tilde{\rho}^{n+1} = \mathcal{S}(\rho^n, V^n)$ be the air density computed with this scheme: note that this is not the "true" density ρ^{n+1}, as computed by the meteorological model. Formulate a correction in order to enforce the mass consistency property.

Solution:

The correction $c^{n+1} = (\rho^{n+1}/\tilde{\rho}^{n+1})\mathcal{S}(c^n, V^n)$ is appropriate. Indeed, if $c^n = \rho^n$, we obtain $c^{n+1} = (\rho^{n+1}/\tilde{\rho}^{n+1})\mathcal{S}(\rho^n, V^n) = \rho^{n+1}$. The drawback of this highly simple approach is that artificial mass can be created. Other algorithms have therefore to be used.

To know more ([138]):

B. Sportisse, D. Quélo, and V. Mallet, *Impact of mass consistency errors for atmospheric dispersion*, Atmos. Env., **41** (2007), pp. 6132–6142

Off-Line Coupling and Mass Consistency The mass consistency property was introduced in Exercise 6.1: a passive tracer that is initially well mixed (its mixing ratio is homogeneous) will remain well mixed.

This property is based on the consistency between the wind velocities used for the computation of ρ and c, respectively. In practice, for off-line coupling, these two fields are not the same ones: the differences may result from interpolations, from different discretization and from different numerical time integration schemes in the meteorological model (that computes density and wind velocity) and in the CTM (that computes concentrations).

The impact for the numerical simulation of concentrations can be important, especially for accidental releases. The advection process is indeed the leading process in this case. We refer to Exercise 6.7 for a simple approach to satisfy the mass consistency property.

6.3 Numerical Simulation of the General Dynamic Equation for Aerosols (GDE)

The numerical simulation of aerosol dynamics (Chap. 5) is the most challenging numerical issue for a CTM. We briefly present the basis of the classical algorithms in this section.

The numerical difficulties are related to the following items:

- there is a wide range of time scales and of sizes (for example, the aerosol diameters range from a few nanometers to a few micrometers);
- the resulting models have a large dimension (given by the product of the number of chemical species in the particulate phase by the number of variables used for the description of the size distribution, for a given aerosol family – there is one aerosol family only per single internal mixing);
- the processes that govern the evolution of size distribution and chemical composition are nonlinear; thermodynamics can also be discontinuous (phase transitions, hysteresis phenomena for crystallization and deliquescence of liquid/solid aerosols).

6.3.1 Size Distribution Representation

We can distinguish two different approaches (Fig. 6.11):

- the *modal methods* are based on a decomposition of the size distribution into three of four modes, usually log-normal functions of the diameter (Sect. 5.1, Chap. 5): the variables are then the parameters that describe the modes (distribution moments). Note that the distribution form is fixed *a priori*.
- the *size-resolved methods* that solve the GDE, without any *a priori* form for the distribution.

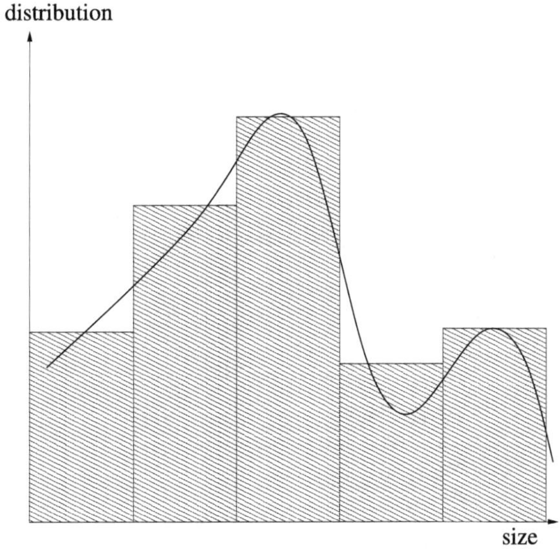

Fig. 6.11 Modal description (*curve*) versus sectional description (*gray boxes*: here 5 sections) of an aerosol distribution as a function of a size variable (mass, volume, diameter)

Among the size-resolved methods, we can also distinguish:

- the *sectional methods* (also referred to as *size-binning methods*). They correspond to the application of finite differences algorithms to the GDE: the size distribution is discretized in *sections* (or *bins*) and the numerical variables are related to averaged or integrated quantities over each section.
- the *variational methods*, for which the size distributions are decomposed along basis functions (finite element method or spectral collocation method; Problem 6.1).

We have made the choice of investigating the size-resolved methods because the resulting numerical issues are more challenging than for modal methods. We will also focus on sectional methods (the most classical methods, widely used for their "robustness").

The aerosol distributions are discretized into a set of *sections* (or *bins*). Let n_b be the number of sections. The objective is then to calculate Q_i^j and N^j, the integrated values over the section j of the mass distribution function for species X_i and of the number distribution, respectively.

Coagulation and condensation/evaporation are usually split, which motivates a distinct presentation for each process. Nucleation is easy to solve (a simple source term) and is usually added to condensation/evaporation. Exercise 6.8 illustrates the dynamical behavior of the aerosol processes with the help of simple analytical cases.

Exercise 6.8 (Analytical Solutions for the GDE) There are analytical solutions for the GDE in a few academic cases.

1. We consider a constant coagulation kernel, K_0. The size variable is supposed to be the volume, v. The coagulation equation is therefore

$$\frac{\partial n(v,t)}{\partial t} = \frac{1}{2} K_0 \int_0^v n(u,t) n(v-u,t) \mathrm{d}u - K_0 n(v,t) \int_0^\infty n(u,t) \mathrm{d}u.$$

Calculate the evolution of the total aerosol number $N(t)$. Comment.

2. We consider the condensation/evaporation equation for the number distribution as a function of the aerosol mass m,

$$\frac{\partial n(m,t)}{\partial t} + \frac{\partial [I(m) n(m,t)]}{\partial m} = 0, \qquad n(m,0) = n_0(m).$$

The condensation kernel is supposed to be in the form $I(m) = \lambda m^{1/3}$ (namely, the growth rate is a linear function of the aerosol diameter). Solve in the case of evaporation ($\lambda < 0$).
Hint: use the variable $\Psi = I(m) n$ (flux) and then integrate along the characteristic curves defined by the growth rate.

Solution:

1. Upon integration of the coagulation equation, we obtain $\mathrm{d}N(t)/\mathrm{d}t = -(K_0/2)N^2$ since

$$\int_0^\infty \int_0^v n(v-u,t) n(u,t) \mathrm{d}u = \left(\int_0^\infty n(u,t) \mathrm{d}u \right)^2 = N(t)^2.$$

Thus, $N(t) = N(0)/(1 + t/\tau)$ with $\tau = 2/(K_0 N(0))$ a characteristic timescale. Note that coagulation results in a decreasing particle number, as expected. In more complicated analytical cases, the Laplace transform can be used to calculate the solutions.

2. The flux Ψ satisfies

$$\frac{\partial \Psi}{\partial t} = I(m) \frac{\partial n}{\partial t} = -I(m) \frac{\partial \Psi}{\partial m}.$$

The characteristic curves are defined as

$$\frac{\mathrm{d}\bar{m}}{\mathrm{d}t} = I(\bar{m}) = \lambda \bar{m}^{1/3}, \qquad \bar{m}(0) = \bar{m}_0,$$

namely $(\bar{m})^{2/3}(t) = (\bar{m}_0)^{2/3} + 2\lambda t/3$. It is easy to check that Ψ is conserved along the characteristic curves. Thus, $\bar{m}_0^{1/3} \times n_0(\bar{m}_0) = m^{1/3} \times n(m,t)$ with $\bar{m}_0$ the initial condition of the characteristic curve with mass m at time t. This gives

$$n(m,t) = n_0((m^{2/3} - 2\lambda t/3)^{3/2}) \left(\frac{(m^{2/3} - 2\lambda t/3)^{3/2}}{m} \right)^{1/3}.$$

The calculation is more complicated in the case of condensation because it requires a boundary condition at $m = 0$ (in practice given by the nucleation rate).

6.3.2 Coagulation

Partitioning Coefficients For the sake of clarity, we focus on the coagulation equation for the number distribution (the so-called Smoluchowski equation, [149]). Upon integration of (5.7) (Chap. 5), we obtain

$$\frac{\mathrm{d}N^j}{\mathrm{d}t}(t) = \frac{1}{2}\sum_{j_1=1}^{j}\sum_{j_2=1}^{j} X^j_{j_1 j_2} N^{j_1} N^{j_2} - N^j \sum_{j_1=1}^{n_b} X_{j j_1} N^{j_1}. \tag{6.59}$$

The coefficients $X^j_{j_1 j_2}$ are the key components of the sectional methods for coagulation. They describe the fraction of aerosols that are in section j, and that result from the coagulation of sections j_1 and j_2, (Fig. 6.12). $X_{j_1 j_2}$ is defined by summing over all sections: $X_{j_1 j_2} = \sum_{j=1}^{n_b} X^j_{j_1 j_2}$.

We refer to [33] for a synthesis of the existing methods and a rigorous justification (the methods are usually rather empirical). A key point is the closure scheme used to form the distribution function inside each section.

Time Integration In a second step, the resulting system of ODE is integrated with a time integration scheme. There are no specific numerical difficulties and explicit methods can be used.

6.3.3 Condensation and Evaporation

The condensation/evaporation equation corresponds to a hyperbolic problem, similar to the advection equation (the spatial coordinates are replaced by size variables). The numerical simulation of this process is the most challenging step for the GDE because there is a wide range of timescales (condensation onto the fine aerosols is a fast process, Exercise 6.9).

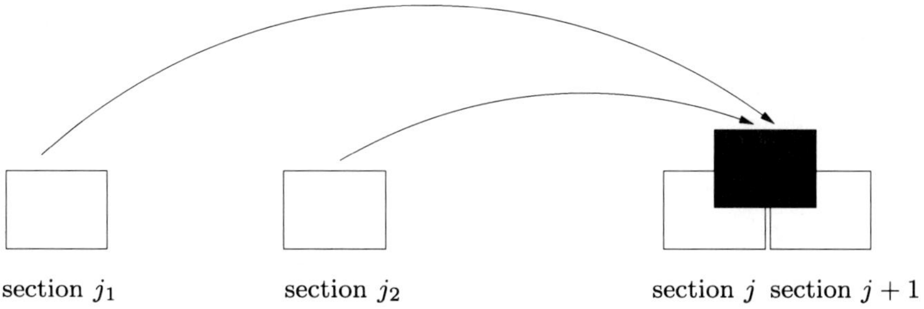

Fig. 6.12 Coagulation of aerosols from sections j_1 and j_2. The resulting aerosols (*in black*) have to be partitioned among the fixed sections that define the discretization of the size distribution functions

As for advection, we can distinguish two numerical approaches:

- the Eulerian methods are based on a *fixed* discretization of the size distribution; the mass transfer fluxes have to be computed at the interface between sections.
- the growth of a given section is tracked in Lagrangian methods.

Due to the rather low number of affordable sections in a three-dimensional CTM, the Eulerian methods are usually not advocated because they may result in a large numerical diffusion (see Exercise 6.6). Hence, we focus on Lagrangian methods.

A popular method is the so-called *moving sectional algorithm* ([66, 75]), that can be interpreted as a Lagrangian approach with specific numerical tricks. A difficult issue is the mapping of the (calculated) Lagrangian distribution onto the fixed sections required for the transport of aerosols inside a CTM.

From the definition of the growth rate in (5.15) (Chap. 5), we obtain a system of EDOs that gives the time evolution of the Lagrangian variables, for example the mass distributions for the n_s chemical species in the particulate phase,

$$\frac{\mathrm{d}Q_i^j}{\mathrm{d}t} = N^j I_i(Q_1^j, \ldots, Q_{n_s}^j), \tag{6.60}$$

where i labels chemical species and j sections.

The growth rate depends on all the chemical species through the thermodynamic equilibrium at the aerosol surface. Solving the equilibrium conditions is a difficult point, which results in a heavy computational burden in the computation of I_i: as a result, the computational cost of a multiphase CTM is mainly related to the computational cost of the thermodynamic module for aerosols.

This system of EDOs is stiff due to the wide range of timescales associated with the different aerosol sizes (Exercise 6.9). Using implicit methods is not necessarily appropriate because the calculation of the Jacobian matrix may be not affordable (this would require a large number of calls to the thermodynamic module). An alternative approach is based on the so-called *hybrid methods*. Similar to the QSSA methods used for chemical kinetics, the fast part of the aerosol variables (namely the variables related to the finest aerosols) are supposed to be at equilibrium while the dynamic mass transfer is solved for the slow part (the variables associated to the remaining part of the aerosol distribution). Note that an explicit solver can then be used (the fast dynamics have been removed).

Exercise 6.9 (Characteristic Timescales for Condensation) Equation (5.19) of Chap. 5 defines the growth rate. Estimate the magnitude of the resulting characteristic timescales as a function of the aerosol size.

Solution:

For a given chemical composition, $I_i(d_p) \sim d_p \times f(Kn, \alpha)$. If $K_n \gg 1$, this yields $I_i(d_p) \sim d_p^2$, while if $K_n \ll 1$, $I_i(d_p) \sim d_p$. We define an approximated timescale by $\tau \simeq m_i(d_p)/I_i$ with $m_i(d_p)$ the mass of species X_i inside aerosols of diameter d_p. As $m_i(d_p) \sim d_p^3$, we obtain $\tau \sim d_p$ for $K_n \gg 1$, and $\tau \sim d_p^2$ for $K_n \ll 1$. Mass transfer is therefore very fast for the finest aerosols, which quickly reach the thermodynamic equilibrium.

To know more ([31]):
E. DEBRY AND B. SPORTISSE, *Reduction of the condensation/evaporation dynamics for atmospheric aerosols: theoretical and numerical investigation of hybrid methods*, J. Aerosol Sci., **37** (2006), pp. 950–966

6.4 State-of-the-Art Modeling System

6.4.1 Forward Simulation

From now on, the CTM is supposed to be built, which means that the physical parameterizations and the numerical algorithms have been chosen. The CTM can be envisioned as an input/output function (a "black box"): the outputs y (concentrations, deposition fluxes, etc.) are computed from the inputs x (parameterization data, initial conditions, boundary conditions, etc.) with

$$y = F(x). \tag{6.61}$$

This defines the so-called *forward mode* of the CTM: for a given input x, we compute the output y.

6.4.2 Uncertainties

If the inputs were accurate (that is to say, if the physical parameterizations and the numerical algorithms were accurate), it could be sufficient to use the CTM only in the forward mode. In practice, the CTMs are characterized by many uncertainties (Table 6.4):

- the input data have often a poor accuracy (a limiting step is the quality of the inventory emissions);
- the meteorological fields have their own uncertainties, which will affect in turn the CTM;
- the subgrid parameterizations (related to the wide range of spatial scales) induce errors that are usually not controlled;
- due to the large dimension, the numerical discretization is often coarse (especially for aerosol dynamics);
- the software systems also contain bugs.

More and more complicated processes should be included in a CTM (especially due to model couplings), which requires the use of finer and finer data, that are usually poorly accurate. Formally, the ideal modeling system is characterized by a trade-off between the data quality and the model complexity (Fig. 6.13). Note that the data are related not only to the input data but also to the observational data, to be used for the model validation and for data assimilation (see below).

Table 6.4 Typical uncertainties of the CTM inputs at the regional scale. Source: [51]

Data	Relative uncertainties
Cloud attenuation	±30%
Dry deposition velocity (O_3 and NO_2)	±30%
Boundary conditions (O_3)	±20%
Anthropogenic emissions	±50%
Biogenic emissions	±100%
Photolysis rate	±30%

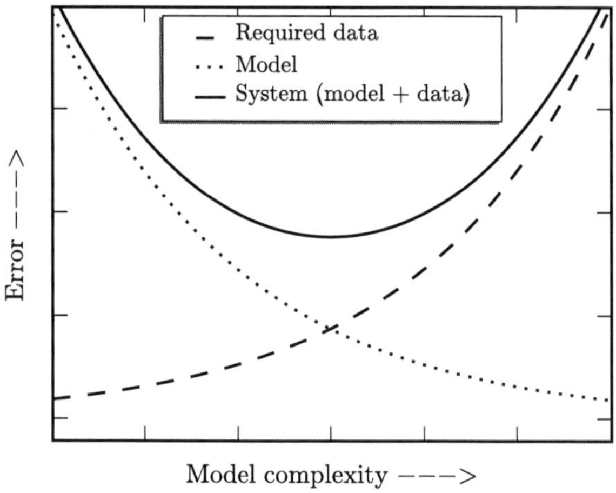

Fig. 6.13 Evolution of model errors and of data errors as a function of the model complexity (resolution). A complicated model is accurate but it requires the use of highly detailed data (that are therefore uncertain in practice); a simple model is poorly accurate but it requires easy-to-obtain data (usually accurate). The global error is evaluated by summing the model error and the data error

6.4.3 Advanced Methods

6.4.3.1 Sensitivity Analysis

We consider a scalar input with continuous values, typically a parameter used in a parameterization or a component of initial or boundary conditions. The uncertainties can be evaluated by sensitivity analysis. The principle is to compute the vector of partial derivatives of outputs with respect to the input data,

$$s = \frac{\partial F}{\partial x}. \tag{6.62}$$

The model associated to $\partial F/\partial x$ is usually referred to as the *linear tangent model.* For a *small* perturbation applied to a scalar input data x (say $\delta x \in \mathbb{R}$), we can calculate the propagation resulting in the outputs, $s\delta x$. For a nonlinear model, this approximation is only valid in the vicinity of the value x at which the partial derivative is computed.

Similarly, it may be useful to search for the impact of perturbations applied to a *set* of input data on a given scalar output. For example, the sensitivity of the mean mercury concentration with respect to the mercury emissions is shown in Fig. 6.14.

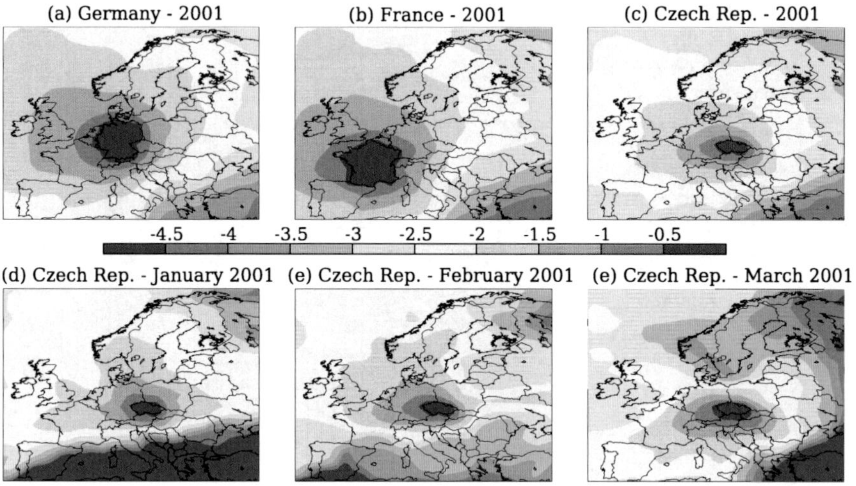

Fig. 6.14 Map of the sensitivity ($\log_{10}$-scale) of the ground averaged mercury concentration with respect to the mercury emissions. The averaged concentration is computed for a given domain and over a given time interval that are indicated for each case. Credit: Yelva Roustan and Marc Bocquet, CEREA. Source: [123]

The computations are carried out with the so-called *adjoint model* (Problem 6.2). Such studies are of course relevant to investigate transboundary pollution: the objective is indeed to express the dependence of targets on emissions. A source/receptor matrix can be generated from the sensitivity coefficients $\partial F/\partial x$, with F the concentration vector (or the vector of deposition fluxes) and x the emission vector.

6.4.3.2 Monte Carlo Methods

Monte Carlo methods are widely used in order to evaluate the global uncertainties in a continuous parameter. Let $P(x)$ be the *probability density function* (PDF) of the uncertain input x. The averaged output value is approached by $\langle y \rangle = \int P(x)\mathrm{d}x$ with a sampling of x by N values $(x_i)_i$,

$$\langle y \rangle \simeq \frac{1}{N}\sum_{i=1}^{N} F(x_i). \tag{6.63}$$

More generally, we can compute the PDF associated to y (Fig. 6.15). The convergence rate can be estimated by $\mathrm{Var}(y)/\sqrt{N}$ with $\mathrm{Var}(y)$ the variance of y, to be viewed as a random variable. As a result, a large number of sampling points have to be computed. The convergence is thus rather slow (even if variance reduction methods can be used in order to quicken the convergence).

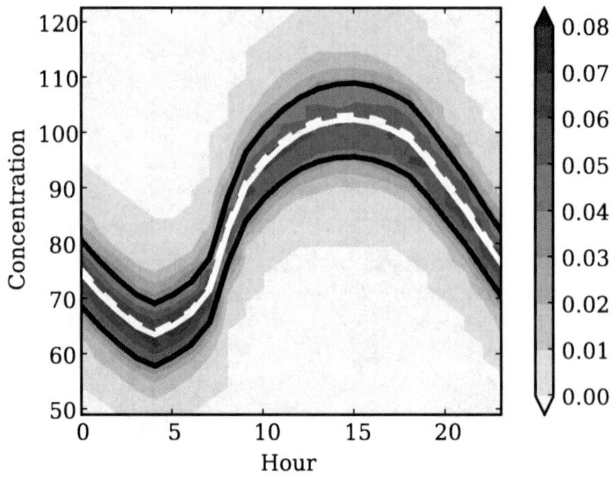

Fig. 6.15 Evolution of the PDF associated to the averaged values of ozone (over one day) over Europe (summer 2001). The PDF is computed by a Monte Carlo method applied to a set of continuous input parameters (with $N = 800$ simulations). Simulation with the POLYPHEMUS system; credit: Vivien Mallet, CEREA

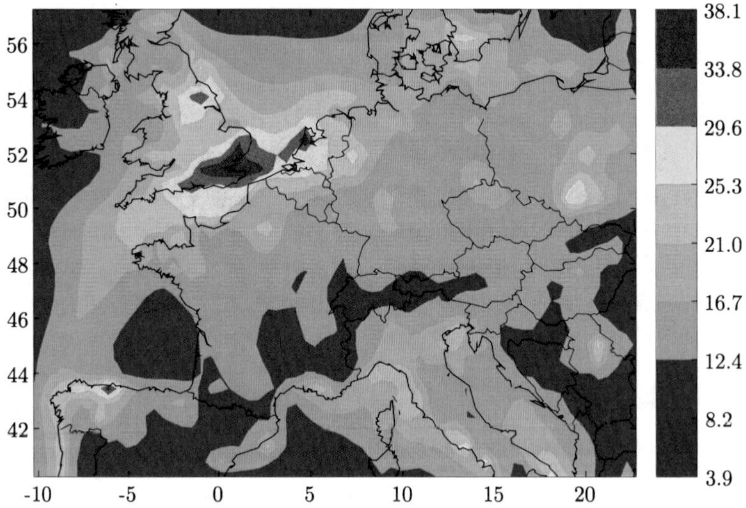

Fig. 6.16 Variance for the ozone concentration (relative uncertainty, in %) in a multi-model configuration of the POLYPHEMUS system (summer 2001). An ensemble of 48 model configurations is used. Credit: Vivien Mallet, CEREA

6.4.3.3 Ensemble Simulation: Multi-Model Ensemble

In practice, there exist a few alternative parameterizations for a given process. There is no *a priori* motivation for selecting one parameterization rather than another. The parameterizations can generate different inputs $(x_i)_i$ or define different model configurations $(F_i)_i$.

A multi-model system is able to provide an *ensemble* of outputs, $(F(x_i))_i$, $(F_i(x))_i$ or $(F_i(x_j))_{i,j}$. This gives an estimation of the spread in the outputs and of the uncertainty propagation (Fig. 6.16).

6.4.3.4 Ensemble Forecast

For forecast purposes, it is desired to estimate not only the uncertainties but also the best available concentrations. *Ensemble forecast methods* are combinations of N model outputs, written as $(y_i)_{i=1,N}$ (Fig. 6.17). Many approaches can be used.

- The simplest method is the *ensemble mean* with $y = (\sum_{i=1}^{N} y_i)/N$.
- For applications related to accidental releases, the median value (chosen among the outputs ranked in increasing order) is often used.
- A more sophisticated and more rigorous approach is based on a linear combination of the model outputs,

$$y = \sum_{i=1}^{N} \alpha_i y_i . \qquad (6.64)$$

The weighting vector α is estimated by comparing the resulting forecast with the available observations: this can be performed by minimizing the cost function that accounts for a deviation from the past observations (the function is not detailed here). The ensemble forecast will favor the "good" models, to be selected when more and more observations become available. Such methods can dramatically improve the forecast quality, especially for photochemistry. They can also be viewed as examples of data assimilation techniques (see next section).

To know more ([91]):
V. MALLET AND B. SPORTISSE, *Ensemble-based air quality forecasts: a multi-model approach applied to ozone*, J. Geophys. Res., **111** (2006), p. 18302

6.4.3.5 Data Assimilation and Inverse Modeling

Background The current state-of-the-art modeling systems include *data assimilation* procedures. Indeed, there are other information sources next to the numerical

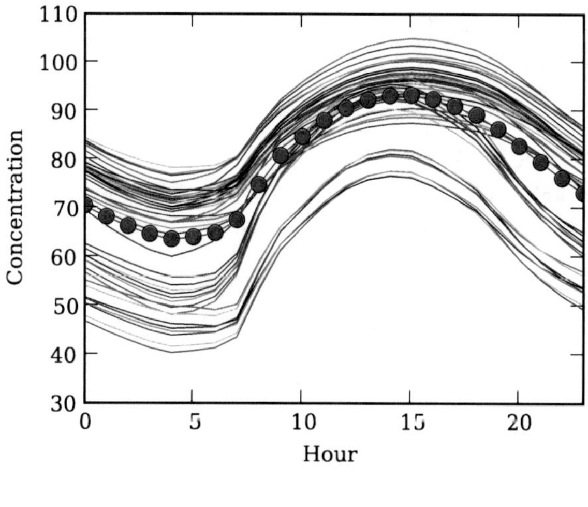

Fig. 6.17 *Ensemble* simulation of ozone with 48 models. Computed with the POLYPHEMUS system (averaged daily evolution, summer 2001). The *red points* stand for the ensemble mean. Credit: Vivien Mallet, CEREA

models: monitoring networks provide observational data. The monitoring networks can be ground networks devoted to air quality, satellite sensors devoted to observation of the atmospheric composition or wifi urban networks in the near-future (with thousands of sensors).

A data assimilation method enables coupling between models and observations. It can be interpreted as the application to the model outputs of a constraint provided by observational data. Similarly, it can be viewed as an interpolation of measurements (that do not give complete information: there are not enough sensors) under the model constraints (that respect physical laws). Exercise 6.10 illustrates the most basic algorithm, namely the so-called optimal interpolation method.

Exercise 6.10 (Optimal Interpolation Method) Let y_m and obs be the values computed by the model and given by the observations, respectively. Both values are uncertain: we write σ_m^2 and σ_o^2 the estimated variances, for y_m and obs respectively. Thus, $1/\sigma_\star^2$ can be interpreted as the confidence given to the value under consideration. In order to estimate the concentration y, we want to minimize the *cost function* J,

$$J(y) = \frac{1}{2\sigma_m^2}(y - y_m)^2 + \frac{1}{2\sigma_o^2}(y - obs)^2.$$

J describes the discrepancy between both information sources, weighted by their confidence. Estimate the so-called *analysis*, y_a, defined as the value that realizes the minimum of J. Comment.

Solution:

J is a convex function whose minimum satisfies $\partial J/\partial y = 0$, namely

$$y_a = \frac{\frac{y_m}{\sigma_m^2} + \frac{obs}{\sigma_o^2}}{\frac{1}{\sigma_m^2} + \frac{1}{\sigma_o^2}}.$$

It is straightforward to comment on the resulting analysis for different magnitudes of the variances. Moreover, this approach has a probabilistic interpretation (not detailed here). Note that the solution quality is given by the ability to compute a minimum of J: when the minimum is "flat", the solution quality is weak (Fig. 6.18). Note that the method requires inversion of the Hessian matrix of J.

Inverse Modeling A similar application is provided by *inverse modeling* (also referred to as *parameter identification*). The objective is then to estimate an *uncertain* input parameter. The typical example is the inverse modeling of emission fluxes in order to lower the uncertainties of an emission inventory. Other applications are to check that a commitment to emission reduction is met, or to localize sources of accidental releases that are detected by a monitoring network.

The simplest case is the linear case $y = Ax$, where $x \in \mathbb{R}^p$ corresponds to the input data. Let H be the *observation operator*, supposed to provide observations $obs = Hy$, defined as a linear function of the state vector y. For a ground sensor, H is associated to time and space sampling for the measurements of a given chemical

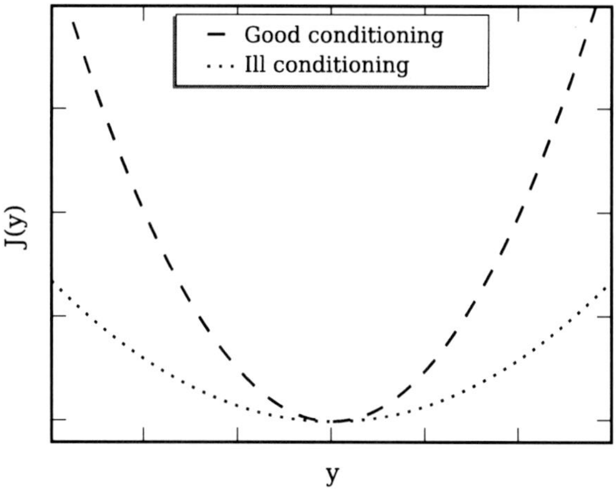

Fig. 6.18 Conditioning of the cost function J with respect to inversion. The conditioning is related to the properties of the Hessian matrix of J (second derivative)

species. It is usually not possible to measure all the components of the state vector and $obs \in \mathbb{R}^m$ with $m \ll p$.

In case $m = p$ and when there is no observational error, we obtain $obs = HAx$: the inversion of this linear algebraic system would give a value for x. In practice, the system is underdetermined and the observational errors have to be taken into account. Let x_b be so-called *background* value for x: it is a first guess of the true value. Similarly to data assimilation, the solution is given by the minimization of the cost function

$$J(x) = \frac{1}{2}(obs - HAx)^T R^{-1}(obs - HAx) + \frac{1}{2}(x - x_b)^T B^{-1}(x - x_b), \quad (6.65)$$

where R and B stand for the error covariance matrices for the observation and the background, respectively. For example, R may be a diagonal matrix with coefficients σ_o^2. The transpose vector of the vector M is M^T.

This method is usually referred to as the *3D-Var method*; "3D" refers to the fact that only space is taken into account (see below for time).

Such methods are used for the inversion of methane emissions (for example [14]), of CO ([23] at the regional scale, [13] at the global scale), of NO_x ([116], Fig. 6.20), of volatile organic compounds ([22]), etc. In the case of accidental releases, we have usually to deal with point-source emissions, contrary to the previous cases (diffuse emissions). A related issue may be the *localization problem* (see for instance Fig. 6.21), especially for radionuclides.

In the future, numerical models will be increasingly used for such applications.

Time Evolution This framework can be generalized to the case of time-dependent systems. Let $(obs_i)_i$ and $(y(t_i))_i$ be the sequence of observed values and of computed values at times $(t_i)_i$, respectively. We write H_i, the observation operator at time t_i, that maps the simulated values to the observations.

We usually split the *sequential methods*, based on estimation theory (Kalman filtering, Fig. 6.19), from the *variational methods*, based on minimization of the

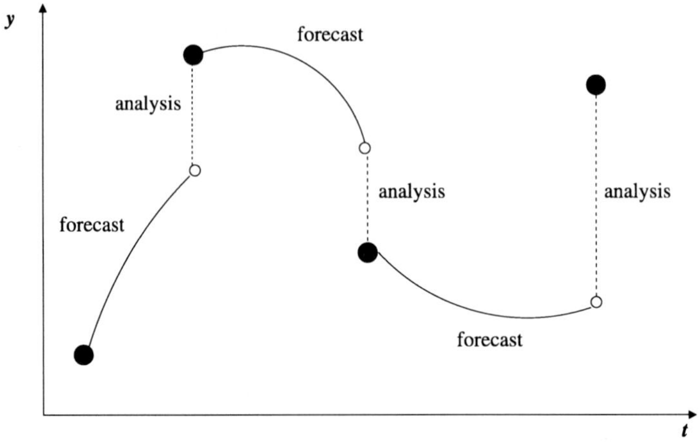

Fig. 6.19 Data assimilation with a sequential method. When an observation is available, the *analysis* is computed so that the state is corrected (with a method similar to Exercise 6.10). The analysis is thereafter used as the initial condition for the model forecast

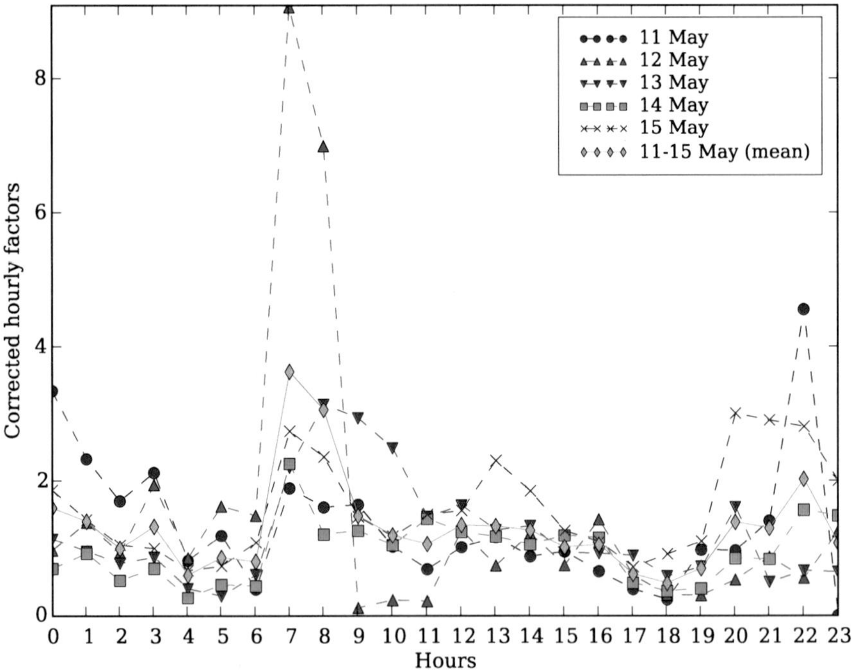

Fig. 6.20 Correction of the hourly emission factors for NO_x with inverse modeling (air quality simulation over Lille in northern France). In the reference emission inventory, the factor is equal to 1. Source: [116]

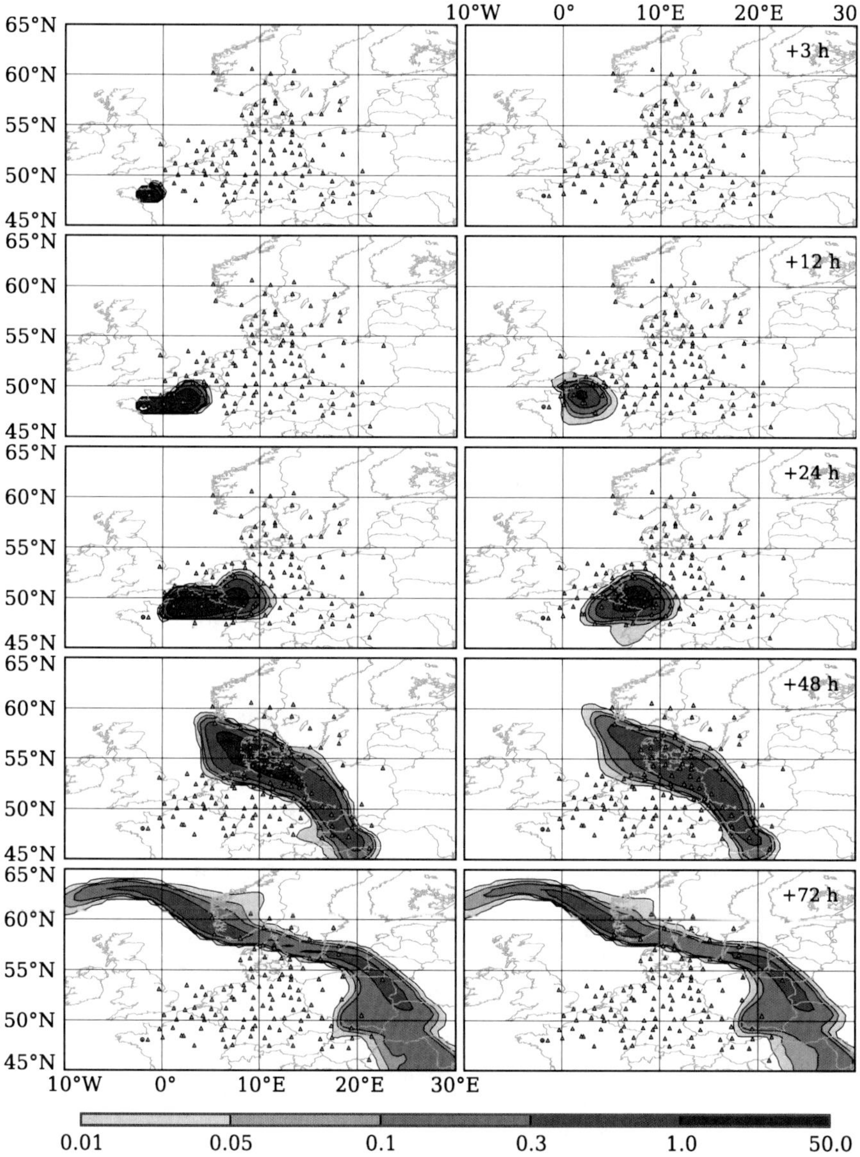

Fig. 6.21 Source reconstruction from the observational data given by a monitoring network, and the resulting forecast of the pollutant dispersion. The observational data are issued from the ETEX campaign (*European Tracer EXperiment*, 1994, [101]): a tracer was emitted near Rennes (in Brittany, France). *Left*: plume simulation with the true emission. *Right*: plume simulation after source reconstruction from the real-time measured data (stations indicated by triangles). In this case, about ten hours of observation are required for localizing the emission source. The unit is $\mathrm{ng\,m^{-3}}$. Credit: Marc Bocquet, CEREA. Source: [15]

cost function over a time interval (an ensemble of observations is taken into account simultaneously).

For example, the parameter to be controlled may be the emission flux x. We assume that the numerical model provides a sequence of values $y(t_{i+1}) = F(y(t_i), x)$. A variational formulation (defining the *4D-Var method*) is based on the minimization of the cost function

$$J(x) = \frac{1}{2}\sum_{i=1}^{n}[obs_i - H_i y(t_i)]^T R_i^{-1}[obs_i - H_i y(t_i)] + \frac{1}{2}(x - x_b)^T B^{-1}(x - x_b), \tag{6.66}$$

where the notations are coherent with the previous cases. The key point is that the sequence of simulated values (y_i) is constrained by the numerical model and depends on x. The cost function $J(x)$ is usually minimized with a gradient-like method. The simplest method is to calculate a sequence $x^{k+1} = x^k - \alpha_k \nabla_x J(x^k)$ ($\alpha_k > 0$ is a numerical parameter): the solution is then obtained after convergence (for k large enough).

The computation of $\nabla_x J$ is unfortunately a challenging issue, as compared to the case of optimal interpolation (Exercise 6.10). We refer to Problem 6.2 for the introduction of a powerful method, based on the so-called *adjoint model.*

6.4.4 Model-to-Data Comparisons

The numerical models are extensively compared to observational data, which should be chosen as various as possible (spatial areas, time periods, chemical or radiative outputs; ground stations, vertical profiles, airborne sensors, satellite data). Notice that it is more relevant to refer to model-to-data comparisons rather than to model *validation*. For example, there are only a limited number of observed chemical species: aerosols are usually measured through a bulk mass (PM_{10}); most of the observations are at ground (while the fields are three-dimensional fields) and are not necessarily well sampled (Fig. 6.22), etc.

The evaluation of a model performance has been strictly defined, especially by the US EPA (*Environmental Protection Agency*). Statistical indicators are used, depending on the targets and on the application:

- for an accidental release (point-source emission of a pollutant not supposed to be in the atmosphere), the indicators will focus on the model's ability to forecast the *arrival times* of the pollutant plume or the maximum values.
- for photochemistry, the concentration fields are much more diffuse and numerous observational data may be available. The classical indicators are based on statistical criteria (Table 6.5). For example, the recommendations of the US EPA focus on the limit values for the so-called MNBE (at $\pm 5\%$), MNGE (at $\pm 30\%$) and UPA (at $\pm 15\%$) for ozone concentrations that are not above a given threshold (typically from 40 to 60 ppb). Note that the results can be highly sensitive to the choice of the monitoring network (Table 6.6 and Table 6.7). A key point is the *representativity* of a monitoring station: actually, a point measurement has to be compared with a numerical output, *averaged* over a grid cell.

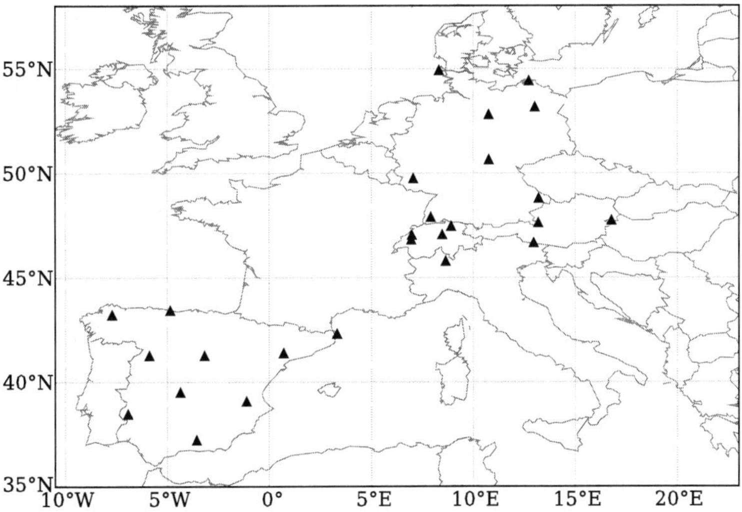

Fig. 6.22 Localization of the EMEP stations for PM_{10}. Credit: Yelva Roustan, CEREA

Table 6.5 A few statistical indicators to compare observed and simulated time sequences with n elements. The simulated values are (y_i) while the corresponding observational data are (obs_i). The averaged value for the field Ψ is denoted as $\overline{\Psi}$. Source: [124]

Indicator	Notation	Formula
Root Mean Square Error	RMSE	$\sqrt{\frac{1}{n}\sum_{i=1}^{n}(y_i - obs_i)^2}$
Bias	Bias	$\frac{1}{n}\sum_{i=1}^{n}(y_i - obs_i)$
Mean Fractional Bias	MFB	$\frac{1}{n}\sum_{i=1}^{n}\frac{(y_i-obs_i)}{(y_i+obs_i)/2}$
Mean Fractional Error	MFE	$\frac{1}{n}\sum_{i=1}^{n}\frac{\lvert y_i-obs_i\rvert}{(y_i+obs_i)/2}$
Correlation	Corr	$\frac{\sum_{i=1}^{n}(y_i-y)(obs_i-\overline{obs})}{\sqrt{(y_i-\overline{y})^2}\sqrt{(obs_i-\overline{obs})^2}}$
Mean Normalized Bias Error	MNBE	$\frac{1}{n}\sum_{i=1}^{n}\frac{y_i-obs_i}{obs_i}$
Mean Normalized Gross Error	MNGE	$\frac{1}{n}\sum_{i=1}^{n}\frac{\lvert y_i-obs_i\rvert}{obs_i}$
Unpaired Peak Prediction Accuracy	UPA	$\frac{y_{max}-obs_{max}}{obs_{max}}$

6.4.5 Applications

The air quality numerical models are used in an increasing number of applications. Most of the problems presented in this book (for example illustrated by exercises or problems) are investigated by numerical simulations. Beyond process studies (scientific understanding of the underlying processes), the models can be used for *forecast* and *impact studies*.

Table 6.6 Model-to-data comparisons for the POLYPHEMUS system over Europe in 2001 for a few gas-phase species: number of stations (it depends on the monitoring network: EMEP and Airbase over Europe, or BDQA over France), averaged measured value ($\mu g\,m^{-3}$), averaged simulated value ($\mu g\,m^{-3}$), RMSE ($\mu g\,m^{-3}$), correlation (%), MFB (%), MFE (%). Source: [126]

Species	Network	Stations	$\overline{obs}$	$\overline{y}$	RMSE	Corr	MFB	MFE
NO_2	EMEP	35	7.5	9.0	5.7	50.0	22	59
	AirBase	1049	23.9	15.3	14.2	57.2	−37	62
	BDQA	75	22.1	13.4	14.6	55.9	−47	70
NH_3	EMEP	3	7.4	6.3	5.5	28.8	12	52
	AirBase	7	12.7	6.6	13.7	27.3	−23	101
HNO_3	EMEP	7	0.7	1.3	1.4	26.2	40	89
SO_2	EMEP	43	2.0	5.3	4.9	46.7	97	106
	AirBase	992	6.5	7.3	6.6	44.2	25	70
	BDQA	10	7.8	6.8	6.5	35.7	−13	59

Table 6.7 Statistical indicators similar to Table 6.6 for aerosols and related species

Species	Network	Stations	$\overline{obs}$	$\overline{y}$	RMSE	Corr	MFB	MFE
PM_{10}	EMEP	26	16.9	15.4	12.5	55.1	−9	50
	AirBase	537	25.4	15.2	17.0	44.9	−45	59
	BDQA	23	19.8	15.5	9.6	57.7	-27	41
$PM_{2.5}$	EMEP	17	12.6	8.3	8.6	54.4	−40	61
Sulfate	EMEP	57	2.5	2.0	1.7	55.6	−6	50
	AirBase	11	1.9	2.3	1.7	49.4	39	65
Nitrate	EMEP	14	2.6	4.0	3.1	41.3	30	75
	AirBase	8	3.5	4.3	2.7	71.7	6	54
Amm.	EMEP	9	1.8	2.0	1.3	51.9	19	49
	AirBase	8	1.8	2.0	0.9	74.4	14	36
Sodium	EMEP	3	1.3	2.4	2.2	62.8	47	68
Chloride	AirBase	7	0.9	3.1	3.5	69.8	83	102

- Forecast, especially for accidental releases.
  The models are then supposed to be *robust*, to use small databases and to run quickly. These requirements explain that the Gaussian models are still widely used for this application. For the forecast of photochemistry, the CTMs have to be *tuned* so that a few forecast targets (such as the ozone peaks) are well reproduced. The assessment of uncertainties and the use of data assimilation methods are also frequent: a forecast system is systematically based on a strong coupling between numerical models and monitoring networks (Fig. 6.21).
  Models can also provide decision-support systems, for example for the design of a monitoring network devoted to air quality or to accidental releases (this is sometimes referred to as *network design*). The numerical model is used to gener-

ate virtual situations: the optimal network is then built by minimizing a function that evaluates its performance (for detection or for data assimilation).
An illustrative example is provided by the *Comprehensive Nuclear Test Ban Treaty* (CTBT, 1996). Banning nuclear tests requires deployment of a monitoring network, so that the emission of radionuclides should be detected and the emission source localized ([158]). For another application, Fig. 6.23 shows optimized monitoring networks for the French nuclear power plants. The security and financial constraints are of course prevailing issues in such a context.

- Impact studies.
For impact studies, the numerical models are supposed to simulate long-term atmospheric compositions. Moreover, they need to describe not only the concentrations but also the sensitivity of the concentrations to emissions. This requirement motivates the use of detailed models, including all physical processes. Note that it is not obvious that the models used for forecasting are appropriate since they are often tuned to a small set of targets.
An increasing number of dispersion models are embedded in *integrated modeling* systems that provide an assessment (for instance with a monetary cost) of different scenarios of future economic activities. The typical example is the RAINS

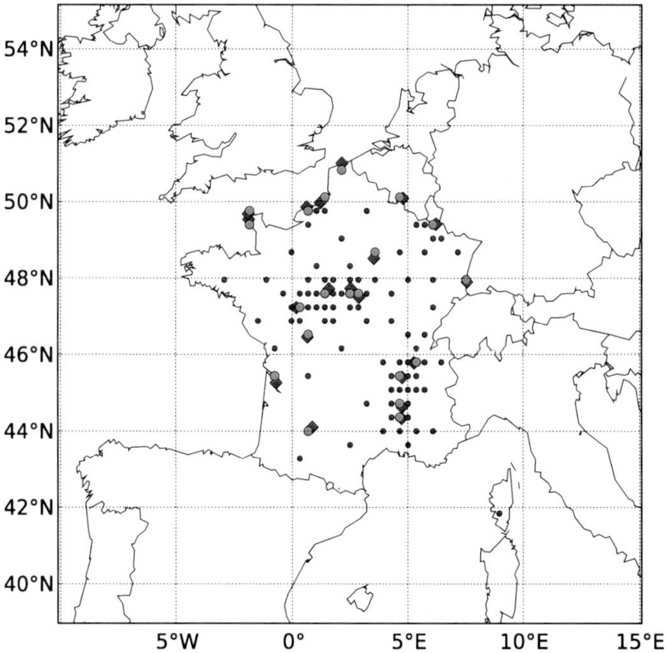

Fig. 6.23 Optimization of a monitoring network for the French nuclear power plants. A network of 100 stations is built from the minimization of a performance function (the algorithm is simulated annealing): the 20 *rhombuses* stand for the nuclear power plants, the 20 *blue circles* stand for the "local" stations, the 80 *dark blue circles* are the other stations (chosen by the optimization procedure). The virtual accidental releases are simulated with the POLYPHEMUS system. Credit: Rachid Abida and Marc Bocquet, CEREA

model (*Regional Air pollution INformation and Simulation*; now GAINS, *Greenhouse Gas and Air Pollution Interactions and Synergies*; [7]), used by the European Union for the CAFE process (see our Introduction). The issue is then to develop *reduced models*, accurate enough, so that they can replace the detailed physical models with a lower computational burden and for long-term studies.

6.5 Next-Generation Models

Air quality models are powerful tools that are increasingly used and improved. In the short term, at least five items should deeply modify the numerical models.

In the near future, a first trend will be an increase of spatial resolution. At the continental scale, the current model resolution ranges from 10 to 100 kilometers and will go down to 1–10 kilometers. This point is of course related to the improvement of meso-scale meteorological models and of emission inventories. This will induce many new open questions at the numerical level, at the parameterization level and at the data level. One may even wonder if a CTM will not use meshes similar to those used for CFD (with millions of cells and unstructured adaptative grids). This will require the use of HPCN (High Performance Computing and Networking).

The second point is related to improvement of "chemical" resolution. The complexity of physical models, especially for aerosols, will increase: a better understanding of Secondary Organic Aerosols (SOAs) implies a better representation of organic species (more variables, [46]), a coupling between organic and inorganic thermodynamics (more computational burden, [112]) and new processes (surface heterogeneous chemistry, polymerization, etc.). Apart from modeling issues, appropriate numerical algorithms will thus be required.

A third point is related to model couplings and feedbacks. This concerns for instance coupling between radiative transfer models, CTMs and meso-scale models for the radiative behavior of the atmosphere. On-line coupling will be the standard approach. Another example is provided by detailed multi-media models in the estimation of environmental impacts through multi-media models.

One must keep in mind that the final use of such models always needs a drastic reduction of CPU costs. Finding the appropriate trade-off between highly detailed physical models (the "arms race") and tailored models that can be used in integrated modeling for impact assessment will still remain a key issue. The current situation is that one *often* uses low-level models and *sometimes* detailed models for such applications. Models of intermediate complexity or algorithms for using detailed models are therefore of great interest. The fourth point is consequently the search for reduced models to be used for integrated modeling and to be derived with a rigorous basis.

The last point is related to uncertainties. All-in-one models, tuned for a small set of target species, should be replaced by platforms able to generate a large set of models and model configurations, not only restricted to a small set of models that are based on the same assumptions. A key issue will be the ability to compute probability density functions for most outputs, in order to evaluate the *atmospheric risks*.

Problems Related to Chap. 6

Problem 6.1 (Variational Methods for the GDE) This problem aims at introducing the variational methods used for solving the GDE. We focus on the coagulation equation for the number distribution function,

$$\begin{aligned}\frac{\partial n}{\partial t} &= \frac{1}{2}\int_0^m K(\tilde{m}, m-\tilde{m}) n(\tilde{m}, t) n(m-\tilde{m}, t) \mathrm{d}\tilde{m} \\ &\quad - n(m,t)\int_0^m K(m, \tilde{m}) n(\tilde{m}, t)\mathrm{d}\tilde{m} \\ &\triangleq f(n, m, t),\end{aligned}$$

where the nucleation threshold, m_0, is supposed to be equal to 0.

Let V be a space of functions of the variable m, with an appropriate scalar product $\langle .,. \rangle$. A weak formulation of the coagulation equation is obtained by calculating the scalar product with a *test function* v:

$$\forall v \in V, \quad \left\langle \frac{\mathrm{d}n}{\mathrm{d}t}, v \right\rangle = \langle f(n, m, t), v \rangle.$$

V is chosen so that all the integrals are well defined.

The numerical solution is defined as a *projection* of the exact solution onto a finite-dimensional subspace of V, V_N (N stands for dimension). Let $(L_1(m), \ldots, L_N(m))$ be a basis of functions for V_N. Hence, the numerical solution can be computed as

$$\Pi n(m,t) = \sum_{j=1}^{N} n^j(t) L_j(m).$$

If the functions $L_j(m)$ are piecewise polynomial functions, the method is referred to as *a finite element method*; if the functions are defined on $[0, \infty[$, the method is said to be a *spectral method*.

1. We investigate the so-called *spectral collocation method*. Let $(m_i)_{i=1,N}$ be a mass discretization. The basis functions are the Lagrange polynomials: L_j is of order N and satisfies $L_j(m_i) = \delta_{ij}$. Formulate the system of EDOs met by $(n^j)_j$. Hint: choose the test functions in order to compute the *degrees of freedom* n^j.
2. Conclude for time integration. Propose a generalization for the whole GDE.

Solution:

1. The *test* functions v to be used for calculating the *degrees of freedom* $n^j(t)$ are the Dirac functions $v_i(m) = \delta(m - m_i)$. This gives

$$\begin{aligned}\frac{\mathrm{d}n^i}{\mathrm{d}t} &= \sum_{j_1=1}^{N}\sum_{j_2=1}^{N} n^{j_1}(t) n^{j_2}(t) \underbrace{\left[\frac{1}{2}\int_0^{m_i} K(\tilde{m}, m_i - \tilde{m}) L_{j_1}(\tilde{m}) L_{j_2}(m_i - \tilde{m})\,\mathrm{d}\tilde{m}\right]}_{A^i_{j_1 j_2}} \\ &\quad - n^i(t) \underbrace{\left[\sum_{j_1=1}^{N}\int_0^{m_i} K(m_i, \tilde{m}) L_{j_1}(\tilde{m}, t)\,\mathrm{d}\tilde{m}\right]}_{B_{i j_1}}.\end{aligned}$$

With a vectorial notation (here, n stands for the vector of components n^i), this equation can be written as $\mathrm{d}n/\mathrm{d}t = A(n)n - \mathrm{diag}(Bn)n$ where $[A(n)n]_i = n^T A^i n$. The matrices $(A^i)_i$ and B can be computed in a preprocessed step.

2. The time integration of the resulting system of EDOs can be carried out with an appropriate scheme. The method is easily generalized to the whole GDE. This can avoid the use of a splitting method for the three processes (coagulation, condensation and evaporation, nucleation). Due to the wide range of timescales for condensation, the time integration scheme must be implicit.
   This framework is more powerful than the sectional methods since it provides a functional description of $n(m,t)$: on the contrary, a sectional method computes only values at the discretization points $(m_i)_i$ (and not at any m).

To know more ([32]):

E. DEBRY AND B. SPORTISSE, *Numerical simulation of the General Dynamics Equation (GDE) for aerosols with two collocation methods*, Appl. Numer. Math., **457** (2007), pp. 885–898

Problem 6.2 (Adjoint Model for the Dispersion Equation) This problem illustrates the notion of an *adjoint model* for the linear case. We consider the dispersion of a passive tracer (let us say a radionuclide). The time evolution of the mixing ratio is governed by

$$\frac{\partial C}{\partial t} + V\nabla C = \frac{1}{\rho}\mathrm{div}(\rho K \nabla C) - \lambda C + S,$$

where S is a point source term. The coefficient λ describes all the linear loss terms (scavenging and filiation). We suppose that $C(t=0)=0$ and $\nabla C = 0$ along the boundaries of the spatial domain Ω. The time interval is $[0, T]$.

The solution C depends on the source term S and is written as C_S. This defines the so-called *forward model*: a given source term generates a mixing ratio field. Let J be a function of the source term defined by

$$J(S) = \iint e(x,t)\rho(x,t)C_S(x,t)\,\mathrm{d}x\mathrm{d}t,$$

with $e(x,t)$ a sampling function (with respect to time and space). The integration is carried out over $\Omega \times [0,T]$. In practice, J represents the observation of the mixing ratio at a given point or the averaged value.

1. Prove that J is a linear function of S.
2. It is desired to calculate the sensitivity of J with respect to the source term (e.g. for inverse modeling). Formulate the equation satisfied by $\nabla_S J$, defined for any perturbation of the source term δS, by

$$J(S+\delta S) = J(S) + \iint \rho(x,t)\nabla_S J\,\delta S(x,t)\,\mathrm{d}x\mathrm{d}t$$

   Is the resulting algorithm efficient?

3. We start with a simple formal calculation in order to introduce a more powerful approach. Let $\langle .,. \rangle$ be the scalar product defined by the air density ρ: $\langle f, g\rangle = \iint \rho f g \,\mathrm{d}x\mathrm{d}t$.
   The dispersion model generates a linear operator M to be applied to the source term S, so that $C_S = MS$. Thus,

$$J(S) = \langle MS, e\rangle = \langle S, M^T e\rangle,$$

   where the definition of the transpose matrix (the *adjoint* matrix) is used. Upon identification, $\nabla_S J = M^T e$ since $J(S) = \langle S, \nabla_S J\rangle$, which provides a direct computation of the gradient.
   The difficult point is to give a meaning for the *adjoint model* $e \to M^T e$. Let $C^\star$ be a spatial field (to be made precise below). Upon integration of the dispersion equation multiplied by $\rho C^\star$, formulate an evolution equation for $C^\star$ (the so-called *adjoint variable*) so that

$$J(S) = \langle C^\star, S\rangle = \iint \rho C^\star S \,\mathrm{d}x\mathrm{d}t.$$

   Hints: use integration by parts and anneal all the terms related to boundary conditions (justified by conditions to be applied to $C^\star$).
4. Comment. Is it possible to reuse the initial model to compute $C^\star$?

Solution:

1. C_S is a linear function of S. This implies the linearity of J.
2. We easily obtain

$$\iint \rho(x,t) \nabla_S J \,\delta S(x,t) \,\mathrm{d}x\mathrm{d}t = J(S+\delta S) - J(S)$$
$$= \iint \rho(x,t)\, e(x,t)\, C_{\delta S}(x,t) \,\mathrm{d}x\mathrm{d}t,$$

   with $C_{\delta S} = C_{S+\delta S} - C_S$ governed by the linear dispersion equation.
   This approach is not efficient because this does not provide $\nabla_S J$. It is only possible, for any perturbation δS, to get $\iint \rho \nabla_S J \,\delta S \,\mathrm{d}x\mathrm{d}t$, which requires that we solve the model for any perturbation. Upon spatial discretization, with n grid cells at ground (this means n point source emissions), this requires n model runs for computing $\nabla_S J$ (one model run per gradient component)! This is not affordable in practice.
3. Upon integration by parts, the equation yields

$$\langle C^\star, S\rangle = \iint \rho C^\star \Big(\frac{\partial C}{\partial t} + V\nabla C - \frac{1}{\rho}\mathrm{div}(\rho K \nabla C) + \lambda C \Big) \mathrm{d}x\mathrm{d}t$$
$$= \iint \Big(-\frac{\partial(\rho C^\star)}{\partial t} C - \mathrm{div}(\rho V C^\star) C - C\mathrm{div}(\rho K \nabla C^\star) + \rho\lambda C^\star C \Big) \mathrm{d}x\mathrm{d}t,$$

   thus

$$\langle C^\star, S\rangle = \iint -\underbrace{\Big(\frac{\partial \rho}{\partial t} + \mathrm{div}(\rho V) \Big)}_{=0} C^\star C \,\mathrm{d}x\mathrm{d}t$$

$$+\iint \rho C\Big(-\frac{\partial C^\star}{\partial t} - V\nabla C^\star - \frac{1}{\rho}\mathrm{div}(\rho K\nabla C^\star) + \lambda C^\star\Big)\,\mathrm{d}x\mathrm{d}t$$

$$= \Big\langle C, -\frac{\partial C^\star}{\partial t} - V\nabla C^\star - \frac{1}{\rho}\mathrm{div}(\rho K\nabla C^\star) + \lambda C^\star\Big\rangle.$$

The terms related to initial and boundary conditions can be omitted if $C^\star(t=T)=0$ and $\nabla C^\star = 0$ along the domain boundaries. As $J(S) = \langle C, e\rangle$, we get $J(S) = \langle C^\star, S\rangle$ if $C^\star$ is governed by

$$-\frac{\partial C^\star}{\partial t} - V\nabla C^\star = \frac{1}{\rho}\mathrm{div}(\rho K\nabla C^\star) - \lambda C^\star + e(x,t).$$

This is a PDE with a backward time integration (which justifies the condition at $t = T$). This equation gives a meaning to $M^T e$, defined as $C^\star$.

4. The dispersion equation for $C^\star$ can be solved by using the initial model: the time sequence has to be reversed for the forcing fields, the sign of the wind velocity has to be reversed and the source term is $e(x,t)$. Thus, the solution is $C^\star$.
The sensitivity analysis is easy to carry out because only *one* model run is required for computing $\nabla_S J = C^\star$. This result does not depend on the dimension of S!
This approach can be extended to the nonlinear case. The adjoint model is a powerful approach to be used for computation of the gradient of a scalar value with respect to input data fields (gridded values). Note that this can be used for computing the gradient of the cost function in variational data assimilation methods. Building the *adjoint code* from a nonlinear code is not easy. *Automatic differentiation techniques* are a promising tool.

To know more ([95, 55]):

G. Marchuk, *Mathematical models in environmental problems*, vol. 16, North-Holland, 1986

F. Hourdin and J.-P. Issartel, *Sub-surface nuclear tests monitoring through the CTBT* ^{133}Xe *network*, Geophys. Res. Lett., **27** (2000), pp. 2245–2248

Appendix 1
Units, Constants and Basic Data

Table A.1 Systeme International (SI) units and conversion factors

Quantity	Unit	Definition	Conversion
Length	Meter	m	
Mass	Kilogram	kg	
Time	Second	s	
Temperature	Kelvin	K	0°C = 273.15 K
Force	Newton	N kg m s^{-2}	1 dyn = 10^{-5} N
Pressure	Pascal	Pa N m^{-2}	1 atm = 1.013×10^5 Pa
Energy	Joule	J kg m^2 s^{-2}	
Power	Watt	W J s^{-1}	1 cal = 4.19 J

Table A.2 Physical constants

	Name	Value
$\mathcal{A}_v$	Avogadro's number	6.022×10^{23} molecule mol^{-1}
G	universal gravitational constant	6.67×10^{-11} m^3 kg^{-1} s^{-2}
R	universal gas constant	8.314 J mol^{-1} K^{-1}
σ	Stefan-Boltzmann constant	5.67×10^{-8} W m^{-2} K^{-4}
c	speed of light in vacuum	3.0×10^8 m s^{-1}
h	Planck's constant	6.63×10^{-34} J s
k_B	Boltzmann's constant ($R/\mathcal{A}_v$)	1.38×10^{-23} J K^{-1}

Table A.3 A few basic data for Earth, air and water

	Name	Value
M_t	mass of the earth	6×10^{24} kg
R_t	radius of the earth	6370 km
S	solar constant	1368 W m^{-2}
g	gravity at surface of earth	9.81 m^{-2}
r	mean earth-sun distance	1.496×10^{11} m
c_p	specific heat of dry air at constant pressure (P)	1005 J kg^{-1} K^{-1}
ν_{air}	dynamic air diffusivity	1.6×10^{-5} m^2 s^{-1}
$c_{p,v}$	specific heat of water vapor at constant pressure	1952 J kg^{-1} K^{-1}
L_v	latent heat of water vaporization at 0 K	2.5×10^6 J kg^{-1}

Table A.4 Atomic weights of a few elements and molar data of a few molecules. The masses, M, are expressed in g mol^{-1}

	Name	M		Name	M
H	hydrogen	1	F	fluorine	19
C	carbon	12	S	sulfur	32
N	nitrogen	14	Cl	chlorine	35.5
O	oxygen	16	Ar	argon	40
OH	hydroxyl	17	HO_2	hydroperoxyl	33
CO	carbon monoxide	28	CO_2	carbon dioxide	44
NO	nitrogen monoxide	30	NO_2	nitrogen dioxide	46
O2	molecular oxygen	32	O_3	ozone	48
SO	sulfur monoxide	48	SO_2	sulfur dioxide	64
CH_4	methane	16	HNO_3	nitric acid	63
NH_3	ammonia	17	H_2SO_4	sulfuric acid	98
H_2O	water	18	$CFCl_3$	CFC-11	137.5
HONO	nitrous acid	47	CF_2Cl_2	CFC-12	121

Table A.5 Multiplying prefixes and definition of angström

Prefix	Symbol	Multiplying factor
tera	T	10^{12}
giga	G	10^9
mega	M	10^6
micro	μ	10^{-6}
nano	n	10^{-9}
pico	p	10^{-12}
angström	A	10^{-10} m

References

[1] *EMEP Assessment Report*, 2004. Chapter 2: Sulphur.

[2] *Mise en œuvre de la stratégie communautaire de réduction des émissions de* CO_2 *des voitures*. Communication de la Commission au Conseil et au Parlement Europen, 2005. COM(2005) 269.

[3] *An assessment of Tropospheric Ozone pollution: a North American perspective*. NARSTO Assessment, 2006. Available at http://www.narsto.org.

[4] La qualité de l'air en Île-de-France en 2006, 2007. Rapport Airparif.

[5] *Scientific assessment of ozone depletion: 2006*. Chapter 8: Halocarbon Scenarios, Ozone Depletion Potentials, and Global Warming Potentials, 2007. World Meteorological Organization.

[6] ACEA, *European Automobile Industry United in Approach towards Further Reducing* CO_2 *Emissions from cars*, June 2007.

[7] M. Amann et al., *RAINS Review 2004*. Report International Institute for Applied Systems Analysis, 2004.

[8] M. Andreae, *World Survey of Climatology: Future Climates of the World*, vol. XVI, Elsevier, 1995, ch. Climatic effects of changing atmospheric aerosol levels.

[9] S. Arrhnius, *On the influence of carbonic acid in the air upon the temperature of the ground*, Philos. Mag., **41** (1896), pp. 237–276.

[10] S. Arya, *Air Pollution Meteorology and Dispersion*, Oxford University Press, 1999.

[11] B. Aumont, S. Szopa, and S. Madronich, *Modelling the evolution of organic carbon during its gas-phase tropospheric oxidation: development of an explicit model based on a self generating approach*, Atmos. Chem. Phys., **5** (2005), pp. 2497–2517.

[12] J. Baker and J. Thornes, *Solar position within Monet's Houses of Parliament*, Proc. Roy. Soc. A, **462** (2006), pp. 3775–3788.

[13] P. Bergamaschi, R. Hein, M. Heimann, and P. Crutzen, *Inverse modeling of the global CO cycle. 1. Inversion of CO mixing ratios*, J. Geophys. Res., **105** (2000), pp. 1909–1927.

[14] P. Bergamaschi, M. Krol, F. Dentener, A. Vermeulen, F. Meinhardt, R. Graul, M. Ramonet, W. Peteres, and E. Dlugokencky, *Inverse modelling of national and European* CH_4 *emissions using the atmospheric zoom model TM5*, Atmos. Chem. Phys., **5** (2005), pp. 2431–2460.

[15] M. Bocquet, *High resolution reconstruction of a tracer dispersion event: application to ETEX*, Quart. J. Roy. Meteor. Soc., **133** (2007), pp. 1013–1026.

[16] D. Bolton, *The computation of equivalent potential temperature*, Mon. Wea. Rev., **108** (1980), pp. 1046–1053.

[17] P. Bousquet, D. Hauglustaine, P. Peylin, C. Carouge, and P. Ciais, *Two decades of OH variability as inferred by an inversion of atmospheric transport and chemistry of methyl chloroform*, Atmos. Chem. Phys., **5** (2005), pp. 2635–2656.

[18] G. Brasseur, J. Orlando, and G. Tyndall, *Atmospheric Chemistry and Global Change*, Oxford University Press, 1999.

[19] M. Brauer, C. Avila-Casado, T. Fortoul, S. Vedal, B. Stevens, and A. Churg, *Air pollution retained particles in the lung*, Env. Health Perspect., **109** (2001), pp. 1039–1043.

[20] P. Brimblecombe, *The Big Smoke: A History of Air Pollution in London Since Medieval Times*, Routledge Kegan & Paul, 1987.

[21] N. Carslaw, *A new detailed chemical model for indoor air pollution*, Atmos. Env., **41** (2007), pp. 1164–1179.

[22] M. Chang, D. Hartley, C. Cardelino, and W.-L. Chang, *Inverse modeling of biogenic emissions*, Geophys. Res. Lett., **23** (1996), p. 3007.

[23] M. Chang, D. Hartley, C. Cardelino, D. Haas-Laursen, and W.-L. Chang, *On using inverse methods for resolving emissions with large spatial inhomogeneities*, J. Geophys. Res., **102** (1997), pp. 16023–16036.

[24] R. Charlson, J. Langner, H. Rodhe, C. Leovy, and S. Warren, *Perturbation of the northern hemisphere radiative balance by backscattering from anthropogenic sulfate aerosols*, Tellus, **43AB** (1991), pp. 152–163.

[25] P. Crutzen *New directions the growing urban heat and pollution "island" effect—impact on chemistry and climate*, Atmos. Env., **38** (2004), pp. 3339–3340.

[26] P. Crutzen, *Albedo enhancement by stratospheric sulfur injections: a contribution to resolve a policy dilemma*? Climatic Change, **77** (2006), pp. 211–219.

[27] P. Crutzen and J. Birks, *The atmosphere after a nuclear war. Twilight at noon*, Ambio, **11** (1982), pp. 114–125.

[28] P. Crutzen, A. Mosler, K. Smith, and W. Winiwarter, N_2O *release from agro-biofuel production negates global warming reduction by replacing fossil fuels*, Atmos. Chem. Phys. Discuss., **7** (2007), pp. 11191–11205.

[29] B. Dahneke, *Theory of Dispersed Multiphase Flow*, Academic Press, New York, 1983.

[30] G. De Moor and P. Veyre, *Les bases de la météorologie dynamique*. Cours École Nationale de la Météorologie, 1991.

[31] E. Debry and B. Sportisse, *Reduction of the condensation/evaporation dynamics for atmospheric aerosols: theoretical and numerical investigation of hybrid methods*, J. Aerosol Sci., **37** (2006), pp. 950–966.

[32] E. Debry and B. Sportisse, *Numerical simulation of the General Dynamics Equation (GDE) for aerosols with two collocation methods*, Appl. Numer. Math., **457** (2007), pp. 885–898.

[33] E. Debry and B. Sportisse, *Solving aerosol coagulation with size-binning methods*, Appl. Numer. Math., **47** (2007), pp. 1008–1020.

[34] P. Denton, *Puffs of smoke, puffs of praise: reconsidering John Evelyn's Fumifugium* (1661), Can. J. Hist., (2000).

[35] B. Dousset and F. Gourmelon, *Satellite multi-sensor data analysis of urban surface temperatures and landcover*, J. Photogramm. Remote Sens., **58** (2003), pp. 43–54. Available at http://www.satlab.hawaii.edu/papers/jprs/.

[36] B. Dousset, F. Gourmelon, and E. Mauri, *Application of satellite remote sensing for urban risk analysis: a case study of the 2003 extreme heat wave in Paris*, in Proceedings of the 2007 Urban Remote Sensing Joint Event, I.E.E.E., 2007. Available at http://www.satlab.hawaii.edu/papers/noref/Dousset-URS-07.pdf.

[37] A. Energy and Environment, *Evaluation of the national plans submitted in 2006 under the National Emission Ceilings Directive 2001/81/EC*, April 2007. Interim report to the European Commission. Version 1.7.

[38] J. Farman, B. G. Gardiner, and J. Shanklin, *Large losses of total ozone in Antarctica reveal seasonal* ClO_x/NO_x *interactions*, Nature, **315** (1985), pp. 207–210.

[39] A.-M. Fjaeraa, *Data Report 2004. Acidifying and eutrophying compounds*, tech. rep., EMEP, 2006.

[40] P. Francis, J. Taylor, P. Hignett, and A. Slingo, *On the question of enhanced absorption of solar radiation by clouds*, Quart. J. Roy. Meteor. Soc., **123** (1997), pp. 419–434.

[41] N. FUCHS, *Mechanics of Aerosols*, Pergamon, New York, 1964.

[42] J. GARRAT, *The Atmospheric Boundary Layer*, Cambridge University Press, 1992.

[43] R. GOODY AND Y. YUNG, *Atmospheric Radiation. A Theoretical Basis*, Oxford University Press, 1986.

[44] E. GORHAM, *Robert Angus Smith, F.R.S. and Chemical Climatology*, Notes and Records Roy. Soc. Lond., **36** (1982), pp. 267–272.

[45] S. GREENFIELD, *Rain scavenging of radioactive particulate matter from the atmosphere*, J. Meteorol., **14** (1957), pp. 115–125.

[46] R. GRIFFIN, D. DABDUB, AND J. SEINFELD, *Secondary organic aerosol 1. Atmospheric chemical mechanism for production of molecular constituents*, J. Geophys. Res., **107** (2002).

[47] D. GROSJEAN, *Secondary organic aerosol: identification and mechanisms of formation*, in Carboneous particles in the Atmosphere Conference, NSF, March 1978. Berkeley, California.

[48] G. GUEROVA, I. BEY, J. ATTIÉ, J. CUI, AND M. SPRENGER, *Impact of transatlantic transport episodes on summertime ozone in Europe*, Atmos. Chem. Phys., **6** (2006).

[49] J. GUIBET, *Carburants et moteurs. Technologies, energie, environnement*, vol. 1, Editions TECHNIP, 1997.

[50] J. GUIBET, *Carburants et moteurs. Technologies, energie, environnement*, vol. 2, Editions TECHNIP, 1997.

[51] S. R. HANNA, J. C. CHANG, AND M. E. FERNAU, *Monte Carlo estimates of uncertainties in predictions by a photochemical grid model (UAM-IV) due to uncertainties in input variables*, Atmos. Env., **32** (1998), pp. 3,619–3,628.

[52] K. HANSEN, J. CHRISTENSEN, J. BRANDT, L. FROHN, AND C. GEELS, *Modelling atmospheric transport of α-hexachlorocyclohexane in the Northern Hemisphere with a 3-D dynamical model: DEHM-POP*, Atmos. Chem. Phys., **4** (2004), pp. 1125–1137.

[53] M. HESS, P. KOEPKE, AND I. SCHULT, *Optical properties of aerosols and clouds: the software package OPAC*, Bull. Amer. Meteor. Soc., **79** (1998), pp. 831–844.

[54] J. HOLTON, *An Introduction to Dynamic Meteorology*, Academic Press, 1992.

[55] F. HOURDIN AND J.-P. ISSARTEL, *Sub-surface nuclear tests monitoring through the CTBT ^{133}Xe network*, Geophys. Res. Lett., **27** (2000), pp. 2245–2248.

[56] R. HUDMAN, ET AL., Ozone production in transpacific Asian pollution plumes and implications for ozone in air quality in California, *J. Geophys. Res.*, **109** (2004).

[57] R. HUSAR, J. PROSPERO, AND L. STOWE, *Characterization of tropospheric aerosols over the oceans with the NOAA advanced very high resolution radiometer optical thickness operational product*, J. Geophys. Res., **102** (1997), pp. 16889–16909.

[58] D. JACOB, *Introduction to Atmospheric Chemistry*, Princeton University Press, 1999.

[59] D. JACOB, *Heterogeneous chemistry and tropospheric ozone*, Atmos. Env., **34** (2000), pp. 2131–2159.

[60] D. JACOB, M. PRATHER, S. WOFSY, AND M. MCELROY, *Atmospheric distribution of ^{85}Kr simulated with a general circulation model*, J. Geophys. Res., **22** (1987), pp. 6614–6626.

[61] D. J. JACOB, J. LOGAN, AND P. MURTI, *Effect of rising Asian emissions on surface ozone in the United States*, Geophys. Res. Lett., **26** (1999), pp. 2175–2178.

[62] M. JACOBSON, *A physically-based treatment of elemental carbon optics: implications for global direct forcing of aerosols*, Geophys. Res. Lett., **27** (2000), pp. 217–220.

[63] M. JACOBSON, *Strong radiative heating due to the mixing state of black carbon in atmospheric aerosols*, Nature, **409** (2001).

[64] M. JACOBSON, *Effects of ethanol (E85) versus gasoline vehicles on cancer and mortality in the United States*, Environ. Sci. Tech. (2007).

[65] M. JACOBSON, J. SEINFELD, G. CARMICHAEL, AND D. STREETS, *The effect on photochemical smog of converting the US fleet of gasoline vehicles to modern diesel vehicles*, Geophys. Res. Lett., **31** (2004).

[66] M. JACOBSON AND R. TURCO, *Simulating condensational growth, evaporation and coagulation of aerosols using a combined moving and stationary size grid*, Aerosol Sci. Technol., **22** (1995), pp. 73–92.

[67] M. Z. JACOBSON, *Fundamentals of Atmospheric Modeling*, Cambridge University Press, New York, 1998.

[68] R. JAENICKE, *Physical and Chemical Properties of the Air*, vol. XV, Springer Verlag, 1988, ch. Properties of atmospherie modeling.

[69] J. JEANS, *The Dynamical Theory of Gases*, Cambridge University Press, 1985.

[70] M. JENNER, *The politics of London Air John Evelyn's Fumifugium and the Restoration*, Hist. J., **38** (1995), pp. 525–551.

[71] L. JESTIN, *Communication personnelle*. Directeur Technique EDF Polska, 2006.

[72] J. JONSON, D. SIMPSON, H. FAGERLI, AND S. SOLBERG, *Can we explain the trends in European ozone levels*? Atmos. Chem. Phys., **6** (2006), pp. 51–66.

[73] M. KANAKIDOU AND P. CRUTZEN, *The photochemical source of carbon monoxide: importance, uncertainties and feedbacks*, Chemosphere: Glob. Change Sci., **1** (1997), pp. 91–109.

[74] KIEHL AND TRENBERTH, *Earth's annual global mean energy budget*, Bull. Amer. Meteor. Soc., **78** (1997), pp. 197–208.

[75] Y. KIM AND J. SEINFELD, *Simulation of multicomponent aerosol condensation by the moving sectional method*, J. Colloid Interface Sci., **135** (1990), pp. 185–199.

[76] D. KITTELSON, J. JOHNSON, W. WATTS, Q. WEI, M. DRAYTON, D. PAULSEN, AND N. BUKOWIECKI, *Diesel aerosol sampling in the atmosphere*, Technical paper series 2000-01-2212, SAE (Society of Automative Engineers), 2000.

[77] L. KLEINMAN *Low and high* NO_x *tropospheric photochemistry*, J. Geophys. Res., **99** (1994), pp. 16831–16838.

[78] P. KREY AND B. KRAJEWSKI, *Comparison of atmospheric transport model calculations with observations of radioactive debris*, J. Geophys. Res., **75** (1970), pp. 2901–2908.

[79] J. KRISTJANSSON, J. KRISTIANSEN, AND E. KAAS, *Solar activity, cosmic rays, clouds and climate: an update*, Adv. Space Sci. (2004).

[80] R. LAMB, *Note on the application of K-Theory to diffusion problems involving nonlinear chemical reactions*, Atmos. Env., **7** (1973), pp. 257–263.

[81] L. LANSER AND J. VERWER, *Analysis of operator splitting for advection-diffusion-reaction problems from air pollution modelling*, in Proceedings 2nd Meeting on Numerical Methods for Differential Equations, Coimbra, Portugal, February 1998.

[82] S. LARSON, G. CASS, K. HUSSEY, AND F. LUCE, *Visibility model verification by image processing techniques*. Final report to the California Air Resources Board, 1984.

[83] M. LAWRENCE, P. JÄCKEL, AND R. VON KUHLMANN, *What does the global mean OH concentration tell us*? Atmos. Chem. Phys. Discuss., **1** (2001), pp. 43–74.

[84] J. LELIEVELD ET AL., *Watching over tropospheric hydroxyl (OH)*, Atmos. Env., **40** (2006), pp. 5741–5743.

[85] G. LESINS, P. CHYLEK, AND U. LOHMANN, *A study of internal and external mixing scenarios and its effect on aerosol optical properties and direct radiative forcing*, J. Geophys. Res., **107** (2002).

[86] I. LEVIN AND V. HESSHAIMER, *Refining of atmospheric transport model entries by the globally observed passive tracer distributions of* ^{85}*krypton and sulfur hexafluoride* (SF_6), J. Geophys. Res., **101** (1996), pp. 16745–16755.

[87] Q. LI, D. JACOB, I. BEY, P. PALMER, B. DUNCAN, B. FIELD, R. MARTIN, A. FIORE, R. YANTOSCA, P. SIMMONDS, AND S. OLTMANS, *Transatlantic transport of pollution and its effects on surface ozone in Europe and North America*, J. Geophys. Res., **107** (2002).

[88] Z. LI, T. ACKERMAN, W. WISCOMBRE, AND G. STEPHENS, *Have clouds darkened since 1995*? Science, **302** (2003). Letters.

[89] K. LIOU, *Radiation and Cloud Processes in the Atmosphere*, vol. 20, Oxford Monograph on Geology and Geophysics, 1992.

[90] U. LOHMANN AND J. FEICHTER, *Global indirect aerosol effects: a review*, Atmos. Chem. Phys., **5** (2005), pp. 717–737.

[91] V. MALLET AND B. SPORTISSE, *Ensemble-based air quality forecasts: a multi-model approach applied to ozone*, J. Geophys. Res., **111** (2006), p. 18302.

[92] B. MALLET AND B. SPORTISSE, *Introduction to Computational Atmospheric Chemistry: From Fundamentals to Advanced Applications of Chemistry Transport Models*, to appear.

[93] V. MALLET, D. QUÉLO, B. SPORTISSE, M. AHMED DE BIASI, E. DEBRY, I. KORSAKISSOK, L. WU, Y. ROUSTAN, K. SARTELET, M. TOMBETTE, AND H. FOUDHIL, *Technical Note: The air quality modeling system Polyphemus*, Atmos. Chem. Phys. Discuss., **7** (2007), pp. 6,459–6,486.

[94] F. MARANO, *Pollution par les particules atmospheriques: état des connaissances et perspectives de recherche*, Documentation Française, 2005, ch. Impacts sur la santé, aspects toxicologiques.

[95] G. MARCHUK, *Mathematical Models in Environmental Problems*, vol. 16, North-Holland, 1986.

[96] R. P. MASON, W. FITZGERALD, AND F. MOREL, *The biogeochemical cycling of elemental mercury. Anthropogenic influences*, Geochim. Cosmochim. Acta, **58** (1994), pp. 3191–3198.

[97] M. MOLEMAKER AND J. VILA-GUERAU DE ARELLANO, *Control of chemical reactions by convective turbulence in the boundary layer*, J. Atmos. Sci., **55** (1998), pp. 568–579.

[98] B. NEMERY, P. HOET, AND A. NEMMAR, *The Meuse Valley fog of 1930: an air pollution disaster*, Lancet, **357** (2001), pp. 704–708.

[99] S. NEMESURE, R. WAGENER, AND S. SCHWARTZ, *Direct shortwave forcing of climate by the anthropogenic sulfate aerosol: sensitivity to particle size, composition, and relative humidity*, J. Geophys. Res., **100** (1995), pp. 26105–26116.

[100] A. NENES, S. PANDIS, AND C. PILINIS, *ISORROPIA: A new thermodynamic equilibrium model for multicomponent inorganic aerosols*, Aquat. Geochem., **4** (1998), pp. 123–152.

[101] K. NODOP, ed., *ETEX symposium on long-range atmospheric transport, model verification and emergency response*, May 1997. Vienna (Austria).

[102] E. NOSLUND AND L. THANING, *On the settling velocity in a nonstationary atmosphere*, Aerosol Sci. Technol., **14** (1991), pp. 247–256.

[103] O. OF AIR AND U.E. RADIATION, *Performance of selective catalytic reduction on coal-fired steam generating units*. Acid Rain Program, June 1997. Final Report.

[104] T. OKE, *City size and the urban heat island*, Atmos. Env., **7** (1973), pp. 769–779.

[105] T. OKE, *Boundary Layer Climates*, Routledge, 1987.

[106] I.P. ON CLIMATE CHANGE, *Climate Change 2001. IPCC Third Assessment Report. The Scientific Basis*, 2001. WMO and UNEP.

[107] C. PAPASTEFANOU, *Residence time of tropospheric aerosols in association with radioactive nuclides*, Appl. Radiat. Isot., **64** (2006), pp. 93–100.

[108] R. PARK, D. JACOB, N. KUMAR, AND R. YANTOSCA, *Regional visibility statistics in the United States: Natural and transboundary pollution influences, and implications for the Regional Haze Rule*, Atmos. Env., **40** (2006), pp. 5405–5423.

[109] S. PARK, S. ROGAK, W. BUSHE, J. WEN, AND M. THOMSON, *An aerosol model to predict size and structure of soot particles*, Combust. Theory Model., **9** (2005), pp. 499–513.

[110] R. PIELKE, *Mesoscale Meteorological Modelling*, Academic Press, 1984.

[111] N. POISSON, M. KANAKIDOU, AND P. CRUTZEN, *Impact of non-methane hydrocarbons on tropospheric chemistry and the oxidizing power of the global troposphere: 3-dimensional modelling results*, J. Atmos. Chem., **36** (2000), pp. 157–230.

[112] B. PUN, R. GRIFFIN, C. SEIGNEUR, AND J. SEINFELD, *Secondary Organic Aerosol 2. Thermodynamic model for gas/particle partitioning of molecular constituents*, J. Geophys. Res., **107** (2002).

[113] J. PUTAUD, R. DINGENEN, U. BALTENSPERGER, E. BRÜGGEMANN, A. CHARRON, M. FACCHINI, S. DECESARI, S. FUZZI, R. GEHRIG, H. HANSSON, R. HARRISON, A. JONES, P. LAJ, G. LORBEER, W. MAENHAUT, N. MIHALOPOULOS, K. MLLER, F. PALMGREN, X. QUEROL, S. RODRIGUEZ, J. SCHNEIDER, G. SPINDLER, H. BRINK, P. TUNVED, K. TORSETH, E. WEINGARTNER, A. WIEDENSOHLER, P. WAHLIN, AND F. RAES, *A European Aerosol Phenomenology*, tech. rep., Joint Research Centre, Institute for Environment and Sustainability, 2003.

[114] M. Quante, *Urban Climate: an example for a significant anthropogenic modification.* European Research Course on Atmospheres, 2004.

[115] D. Quélo, M. Krysta, M. Bocquet, O. Isnard, Y. Minier, and B. Sportisse, *Validation of the polyphemus system: the ETEX, Chernobyl and Algeciras cases*, Atmos. Env., **41** (2007), pp. 5300–5315.

[116] D. Quélo, V. Mallet, and B. Sportisse, *Inverse modeling of* NO_x *emissions at regional scale over Northern France. Preliminary investigation of the second-order sensitivity*, J. Geophys. Res., **110** (2005).

[117] F. Raes, R. Dingenen, E. Vignati, J. Wilson, J.-P. Putaud, J. Seinfeld, and P. Adams, *Formation and cycling of aerosols in the global troposphere*, Atmos. Env., **34** (2000), pp. 4215–4240.

[118] V. Ramanathan et al., *Indian Ocean Experiment (INDOEX): an integrated analysis of the climate forcing and effects of the great Indo-Asian haze*, J. Geophys. Res., **106** (2001), pp. 28371–28398.

[119] V. Ramanathan and P. Crutzen, *Atmospheric Brown Clouds*, Atmos. Env. (2003), pp. 4033–4035.

[120] E. Reiter, *Stratospheric-tropospheric exchange processes*, Rev. Geophys. Space. Phys., **13** (1975), pp. 459–474.

[121] A. Robock, L. Oman, and G. Stenchikov, *Nuclear winter revisited with a modern climate model and current nuclear arsenals: still catastrophic consequences*, J. Geophys. Res., **112** (2007), p. 13107.

[122] F. Rosner, *The life of Moses Maimonides, a prominent medieval physician*, Einstein Quart. J. Bio. Med., **19** (2002), pp. 125–128.

[123] Y. Roustan and M. Bocquet, *Sensitivity analysis for mercury over Europe*, J. Geophys. Res., **111** (2006).

[124] A. Russell and R. Dennis, *NARSTO critical review of photochemical models and modeling*, Atmos. Env., **34** (2000), pp. 2,283–2,234.

[125] A. Sandu, J. Verwer, M. Van Loon, G. Carmichael, F. Potra, D. Dabdub, and J. Seinfeld, *Benchmarking stiff ODEs solvers for atmospheric chemistry problems I: implicit versus explicit*, Atmos. Env., **31** (1997), pp. 3151–3166.

[126] K. Sartelet, E. Debry, K. Fahey, M. Tombette, Y. Roustan, and B. Sportisse, *Simulation of aerosols and related species over Europe with the Polyphemus system. Part I: model-to-data comparison for year 2001*, Atmos. Env. (2007).

[127] S. Schwartz, *The whitehouse effect: Shortwave radiative forcing of climate by anthropogenic aerosols. An overview*, J. Aerosol Sci., **27** (1996), pp. 359–382.

[128] C. Seigneur, *Current status of air quality modeling for particulate matter*, J. Air Waste Manage. Assoc., **51** (2001), pp. 1508–1521.

[129] S. Seigneur, P. Karamchandani, K. Lohman, and K. Vijayaraghavan, *Multiscale modeling of the atmospheric fate and transport of mercury*, J. Geophys. Res., **106** (2001), pp. 27795–27809.

[130] J. Seinfeld and S. Pandis, *Atmospheric Chemistry and Physics*, Wiley-Interscience, 1998.

[131] N. Selin, D. Jacob, R. Park, R. Yantosca, S. Strode, L. Jaegle, and D. Jaffe, *Chemical cycling and deposition of atmospheric mercury: global constraints from observations*, J. Geophys. Res., **112** (2007).

[132] F. Selsis, *Évaporation plantaire*. In *Formation plantaire et exoplantes*, École CNRS de Goutelas XXVIII, 2005. Édit par J.L. Halbwachs, D. Egret et J.M. Hameury.

[133] C. Sorensen, *Light scattering by fractal aggregates: a review*, Aerosol Sci. and Technol., **35** (2001), pp. 648–687.

[134] B. Sportisse, *An analysis of operator splitting techniques in the stiff case*, J. Comp. Phys., **161** (2000), pp. 140–168.

[135] B. Sportisse, *A review of parameterizations for modeling dry deposition and scavenging of radionuclides*, Atmos. Env., **41** (2007), pp. 2683–2698.

[136] B. Sportisse and R. Djouad, *Reduction of chemical kinetics in air pollution modelling*, J. Comp. Phys., **164** (2000), pp. 354–376.

[137] B. Sportisse and R. Djouad, *Mathematical investigation of mass transfer for atmospheric pollutants into a fixed droplet with aqueous chemistry*, J. Geophys. Res., **108** (2003), p. 4073.

[138] B. Sportisse, D. Quélo, and V. Mallet, *Impact of mass consistency errors for atmospheric dispersion*, Atmos. Env., **41** (2007), pp. 6132–6142.

[139] R. Stull, *An Introduction to Boundary Layer Meteorology*, Kluwer Academic Publishers, 1988.

[140] H. Svensmark and E. Friis-Christensen, *Variation of cosmic ray flux and global cloud coverage: a missing link in solar-climate relationships*, J. Atmos. Solar-Terr. Phys., **59** (1997), pp. 1225–1232.

[141] G. Thomas and K. Stamnes, *Radiative Transfer in the Atmosphere and Ocean*, Cambridge University Press, 1999.

[142] F. Troussier, *Evolution spatio-temporelle des teneurs en composs organiques volatils en atmosphere urbaine et priurbaine, et contribution de leurs sources*, PhD thesis, Université des Sciences et Technologies de Lille, 2006.

[143] R. Turco, *Air pollution: a Los Angeles case study*, 2003. Teaching notes.

[144] N. R. C. Usa, *Rethinking the Ozone Problem in Regional and Urban Pollution*, 1991.

[145] H. Vehkamki, M. Kulmala, I. Napari, K. Lehtinen, M. Timmreck, A. Noppel, and Laaksonen *An improved parameterization for sulfuric acid-water nucleation rates for tropospheric and stratospheric conditions*, J. Geophys. Res., **107** (2002), p. 4622.

[146] A. Venkatram and J. Pleim, *The electrical analogy does not apply to modeling dry deposition of particles*, Atmos. Env., **33** (1999), pp. 3075–3076.

[147] J. Vila-Guerau de Arellano, A. Dosio, J. Vinuesa, A. Holtslag, and S. Galmarini, *The dispersion of chemically reactive species in the atmospheric boundary layer*, Meteor. Atmos. Phys., **87** (2004).

[148] J. Vila-Guerau de Arellano, P. Duynkerke, P. Jonker, and P. Builtjes, *An observational study on the effects of time and space averaging in photochemical models*, Atmos. Env., **27A** (1993), pp. 353–362.

[149] M. von Smoluchowski, *Versuch einer mathematischen Theorie des koagulationskinetic kolloider Losunggen*, Z. Phys. Chem. (Leipzig), **92** (1917).

[150] J. Wallace and P. Hobbs, *Atmospheric Science: An Introductory Survey*, Academic Press, second ed., 2006.

[151] F. Wania and D. Mackay, *Tracking the distribution of persistent organic pollutants*, Environ. Sci. Tech., **30** (1996), pp. 390–396.

[152] P. Warneck, *Chemistry of the Natural Atmosphere*, Academic Press, 1999.

[153] C. Weschler, *Ozone-initiated reaction products indoors may be more harmful than ozone itself*, Atmos. Env., **38** (2004), pp. 5715–5716.

[154] L. Whitehouse, A. Tomlin, and M. Pilling, *Systematic lumping of complex tropospheric chemical mechanisms: a time-scale based approach*, Atmos. Chem. Phys. Discuss., **4** (2004), pp. 3785–3834.

[155] J. Wieringa, *Representative roughness parameters for homogeneous terrain*, Boundary-Layer Meteor., **63** (1993), pp. 323–364.

[156] E. Wilkins *Air pollution aspects of the London fog of December 1952*, Quart. J. Roy. Meteor. Soc., **80** (2006), pp. 267–271. From the initial article in 1954.

[157] M. Williams *Air pollution policy: 1952–2002*, Sci. Total Env., **334** (2004), pp. 15–20.

[158] G. Wotawa, L. De Geer, P. Denier, M. Kalinowski, H. Toivonen, R. D'Amours, F. Desiato, J.-P. Issartel, M. Langer, P. Seibert, A. Frank, C. Sloan, and H. Yamazawa, *Atmospheric transport modelling in support of CTBT verification—overview and basic concepts*, Atmos. Env., **37** (2003), pp. 2529–2537.

Index

A
ABL, 93
absorption cross section, 54
absorption layer (Chapman's theory), 67
absorption spectrum, 63
accommodation coefficient, 196, 212, 224
accumulation (mode), 185
acid
  nitric (HNO_3), 189, 197, 219
  sulfuric (H_2SO_4), 187–189, 198
acid rains
  history, 3
  NAPAP, 6
  processes, 213
acidic solution, 213
actinic flux, 139
activation, 209
activation energy, 136
adiabatic lapse rate
  dry air, 100
  moist, 103
adjoint model, 274, 279
adjoint PDE, 282
advection schemes, 254
aerodynamic diameter, 182
aerosol column, 181, 223
aerosol deposition, 130
aerosols, 179
  black carbon, 80, 191
  coagulation, 188, 193
  direct effect, 78, 87
  dynamics, 188
  fine, 227
  GDE, 192
aerosols (*cont.*)
  images, 183
  indirect effect, 81
  inorganic, 181
  marine, 182, 222
  maritime, 180
  mineral, 180
  mixing state, 78, 185
  modes, 185
  nucleation, 188, 198, 228
  organic, 181
  residence time, 186
  secondary organic aerosols (SOA), 191
  settling velocity, 182
  size, 184
  stratospheric, 187
air (composition), 17
air parcel, 101
Airparif, 85, 87, 170
Aitken (mode), 185
albedo, 59
  of aerosols, 90
  surface, 62
ammonia (NH_3), 188, 189, 197, 199, 216
ammonium nitrate, 189, 197, 217
ammonium sulfate, 189, 217
antarctic ozone hole, 156
anthropogenic emissions, 22
anticyclone, 116
aqueous-phase chemistry, 214
Arrhenius' law, 136
Arrhenius (Svante), 133

B. Sportisse, *Fundamentals in Air Pollution*,

atmospheric boundary layer
ABL, 93
neutral, 97
stable, 98
unstable, 97
atmospheric dilemma, 12
biofuels, 170
fine aerosols, 227
sulfate aerosols, 79
Avogadro number, 18

B

Beer-Lambert law, 52
biofuels, 170
biogenic emissions, 22
biomass burning, 22, 83, 180
Boussinesq approximation, 117
Box models, 242
branching reaction, 144
breeze (urban), 125
brown cloud, 82
Brunt-Vaisala frequency, 95, 103
buoyancy, 98, 101

C

carbon dioxide (CO_2), 7, 71
and rainwater pH, 213
greenhouse effect, 64, 76
carbon monoxide (CO)
emissions, 159
oxidation, 165, 177
carbonyl sulfide (OCS), 187
CCN (*Cloud Condensation Nuclei*), 190, 203, 204
CFC
chemical reactions, 154
lifetime, 156
Montreal protocol, 7, 41
residence time, 33
CFL condition, 257, 258
Chapman cycle, 151
chemical kinetics, 134
chemical lifetime, 144
VOCs, 146
chemical regimes, 165, 176
chemistry-transport models (CTM), 243
Chernobyl accident, 13
Clean Air Act, 5
climate engineering, 79
closure schemes, 119, 234
cloud, 202
albedo, 89
direct effect, 78, 89
cloud condensation nuclei (CCN), 81
CLRTAP, 6
coagulation, 188, 193, 263
column
of aerosols, 181
of liquid water, 202
ozone, 27, 156
combustion, 159
composition
of aerosols, 181
of dry air, 17
concentration
mass, 19
molecule, 20
condensation and evaporation, 188, 195, 263
conditional stability, 104
constant
Boltzmann, 18
Henry, 211
Planck, 49
Stefan-Boltzmann, 51
universal gas, 18
continuity equation, 114
controversies
cosmic rays, 84
Convention on Long-Range Transboundary Air Pollution (CLRTAP), 6
Coriolis force, 113, 114, 122
crystallization, 197
CTBT, 277
CTM, 231, 243

D

Damköhler number, 236
data assimilation, 269
3D-Var, 271
4D-Var, 274
deliquescence, 81, 197
diffusion
eddy (K_z), 236
molecular (dynamic), 28
direct effect, 78, 87
dissolution, 211
Dobson Unit, 27, 156
DRH, 197
dry deposition, 34, 239, 245
dry deposition (aerosols), 130

E
eddy diffusion (K_z), 244
EKMA, 166
Ekman layer, 97, 120
EMEP, 6, 274
emission reduction
NEC directive, 7
strategy, 167
emissions, 243
carbon, 27
chemical speciation, 23
nitrogen oxides, 22
of aerosols, 180
VOC, 22
emissivity, 52
energy
latent, 70
sensible, 70, 98
turbulent kinetic, 110
ensemble forecast, 268, 269
enthalpy, 99
entrainment layer, 106, 112
entropy, 205
EPA, 5, 274
equations
advection, diffusion, reaction, 232
continuity, 114
diphasic (mass transfer), 212, 223
general dynamic equation for aerosols (GDE), 192
Köhler, 209
Koschmieder (visibility), 85
reactive dispersion, 232
Schrödinger, 49
equivalent PDE, 258
escape velocity, 28
ETEX, 271
ethanol, 170
Eulerian viewpoint, 255
eutrophication, 4
exosphere, 28
explicit schemes, 250
extinction, 59
cross section, 60

F
fall-off reaction, 136
feedbacks, 74, 158
fine particles, 227
finite difference method, 255
first law of thermodynamics, 99
forecast, 269, 276
free atmosphere, 96
friction velocity, 122
Fumigium, 3

G
Gaussian models, 239
Gaussian Puff Model, 241
GDE, 192, 260
analytical solution, 261
sectional methods, 261
variational methods, 279
geo-engineering, 12
geostrophic wind, 107, 113
global warming potential (GWP), 77
Göteborg protocol, 6
gravitational sedimentation, 182, 225
gravitational settling, 33
Greenfield gap, 221
greenhouse effect, 65, 71
GWP, 77
greenhouse gas, 65

H
Hadley's circulation, 31
haze index, 86
Henry's law, 211
heterogeneous reactions, 190, 201
history (of air pollution), 1
homosphere, 28
hydrogen, 28
hydrogen chloride, 148
hydrostatic approximation, 117
hydrostatic equation, 25
hydroxyl (OH), 145
inversion, 175
source, 166
hysteresis, 197

I
ideal gas law, 18
Île-de-France, 170
implicit schemes, 250
indirect effect, 81
indoor air quality, 172
infrared radiation, 52, 65
integrated modeling, 277
Inter-Tropical Convergence Zone (ITCZ), 31, 36

interhemispheric exchange, 36
internal/external mixing, 78, 185, 190
inverse modeling, 269, 270
  OH, 175
inversion layer, 97, 103
ionization, 64
IPCC, 8, 73
ITCZ, 31, 36

J
Jacobian matrix, 250, 251
Jaenicke formula, 186
Jeans' escape, 28
Junge layer, 187

K
K theory, 119
Kelvin effect, 196, 206
kinetic rate, 136
Kirchhoff's law, 54
Köhler equation, 209
krypton, 36

L
Lagrangian Particle model, 241
Lagrangian viewpoint, 255
latent heat, 104, 114, 126
law
  Henry, 211
  Planck, 49
  Raoult, 208
  Stefan-Boltzmann, 51
  Stokes, 182
lead, 6
liquid water content, 202
  radiative properties, 80
low-level jet streams, 98
low-level nocturnal jets, 121
LUC, 239
lumping, 149

M
mass action law, 136
mass consistency, 233, 260
mass of the atmospherere, 26
mass transfer
  aerosols, 195
  clouds, 210, 223
MCF (methyl-chloroform), 175
mean free path, 194
mercury, 37, 266
methane (CH_4), 133, 144, 159
  chemical lifetime, 145
  lifetime of OH, 143
  oxidation, 164
  residence time, 34
Mie scattering, 56, 57
mixing height, 93, 111
mixing layer, 106
mixing ratio, 17
modal (model), 185, 260
model hierarchy, 239
model species, 149
model-to-data comparison, 274
modeling system (state-of-the-art), 265
molar mass of air, 18
Monin-Obukhov length, 124
monitoring network, 8
Monte Carlo method, 267
Montreal Protocol, 7

N
Navier-Stokes Equations, 114
NEC directive, 7
network design, 276
nitrate, 213, 216
nitric acid (HNO_3), 147, 157, 216
nitrogen oxides (NO_x)
  disbenefit, 167
  emissions, 22
  NO_x-limited regime, 165, 176
nuclear winter, 92
nucleation, 188, 198, 228
number
  of Avogadro, 18
  of Loschmidt, 20
numerical diffusion, 258

O
OCS (carbonyl sulfide), 187
off-line coupling, 232, 260
OH, *see* hydroxyl (OH)
Oke's law, 128
on-line coupling, 232
operator splitting, 246
optical depth, 54
optimal interpolation, 270
oxidation, *see* oxidizing power, 142
  chain, 143
  CO, 165

oxidation, *see* oxidizing power (*cont.*)
methane, 164
VOCs, 163
oxidation chain, 163
oxidizing capacity, 142
oxidizing power, 133, 177
oxygen, 139
ozone
artic hole, 156
column, 27
ozonolysis, 153
stratospheric, 150
titration, 171
tropospheric, 159
ozone hole, 156

P
PAN, 148
passive remote sensing, 66
PDE
advection-diffusion-reaction, 233
PDF, 267
persistent organic pollutants, 35
pH, 213
of clouds, 215
photolysis, 135
photon (energy), 49
photostationary equilibrium, 162
Pinatubo Mount (volcan), 79, 80
Planck's law, 49
PM, 182
POP, 35
potential temperature, 99
ppb, ppm, ppt, 17
pressure, 25
partial, 203
saturation vapor, 203
primary species, 21
PSC, 157

Q
QSSA, 145, 253
quantum yield, 139
quasi steady state assumption (QSSA), 145

R
radiance, 47
radiating forcing, 76
radiation
solar, 64
radiation spectrum, 46
radiative budget, 68
radiative energy, 47
radiative energy budget, 69
radiative energy (budget for the Earth)
magnitude, 45
radiative forcing, 87
aerosols, 78
clouds, 78
definition, 73
greenhouse gas, 76
radiative transfer
aerosol, 60
Beer-Lambert law, 52
blackbody emission, 50
emission effective temperature, 68
extinction, 59
Kirchhoff's law, 54
Planck's law, 49
radiation spectrum, 46
radiative transfer equation, 58
scattering, 55
Stefan-Boltzmann law, 51
Wien's displacement law, 51
radical, 143
radioactive decay, 32, 34, 40
radioelements, 34
radon, 40
rainwater composition, 217
rainwater pH, 213
Raoult law, 208
Rayleigh number, 108
Rayleigh scattering, 56
reaction
fall-off, 136
reactions
heterogeneous, 190, 201
reduction of emissions
CFCs, 41
refractive index, 56, 61
reglementation
ACEA agreement, 10
EURO norms, 10
European directives, 6
history, 8
regulation
Clean Air Act, 5
relative humidity, 196
reservoir species, 147

residence time, 35
  aerosols, 40, 186
  trace species, 32
  water, 203
resistance model (dry deposition), 130
respiration (and particles), 180
Reynolds number, 108
Richardson number, 111
richness (of the mixture), 159
roughness height, 123
runaway greenhouse effect, 74

S

scale height, 26
  aerosols, 226
scattering, 55
scattering cross section, 59
scavenging coefficient, 219
sea salts, 180
secondary species, 21
sedimentation, 33
segregation effect, 236
semi-volatile organic compounds, 199
sensible heat, 126
sensitivity analysis, 266
sink species, 147
sky color, 84
  and particles, 85
  blue, 57
  NO_2, 85
smog
  Great London smog, 4
  history, 3
Smoluchowski equation, 263
SNAP, 243
SO_2, *see* sulfur dioxide (SO_2)
SOA, 191
solar constant, 68
solar radiation, 52
solid angle, 47
soot, 191, 227
source/receptor matrix, 267
speciation, 23
species
  trace, 17
specific heat, 99
specific humidity, 19
splitting, 246
stability, 251
standard thermodynamical conditions, 18
Stefan-Boltzmann law, 51
stiffness, 252
stoichiometry, 135
Stokes (law of), 182
stratopause, 26
stratosphere, 23
strontium, 38
sulfate, 213, 215
sulfur dioxide
  and acid rains, 213, 217
  and clouds, 215
  and fine particles, 227
  Great Smog, 4
  oxidation, 189
  residence time, 188
sulfuric acid (H_2SO_4), 35, 227
supersaturation, 204
surface boundary layer, 106, 122
surface tension, 206
surrogate species, 149
SVOCs, 199

T

temperature
  dew point, 204
  emission effective, 68
  potential, 99
  virtual (wet air), 19
temperature inversion, 153
thermal espace, 28
thermal stratification, 98
thermodynamic equilibrium, 189, 196, 264
thermodynamics
  first law, 205
third body, 135
timescales, 30, 250
  atmospheric transport, 30
  CFCs, 35, 156
  chemical lifetime, 144
  interhemispheric exchange, 37
  meteorology, 94
  troposphere/stratosphere transfer, 40
titration of ozone, 162, 170, 171, 237
trace (species), 17
transcontinental transport, 171
transition (energy levels), 48
tropopause, 26
troposphere, 23
turbulence, 106
Twomey effect, 81, 89

U
uncertainties, 265
upwind scheme, 256
urban boundary layer, 129
urban breeze, 125
urban climate, 125
urban heat island, 127

V
visibility, 84
visual contrast, 84
VOC
  and aerosols, 199
  emission reduction, 167
  emissions, 22
  lifetime, 146
VOC (*cont.*)
  lumping, 149
  oxidation, 163
  VOC-limited regime, 165, 177
volcanic emissions, 22, 79, 80, 187
von Karman constant, 123

W
water vapor, 19
  condensation, 202
  greenhouse effect, 65, 74
wet scavenging, 218
Wien's displacement law, 51

Z
Zeldovitch mechanism, 161